Physics and Chemistry of
Mineral
Surfaces

CRC Series in
Chemistry and Physics of
Surfaces
and
Interfaces

Series Editor
John P. LaFemina, Ph.D.

Titles in the Series

Physics and Chemistry of Mineral Surfaces
Patrick V. Brady

Surface Analytical Techniques for Probing Biomaterial Processes
John Davies

Physics and Chemistry of

Mineral
Surfaces

Edited by
Patrick V. Brady, Ph.D.
Sandia National Laboratories
Albuquerque, New Mexico

CRC Press
Taylor & Francis Group
Boca Raton London New York

CRC Press is an imprint of the
Taylor & Francis Group, an **informa** business

CRC Press
Taylor & Francis Group
6000 Broken Sound Parkway NW, Suite 300
Boca Raton, FL 33487-2742

First issued in paperback 2019

ISBN-13: 978-0-8493-8351-9 (hbk)
ISBN-13: 978-0-367-40131-3 (pbk)
Library of Congress Card Number 95-47556

Library of Congress Cataloging-in-Publication Data

Physics and chemistry of mineral surfaces / edited by Patrick
    V. Brady.
        p.   cm. -- (CRC series in chemistry and physics of surfaces
    and interfaces)
    Includes bibliographical references and index.
    ISBN 0-8493-8351-X (alk. paper)
    1. Minerals--Surfaces.   2. Mineralogical chemistry.   3. Surface
chemistry.   I. Brady, Patrick V., 1961–   . II. Series.
QE364.2.S87P48   1996
549'.131--dc20
                                                            95-47556
                                                            CIP

Visit the Taylor & Francis Web site at
http://www.taylorandfrancis.com

and the CRC Press Web site at
http://www.crcpress.com

EDITORIAL ADVISORY BOARD

AIMS AND SCOPE

The aim of this series is to develop a literature devoted to cross-disciplinary, *critical* reviews of surface and interface physics and chemistry. This series will take a broad approach to covering these fields with a particular emphasis on the study of insulating systems including metal oxide, mineral, glass, ceramic, and polymer surfaces and interfaces.

The volumes in this series will comprise edited works, containing chapters authored by different researchers, single author volumes dedicated to singular research topics, and selected conference proceedings. Titles will focus on the fundamentals of surface and interface science, as well as the applications of surface chemistry and analytical techniques.

This book series faces several challenges. The first challenge is to help develop a common language for the researchers in this area that will facilitate the transfer of knowledge and information across the traditional disciplinary boundaries. To meet this challenge, each of the volumes in the series (or each article in a particular volume) will contain a general introduction to the topical area, its nomenclature, and notation. The second challenge is to produce *critical* reviews which not only detail the work that has been performed in a particular subject area, but critically reviews it by defining the outstanding issues in the field and describing how the work that has been performed addresses (or does not address) these issues. Most importantly, the review articles should present a comprehensive, conceptual framework in which the work that has been performed can be understood and interpreted. In doing this, the review will help to define the outstanding issues and shape the focus and future direction of research in this field.

Readership: These reviews would be written by the pioneering experts in the field and be targeted at both graduate students and established researchers just entering the field of insulator surface science as well as researchers in related fields that wish to keep abreast of the latest developments. Moreover, this series can serve to catalyze the information exchange among the researchers within this broad multi-disciplinary research area.

SERIES PREFACE

It is especially gratifying for me to see this volume come to publication. In 1992, when CRC Press and I first agreed to create the *Chemistry and Physics of Surface and Interfaces* series, I had been having discussions with Andy Gratz and Mike Hochella about a collaborative research project to study the surface structure and chemistry of calcite. As we thought about appropriate topics for the series, we immediately thought of a volume that would lay out the broad landscape of mineral surface structure and chemistry from its most fundamental to applied aspects. Andy was very excited about this venture, having pioneered the application of scanning force microscopy to mineral surface and interfacial chemical processes, such as the dissolution of quartz, and he quickly agreed to act as the Editor for this volume. Andy pulled together a wonderful team of contributors, each representing the state-of-the art in their particular field. Tragically, Andy passed away in 1993. He left behind a rich legacy of scientific contributions, for which he was remembered with a special symposium in his honor at the 1994 Annual Meeting of the Geological Society of America, entitled *Frontiers of Mineral Surface Geochemistry: A Symposium in Memory of Andrew J. Gratz (1962–1993)*.

In the interim, I managed to persuade Pat Brady, one of the chapter authors, to take over the duties of Editor. Pat quietly, but confidently, put his own mark on this volume, driving all of the chapter authors (myself included!) to meet their deadlines, and coordinating a peer review process of each of the chapters to ensure the highest quality.

And so, three years after its conception, the volume *Physics and Chemistry of Mineral Surfaces* is born. Along the way, the field of geochemical surface science has blossomed with the application of traditional experimental and computational surface science tools to problems in surface geochemistry. The authors of this volume have all made significant contributions to the definition of this exciting field of study. And through their work, I believe lies the path to a greater understanding of the natural geochemical environment in which we all live.

John P. LaFemina

October 1995
Richland, Washington

PREFACE

Reactive mineral surfaces control the bulk chemical and physical properties of many natural and engineered systems. Mineral surface reactivity is therefore of critical interest to workers in the fields of materials science, applied physics, physical chemistry and the earth and environmental sciences, to name but a few. The aim of *The Physics and Chemistry of Mineral Surfaces* is to first outline atomistic controls on mineral surface structure and reactions; to then apply these concepts to understand sorption, mineral corrosion and growth; and to ultimately consider the role of surfaces in environmental and geochemical processes. Chapters 1 and 2 by Gibson and LaFemina, and Henrich, respectively, explore the structure of mineral surfaces and outline the atomistic origins of surface reactivity. In Chapter 3 Hayes and Katz combine X-ray absorption spectroscopy with surface complexation theory to establish the chemical controls on metal sorption. The surface complexation approach is linked to mineral corrosion and growth by Brady and House in Chapter 4. Chapter 5, by Brady and Zachara, summarizes a number of specific areas in the earth and environmental sciences where mineral surface science is currently being applied.

This book was intended by Andrew J. Gratz, the original editor, to integrate surface physics and chemistry and apply them to macroscopic, sometimes global, processes. My efforts are dedicated solely to Andy's memory. I appreciate the continuing support of the U.S. DOE-BES/Geosciences, the U.S. Nuclear Regulatory Commission, the U.S. National Science Foundation and the Petroleum Research Fund of the American Chemical Society. This work was supported by the United States Department of Energy under contract DE-AC04-94AL85000. Many thanks to Felicia Shapiro and Renée Taub at CRC for helping get this all into shape.

Pat Brady

Albuquerque, New Mexico

THE EDITOR

Patrick V. Brady, Ph.D., was born in Rome, Georgia, in 1961 and went to the University of California at Berkeley on a football scholarship, receiving an A.B. in Geology in 1984; followed by an M.S. (1987) and then a Ph.D. (1989) in Geochemistry at Northwestern University. After a post-doctoral appointment at the Swiss Federal Institute of Technology — Institute of Water Resources and Water Pollution Control (EAWAG) in Dübendorf, Switzerland, and two and a half years as an Assistant Professor at SMU in Dallas, Texas, Brady became a senior member of the Technical Staff at Sandia National Laboratories in Albuquerque, New Mexico in 1992. Brady's specialties are surface chemistry, mineral kinetics, and global change.

CONTRIBUTORS

Patrick V. Brady
Sandia National Laboratories
Albuquerque, New Mexico

Andrew Gibson
Department of Materials and Interfaces
Environmental Molecular Science
Laboratory
Pacific Northwest Laboratory
Richland, Washington

Kim F. Hayes
Department of Civil and Environmental
Engineering
University of Michigan
Ann Arbor, Michigan

Victor E. Henrich
Department of Applied Physics
Yale University
New Haven, Connecticut

William Alan House
Institute of Freshwater Ecology —
River Laboratory
East Stoke, Wareham
Dorset, United Kingdom

Lynn E. Katz
Department of Civil and Environmental
Engineering
University of Maine
Orono, Maine

John P. LaFemina
Pacific Northwest Laboratories
Richland, Washington

John M. Zachara
Pacific Northwest Laboratories
Richland, Washington

CONTENTS

Chapter **1**

STRUCTURE OF MINERAL SURFACES

Andrew S. Gibson and John P. LaFemina

CONTENTS

0-8493-8351-X/96/$0.00+$.50

I. INTRODUCTION

A. Purpose and Scope of Chapter

There is no more fundamental property of a surface or an interface than its atomic geometry.[1,2] The physical arrangement of the atoms controls every aspect of the physics and chemistry of the interface. The goal of this chapter is to review the current state of the art in experimental and theoretical determinations of surface and interface geometry. To a large extent we limit the scope of the discussion to mineral surfaces for which experimental determinations or theoretical predictions have been made, rather than make conjectures concerning the properties of surfaces which may be environmentally or geologically interesting, but about which nothing is really known. We will demonstrate, however, that the atomic structure of mineral surfaces can be qualitatively understood, and predicted, using a set of five simple principles. These principles, based upon fundamental chemistry and physics, have been derived from over 20 years of research on semiconductor surfaces and interfaces,[1-4] and have recently begun to be applied to mineral oxides.[5] These principles offer the opportunity to make predictions for some mineral surfaces that have not, as yet, been studied experimentally.

The range of minerals for which detailed experimental or theoretical information is available, unfortunately, is limited.[6] Experimental limitations derive mainly from the paucity of well-characterized single-crystal materials for study. In addition, the insulating nature of these materials limits the use of charged particle spectroscopies, such as low-energy electron diffraction (LEED),[7] which can be used to provide quantitative information on surface structures. This lack of detailed experimental information has also limited the development of semiempirical and empirical quantum mechanical models; models which, because of the complexity associated with *ab initio* methods, have traditionally been the first to characterize surface relaxations and reconstructions.[2] These difficulties will be explored in more detail in Section II.

In this chapter we will focus on the subset of mineral systems for which the most detailed understanding of surface atomic structures exists. The results of computational studies, at all levels of approximation, and experimental surface structure determinations by LEED,[7] X-ray photoelectron diffraction (XPD),[8] Auger photoelectron diffraction (APD),[8] ion scattering,[9] and scanning probe microscopies[10] will be presented and discussed in the context of the surface structure principles described in detail later in the chapter.

The level of presentation in this chapter is such that readers familiar with fundamental concepts in solid state atomic structure and chemical bonding should have no difficulty. Good undergraduate-level texts include: *Introduction to Solid State Physics* by C. Kittel[11] and *Chemistry in Two Dimensions: Surfaces* by Gabor A. Somorjai.[12] More in-depth treatments of these topics can be found in several excellent graduate-level texts: *Physics at Surfaces* by Andrew Zangwill;[13] *Solids and Surfaces: A Chemist's View of Bonding in Extended Structures* by Roald Hoffmann;[14] *Atomic and Electronic Structure of Surfaces: Theoretical Foundations* by M. Lanoo and P. Friedel;[15] and *Electronic Structure and the Properties of Solids: The Physics of the Chemical Bond* by W.A. Harrison.[16]

B. Basic Nomenclature

It is useful, at this time, to review some basic nomenclature that will be used throughout this chapter. A surface is said to be *relaxed* if it displays the same symmetry as the bulk material. *Reconstructed* surfaces, on the other hand, display a surface symmetry different from the bulk. Relaxed and reconstructed surfaces are designated with the label (n × m), where n and m represent the ratio of the reconstructed to unreconstructed surface translation vectors for the two directions parallel to the surface. Relaxed surfaces, therefore, carry the designation (1×1). In some cases an additional designation of p (for primitive) or c (for centered) is added to more fully reflect the symmetry of the surface unit cell. Finally, if the surface translation vectors for the reconstructed surface are rotated from the surface translation vectors of the unreconstructed (or truncated bulk) surface the designation $R\Theta°$ is added, where Θ is the angle of rotation.

When discussing the mechanisms and driving forces for surface relaxations and reconstructions, we will repeatedly refer to the surface *dangling bond* charge density. This is the excess charge density remaining at the surface in the "dangling" bonds which were used to bind the surface atoms to their, now missing, neighbors in the bulk material. This dangling bond charge density is localized at the surface in *surface states* which can be primarily derived from the surface cations (cation-derived surface states) or surface anions (anion-derived surface states). As will be seen, this is an enormously useful concept, not only on semiconductors, but also in ionic materials.

C. Principles Governing Surface Structures

Although the set of mineral surfaces that have had their surface atomic and electronic structure examined in detail is small, there is much that can be learned from these studies that is generally applicable.[5,6] As stated earlier, there exists a set of simple physical and chemical principles, derived from the study of semiconductor surfaces,[1-4] that can be used to qualitatively understand and predict the surface relaxations and reconstructions which occur at mineral surfaces. In this section each of these principles will be described.

1. Principle 1: Stable Surfaces Are Autocompensated

Typically, bulk materials are stable when the bonding orbitals (or bands) are fully occupied and the antibonding orbitals (or bands) are fully unoccupied. For binary solids, the bonding orbitals (or valence bands) are associated with the most electronegative atoms (or anions), while the antibonding orbitals (or conduction bands) are associated with the electropositive atoms (or cations). By completely filling the valence bands and completely emptying the conduction bands, the material is charge neutral and most stable. At the surface, this stability condition can be recast by requiring the anion-derived dangling bond surface states to be completely filled and the cation-derived dangling bond surface states to be completely empty. In this way the surface remains charge neutral and stable.[16] Such charge neutral surfaces are said to be autocompensated.[17] It must be emphasized that the principle of autocompensation is not identical to satisfying the laws of valence, but is a more demanding constraint. We return to this point below in a discussion of polar surfaces.

One difficulty with mineral surfaces is the large variety of surface stoichiometries available to them.[6] Fortunately, it is possible to discriminate between all possible surface stoichiometries and identify those most likely to be exhibited experimentally by determining which surface stoichiometries are autocompensated. To do this, however, we must learn how to count dangling bond electrons using a method first described by Harrison,[16] and illustrated here with the chalchogenide zincblende ZnS (110) cleavage surface.

In bulk zincblende, the Zn and S atoms are fourfold coordinated. Each Zn atom contributes two valence electrons to the four bonds, or $1/2$ electron per bond. Each S atom contributes six valence electrons to four bonds, or $3/2$ electrons per bond. Each ZnS unit cell then has two electrons per bond ($1/2 + 3/2$) and is stable. The truncated bulk (110) cleavage surface, shown in Figure 1, comprises zigzag rows of ZnS, and the surface unit cell comprises one ZnS formula unit. The surface atoms are now threefold coordinated with one dangling bond. In each unit cell, therefore, there is one Zn dangling bond (with $1/2$ electron) and one S dangling bond (with $3/2$ electrons). By transferring the $1/2$ electron from the Zn dangling bond to the S dangling bond, the anion dangling bond (and anion-derived surface state) is completely filled and the cation dangling bond (and cation-derived surface state) is completely empty. The surface is autocompensated, charge neutral, and stable.

Zincblende (110))

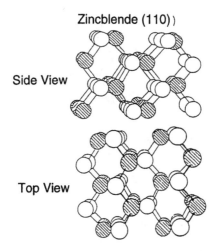

Side View

Top View

Figure 1
Illustration of the ideal (truncated bulk) zincblende structure (110) surface, top and side views. (From Duke, C.B. *Surface Properties of Electronic Materials*. D.A. King, D.P. Woodruff, Eds. Elsevier: Amsterdam, 1988; 69. With permission.)

The (111) surface of the zincblende structure, shown in Figure 2, is not autocompensated.[18] Let us consider the Zn terminated surface. The S atoms in the subsurface layer contribute $3/2$ electron to each of the four bonds with the nearest neighbor Zn atoms. If the surface Zn atoms, which are threefold coordinated, contribute $1/2$ electron to each bond, there will be a surplus of $1/2$ electron per surface Zn atom. This additional electron density will be required to occupy an electronic state of Zn or conduction band character. For this reason the surface is not autocompensated and is likely to be thermodynamically unstable. In the course of this chapter we will see how different surfaces defect or facet at such surfaces in order to become autocompensated. Surfaces such as the (111) zincblende surface are referred to as polar surfaces, since in a classic electrostatic point ion model of the system the surfaces are charged.

ZINCBLENDE (111)

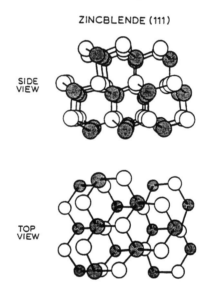

SIDE
VIEW

TOP
VIEW

Figure 2
Schematic illustration of the side and top views of the ideal (truncated bulk) zincblende (111) surface. The surface atoms can either be cations — cation-terminated, usually designated (111) — or anions — anion-terminated, usually designated (-1-1-1). (From Duke, C.B. *Surface Properties of Electronic Materials*. D.A. King, D.P. Woodruff, Eds. Elsevier: Amsterdam, 1988; 69. With permission.)

As is clear from the above examples, all cleavage surfaces, which by definition have stoichiometric numbers of surface anions and cations, are autocompensated. Fracture and growth surfaces, however, can exhibit many possible stoichiometries which can be discriminated by determining which are autocompensated and which are not.

2. Principle 2: Rehybridize the Dangling Bond Charge Density

Once a particular surface stoichiometry has been determined to be stable (i.e., autocompensated), the primary factor in determining the nature of the surface relaxation or reconstruction is the energy that can be gained by rehybridizing the surface dangling bond charge density in response to the reduced atomic coordination at the surface. This can result in a charge transfer between atoms at the surface (as in the zincblende example above), the formation of new bonds between surface atoms (e.g., surface dimerization), or the formation of new bonds between surface atoms and atoms adsorbed to the surface (i.e., adatoms). In each of these instances the chemical bonding at the surface is different from that of the bulk and leads to different chemical and physical properties at the surface.

3. Principle 3: Form an Insulating Surface

By pairing up the electrons in the dangling bonds and opening up a gap between the occupied and unoccupied surface states, an insulating surface is formed. This can be done in several ways (e.g., surface dimerization or adsorption) as described above. The key point, however, is that the formation of an insulating surface stabilizes the occupied surface states while destabilizing the unoccupied surface states, resulting in a net lowering of the surface energy.

4. Principle 4: Conserve Near-Neighbor Bond Lengths

As the surface atoms change their atomic positions in order to minimize the energy of the dangling bond charge density, changes in the local bonding environment (near-neighbor bond lengths and angles) lead to the formation of local strain fields that are energetically unfavorable.[2] Because the force constants for bond bending are smaller than those for bond compression or stretching (by one to two orders of magnitude, typically), changes in near-neighbor bond lengths cost the most energy and, if sufficiently severe, can prevent the surface atoms from moving. Consequently, the relaxation or reconstruction of a surface results from a balance of the energy lowering because of the rehybridization of the dangling bond charge density and the formation of an insulating surface, and the elastic energy cost associated with significantly distorting the local bonding environment. Of course, relaxations and reconstructions that (nearly) conserve near-neighbor bond lengths minimize their elastic energy costs.

Because of the importance of minimizing surface and subsurface strain, the qualitative nature of surface relaxations and reconstructions is determined primarily by the topology, or atomic connectivity, of the surface.[2-4] Surfaces whose topology allows approximately bond-length conserving motions of the surface atoms typically undergo large atomic motion surface relaxations and reconstructions. Conversely, surfaces whose to-pology forbids such motions typically undergo small atomic motion re-laxations.

5. Principle 5: Kinetics Are Important

It is important to remember that the experimentally observed surface is that which is kinetically accessible under the particular surface-process-ing conditions used.[3,4] For cleavage surfaces, the experimentally observed surface is "activationless" in that the activation barrier to relaxation or reconstruction is less than the energy supplied by the cleavage process. This is no guarantee, however, that the observed surface is the thermo-dynamically most stable surface. As an example, the (111) surface of the semiconductor silicon exhibits a (2×1) structure upon cleavage at low temperatures (<360°C), but irreversibly reconstructs to the thermodynam-ically stable (7×7) structure upon heating to 400°C.[2] Therefore, by saying that surface kinetics plays a major role in determining the experimentally accessible surface, we remind ourselves that the surface-processing con-ditions are important. Whatever surface stoichiometry is exhibited, how-ever, must be autocompensated, and the resulting surface atomic geom-etry will be guided by the energy balancing between the rehybridization of the surface dangling bond charge density and the formation of local strain fields.

D. Chapter Organization

This chapter is organized as follows. In Section II a brief review of the experimental and computational methods used to determine surface atomic geometries is presented. Because of the importance of surface topology, or atomic connectivity, in determining the qualitative nature of the surface relaxation or reconstruction, the examination of mineral sur-face structures contained in Section III is organized by crystal lattice type. As stated previously, we limit the discussion to clean surfaces for which some detailed experimental or theoretical information is available. The chapter concludes in Section IV with a discussion of the major issues associated with the atomic structure of mineral surfaces and speculations on promising areas for future research.

II. METHODS OF SURFACE STRUCTURE DETERMINATION

In this section a brief review is given of the most common experimen-tal and theoretical methods for surface structure determination. These descriptions are by no means exhaustive, and references to more complete works on each individual technique are given throughout the section. The purpose of this section is to give the reader a sufficient understanding of the methods so that the evaluation of the structural determinations is understandable. Consequently, the discussion will focus more on describing the fundamental, and relative, strengths and weaknesses of each of the methods rather than on the details of any particular technique.

A. Experimental Methods

There are two major classes of experimental techniques for the determination of surface atomic structure: scanning probe microscopies[10] and electron diffraction methods.[8] In the former, a tip is scanned across the surface and the measurement of the resulting current flow (in scanning tunneling microscopy) or interatomic force (in atomic force microscopy) renders a real-space "image" of the surface atoms. In the latter, the diffraction (or scattering) of electrons by atoms in the surface renders a reciprocal-(or k-)space image of the surface. These are, of course, grossly simplified descriptions of these methods. Yet, they do capture the essential "picture." In the following pages, we present a more accurate description of these methods, their strengths, and their limitations.

1. Scanning Probe Microscopies

Mineral solubility and sorption reactions depend upon the atomic geometry and nanometer-scale morphology of mineral surfaces.[19] Scanning probe microscopies, the scanning tunneling microscope (STM)[20] and the atomic force microscope (AFM),[21] are ideal tools for studying mineral surfaces over the length scales required to understand these phenomena, not only in vacuum, but also in air and aqueous environments. Even more impressive is the capability to study such processes as growth and dissolution in real time.[22] These probes come much closer to direct observation of which sites on a surface are responsible for chemical reactions, as opposed to scattering techniques which require mathematical manipulation for their interpretation.

a. Scanning Tunneling Microscope

In an STM, a bias voltage (ca. 100 mV) is placed between the surface of interest and the tip. The tip is then moved to within several (1 to 4) angstroms of the surface (for atomic resolution), and, depending upon the sign of the voltage bias, a tunneling current (ca. 1 nA) will either flow out of the surface into the tip, or vice versa.[20] By scanning the tip across the surface at a constant height (constant height mode) the tunneling current will modulate. Conversely, the tip can be scanned in such a way as to maintain a constant tunneling current (constant current mode, which is most common) with the height of the tip above the surface modulated. In either case, the surface map of the modulated current (height) contains information about the structure of the surface being scanned. The constant current mode is most widely used in STM experiments, as the measurement of current provides a feedback mechanism which requires that the tip remain positioned within a few angstroms of the surface.[20] The sensitivity of the STM, and the basis of its operation, is that the tunneling current decays exponentially in the distance that separates the surface from the tip.[20]

The exact interpretation of these maps is the subject of many research efforts. What is agreed upon is that the STM images contain *direct* information about the *electronic* structure of the surface, via the interaction of the charge density at the tip with the charge density at the surface. Since it is the geometric arrangement of the atoms at the surface that determines the electronic structure, the STM image then contains *indirect* information about the surface *atomic* structure. To understand the extent to which we are and are not investigating properties of the surface, rather than the properties of the tip-sample interaction, it is useful to consider the theory underlying this instrument.

Consider the following STM experiment. The sample and bulk portion of the tip are described by the crystalline Hamiltonian H_o, while the end of the tip is described by the potential V_{tip}. If we consider a small bias, the current that will flow from the tip to the sample can be calculated using scattering theory:[23]

$$J = \frac{8\pi^2 e}{h} \sum_{i,f} \left| \int \psi_f V_{tip} \psi_i dr \right|^2 \delta(E_f - E_i) \tag{1}$$

In Equation 1, ψ_i describes an electron originating in the tip and moving toward the sample [and is an eigenstate of $(H_o + V_{tip})$], ψ_f describes the electron ending up in the sample (and is an eigenstate of H_o. This treatment includes no temperature effects. The current given by Equation 1 is difficult to compute for realistic Hamiltonians because of the large number of atoms that need to be included to adequately describe both the sample and the tip as a single system. If we assume, however, that it is the electronic structure of the surface that is important and that the tip is spherical and serves only as a source of electrons, both ψ_i and ψ_f can be replaced by ψ_s, the wavefunction for the sample. If we further assume that the potential, V_{tip}, is constant, then Equation 1 reduces to,

$$J = G(R) \left| \psi(E_F r) \right|^2 \tag{2}$$

where $G(R)$ is a function determined by the radius of the spherical tip, E_F is the Fermi energy of the sample, and ψ is the electronic wavefunction with energy E_F. What is special about this formula is that the tunneling current depends only upon the density of states of the sample and, in this approximation, does not depend upon the details of the electronic structure of the tip. Furthermore, the quantity in Equation 2 (referred to as Tersoff-Hamann theory[24]) is amenable to calculations with elaborate descriptions of the electronic structure of the surface.

One major limitation of using STM on mineral surfaces is that the sample must be sufficiently conducting to support the tunneling current. For the majority of minerals, which are highly insulating, this condition severely restricts the use of the STM. The STM has been used, however, with considerable success on semiconducting mineral surfaces,[25-27] and these will be discussed in Section III.

b. Atomic Force Microscope

The importance of the AFM is the ability to resolve isolated atomic defect structures, such as steps or impurity atoms, on both conducting and insulating surfaces.[21] These results were initially obtained on layered materials such as mica and graphite,[28] but the method has been used successfully on ionic materials including NaCl,[29] LiF,[30] PbS,[31] AgBr,[32] and $CaCO_3$.[33] Perhaps most important for the mineral surfaces has been the success of the method under aqueous environments.[34]

In the AFM, the tip is in contact with the surface and attached to a cantilever. As the tip is raster scanned over the surface of the sample, laser light reflected from the cantilever is used to measure deflection of the tip. The microscope may be operated in a mode of constant force by using the measured cantilever deflection with a feedback mechanism, which is analogous to the constant current mode in the STM. Atomic resolution with the AFM is achieved in what is referred to as contact mode, where the force between sample and tip is usually repulsive and between 10^{-8} and 10^{-9} N.[21] Chemical bonding (i.e., overlap of the electron charge density) is responsible for the interaction between sample and tip at the tip-sample separation in this mode of operation and is the basis for the atomic resolution.[21] The AFM is also operated in what is referred to as the non-contact mode, where an attractive interaction between the sample and tip is mediated by the van der Waals force, the exchange and electrostatic forces. Atomic resolution is not observed in this case, which will not be considered in this section. An obvious benefit of the AFM over the STM for mineral surfaces is that the samples are not required to be metals or semiconductors.

A good approximation for the force on an individual atom of the tip (or the substrate) can be calculated using density functional theory, which is discussed in Section II.B. The force is the first derivative of the energy with respect to some displacement, and because of the variational principle in quantum mechanics it may be computed from the ion-electron interaction, but without computing the change in the wavefunction due to the displacement.

Consider what the energy vs. position profile must look like. At small tip-sample separations (<2 to 3 Å), the ion-ion repulsion rapidly becomes larger than the electron-ion attraction for the ion at the end of the tip. The corrugation observed in the AFM is partly due to the fact that the position of this minimum is a function of the lateral position of the tip above the surface, such that the force may be attractive at one lateral coordinate, but repulsive at another. The force variation of the tip-sample system is complicated by the finite size of the tip. While the atom at the apex of the tip may feel strong repulsion, the longer range attractive forces acting on many atoms may complicate interpretation. It is widely believed that the surface of the substrate is deformed over many angstroms.[35] By using elastic constants computed using density functional theory, an elastic continuum model for graphite has been used to predict that at a load of 10^{-9} N, using a cylindrical tip with a radius of 2.75 Å, the maximum layer distortion is approximately 0.5 Å and the healing length, defined as the radius at which the layer distortion decreases to half its maximum value, is approximately 5 Å.[35]

Short-range lateral forces which produce energy losses through energy transfer to shear modes are the main cause of friction when the AFM tip is raster scanned across the surface of the substrate.[36] This effect is both small and reversible when the AFM is operated in the noncontact mode. It has been estimated that these lateral forces in the contact mode are still an order of magnitude smaller than the perpendicular forces because the attractive forces are additive in the perpendicular direction, but tend to cancel in the lateral force.[36] Extensive computer simulations using empirical potentials[37] have been directed at a better understanding of the effects of lateral forces. Surface melting, nano indentation, formation of a connective neck, wetting mechanism, and hysteresis of the retracting tip have been proposed as aspects of atomic force microscopy.[35-37]

Time-resolved measurements with the AFM have been used to study the dissolution of calcite.[38] Dissolution is found to occur because of shallow and flat pits which grow with a constant velocity in the lateral directions. The experiments have determined that the velocities of the two inequivalent pit edges for the calcite structure are different by a factor of 2.5. Such macroscopic information puts important constraints on the types of microscopic processes which determine dissolution and is an important application of the AFM.

2. Electron Diffraction Methods

In this section we will examine the fundamental similarities and differences among the most commonly used electron diffraction methods for surface structure determination: LEED,[7] XPD,[8] and APD.[8]

The electron diffraction, or scattering, methods yield indirect information on the atomic geometry of the surface by measuring how the path of an electron through the solid changes with the electron energy and crystal orientation. In this way, the methods are fundamentally the same. They differ in details, such as the energy of the scattering electron (which controls the mean free path of the electron in the crystal) and whether the electron originates outside the crystal (from an incident beam as in LEED) or inside the solid as the result of a photoexcitation event (as in XPD and APD). In all of these methods, however, the extraction of quantitative surface structural information from the data is done by measuring the scattered electron intensity as a function of crystal orientation and electron energy.[8] A model of the surface structure is composed, and the scattering intensity curves are computed as a function of electron energy and crystal orientation. The surface structure of the model is then varied to obtain the best fit between the computed and measured intensity curves.

The remainder of this section will focus on briefly describing each of these methods individually, characterizing the differences between them, and assessing their relative strengths and weaknesses. For more in-depth treatments of these methods, several excellent reviews are available.[7,8,39]

a.   Low-Energy Electron Diffraction (LEED)

In LEED the incident electrons have an energy that is typically between 20 and 300 eV.[39] At these energies, the interaction between the incident electrons and the solid is strong, and the inelastic mean free path is on the order of several angstroms.[39] Since the elastically backscattered electrons have sampled only the top few atomic layers, LEED experiments are a probe of the surface region. The underlying physics of the elastic interaction between the incident electrons and the solid is the independent electron Schrödinger equation,[40-42]

$$\left(-\nabla^2/2 + V\right)|\psi > = -ihd|\psi > /dt \qquad (3)$$

where the potential operator, V, describes the coulomb interaction, and includes exchange and correlation effects in the mean field, and $|\psi>$ is the wavefunction for the incoming electron. Since Equation 3 is a wave equation, it is reasonable that the electron should exhibit wavelike phenomena, such as reflection, refraction, interference, and diffraction. The elastic backscattering of the LEED electrons is a coherent process, with the backscattered electrons interfering either constructively or destructively to produce a LEED pattern such as the one shown in Figure 3. The bright spots correspond to a direction in space where the electron waves interfere constructively to produce a high intensity of electron flux, and, conversely, the dark regions correspond to destructive interference.

Because LEED is a coherent process, the presence of long-range order in the sample is necessary to observe a diffraction pattern.[8] The symmetry of the LEED pattern is used to determine the symmetry of surface reconstructions. The information on the surface symmetry obtained in a LEED experiment includes the spot intensities, not just the spot positions. While

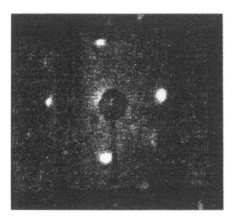

Figure 3
A (1 × 1) LEED pattern from the MgO (001) surface. (From Tran, T.T.; Chambers, S.A. *Appl. Surf. Sci.* 1994, *81*, 161. With permission.)

the translational symmetry of the surface determines spot positions, the point group symmetries, such as the rotation symmetries, affect the intensity of the spots. Under most circumstances the symmetry operations in the LEED pattern are those common to the surface structure and to the incident electron beam, and at normal incidence the spot intensities have the same symmetry as the surface structure. The contents of the unit cell, both the number and positions of the atoms, do not affect the positions of the diffraction spots. It is the spot intensities that can be used to determine detailed information for atomic positions. Once the intensity vs. energy (or I-V) curves have been obtained for each discrete diffraction beam, they must be deconvoluted to obtain information on the positions of the atoms in the surface and near-surface region. Unfortunately, a direct inversion of the data, as is done in X-ray diffraction (XRD), is not possible. In XRD, the data can be directly inverted because the photon undergoes only a single scattering event. The backscattered electrons in a LEED experiment, however, undergo multiple scattering events while in the solid (because of the strong electron-solid interactions) which prevent the direct inversion of the data. In other words, it is not reasonable to make the approximation that a wave scatters from a single site when attempting the solution to the Schrödinger equation (Equation 3). A second reason for the failure of direct inversion techniques is the small amount of data collected in a LEED experiment, typically I-V curves from only a few (six to eight) diffracted beams over a limited range of energy.[8]

Figure 4 shows a comparison between experimental and theoretical I-V curves for the (001) surface of MgO.[43] The experimental curves are averages of symmetrically equivalent beams. The theoretical curves were calculated with a model of the surface consisting of six MgO layers, an outward relaxation of the top layer of ions by 1% of the MgO bond distance, and a differential relaxation of the O anions outward of 5% of the bond distance. The agreement between the experimental and calculated intensity of the LEED beam indicates that the MgO (001) surface is well described by the proposed structural model. Differences between theoretical and experimental curves increase at lower energies, where experimental surface-charging problems were encountered.

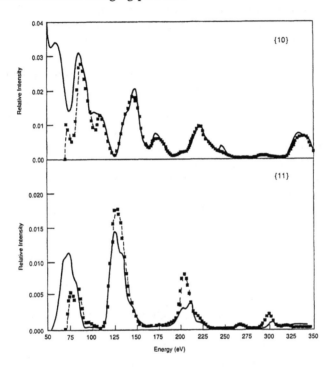

Figure 4

Comparison of the measured and calculated intensity-voltage (I-V) curves for the averaged {10} and {11} beams. The solid lines are the computed intensities for the MgO (001) surface with a 5% rumpling and a 1% relaxation. (From Blanchard, D.L.; Lessor, D.L.; LaFemina, J.P.; Baer, D.R.; Ford, W.K.; Guo, T. *J. Vac. Sci. Tech.* 1991, *A9*, 1814. With permission.)

If the surface geometry is strongly disordered, the direction of the constructive interference from one region of the surface is different from that of other parts of the surface. A diffuse LEED pattern suggests a disordered geometry of surface atoms. Conversely, sharp spots in the LEED suggest a periodic geometry of surface atoms. However, when the LEED pattern of a surface is described as having a particular symmetry, say (1 × 1), this does not necessarily mean that the surface is well ordered with a low density of steps and point defects. It is possible that the 1 × 1 LEED pattern results from subsurface layers of atoms, with the disordered surface layer contributing to a diffuse background. Determining whether a diffuse LEED pattern is a result of such microscopic surface disorder or is a consequence of more macroscopic disorder, such as steps, or the motion of surface atoms due to thermal vibration is appropriately described as qualitative LEED.

In many cases LEED is used to determine the structure of a surface superlattice, which is a relaxed or reconstructed termination of the bulk structure. In some cases, however, the question arises whether the surface unit cell is due to a reconstruction of the clean substrate or whether it is due to a more complicated phenomenon such as an overlayer of different atoms. When addressing such a question, the interpretation of LEED experiments suffers from a lack of atom specificity. Consequently, from an observation of weak fractional order spots, it may not be possible to differentiate between a surface reconstruction and an overlayer of an adsorbate with low atomic number. To overcome this difficulty more-atom-specific techniques, such as XPD or APD,[8] are being applied to mineral surfaces. An additional limitation of surface structural techniques such as LEED arises in epitaxial mineral film growth. The structural sensitivity of LEED and the scanning probe microscopies decreases very quickly for layers beneath the surface.[7] Similarly, bulk-sensitive techniques such as XRD tend to be dominated by signals from deep in the bulk substrate.

The ideal probe for the characterization of thin film overlayers would have element specificity and a probing depth of several tens of angstroms. High-energy APD and XPD fulfill both of these requirements. Since all elements possess unique Auger and photoelectron spectra, it is almost always possible to find a unique high kinetic energy line for one kind of atom in the overlayer that does not interfere with lines from the supporting substrate. The higher kinetic energies increase the penetration depth, allowing more layers to be probed, and simplify interpretation since the probability of multiple elastic scattering events is reduced, and the system can be modeled with a single scattering formalism. This is important, because the solution to the multiple scattering equations would be prohibitively expensive for large systems. These techniques have recently been applied to MgO (001).[44,45]

B. Theoretical Methods

This section describes various computational methods used to determine the atomic structure of insulator and semiconductor surfaces. All of these methods involve evaluation of an expression for the energy of a given geometric arrangement of atoms. The equilibrium surface structure is obtained by variation of the atomic positions in order to minimize the energy expression. The approximations associated with several of the more commonly used energy expressions will be presented and their advantages and disadvantages discussed. The methodologies can be roughly classified into two groups: those that are based on quantum mechanics and those that use empirically determined classic potentials. The quantum mechanical methods include both *ab initio* calculations,[46] which do not include any fitting procedures for the chemical species or geometric arrangement of bonds, and semiempirical tight-binding calculations which rely on the fitting of parameters to known quantities,[16,47] such as those of the bulk material or small molecules. The empirical classic potential models[48] have primarily been developed, and applied to, "ionic" insulators because of the presumed predominance of the coulombic interactions between the ions in the lattice.[49-51]

Theoretical methods have been used to compute the surface energy and the structures of many materials. The surface energy is what determines the shape of crystals and may be used to understand faceting at certain surfaces.[13,52] It is defined as the difference in energy per unit area between a semi-infinite slab with the particular surface termination of interest and the energy of the same number of atoms in the bulk structure. For a cleavage face, such as the rocksalt (001) surface, the energy cost to cleave the crystal (per unit area) is equal to twice the surface energy. The factor of two arises because two faces are created in the cleavage process. The situation becomes more complicated at noncleavage surfaces and surfaces which are defected such that the surface stoichiometry is different from that of the bulk material. For example, a surface may defect to form a periodic array of vacancies. This problem with the surface energy arises because it is no longer sensible to subtract the energy of the stoichiometric bulk material. Instead, it is necessary to make an assumption about the energetics of the final state of the atoms that have been removed from the surface. Usually, it is assumed that the material transforms to its standard state.

Having selected a method for computing the properties of the mineral of interest, the lowest energy structure may be determined by guessing many possible structures and interpolating between them. This algorithm involves taking the derivative of the energy with respect to the lattice positions numerically and is usually slowly convergent. Fortunately, some of the energy functionals presented in this section can be implicitly differentiated with respect to lattice positions, allowing for the direct evaluation of a force on each atom.[53] By moving the atoms in the direction of the force, the structure converges to minimize the energy with relatively much fewer energy function evaluations. Implementations of these methods, which give expressions for atomic forces, have been applied with much greater success to the more complicated surface reconstructions or relaxations.

Different mathematical techniques are used to treat the boundary conditions of a semi-infinite surface. Slab models treat the infinite (in the two dimensions parallel to the surface) two-dimensional nature of the surface properly, but have a finite thickness in the third dimension perpendicular to the surface.[54-56] Green's function techniques treat the semi-infinite nature of the surface exactly (and as a result are the most complex).[57-59] Finally, the cluster methods model the surface with a finite set of atoms.[60] The finite slab and cluster calculations must be checked for convergence with respect to the thickness of the slab and the size of the cluster for each system that is treated. As a general rule, it is low-coordination small bandgap materials, such as the tetrahedrally bonded semiconductors, which require a large number of atoms to obtain an accurate representation of the surface structure. By comparison, the structural details of highly coordinated, large band gap materials are adequately converged with slab thicknesses of only two or three formula units.

This section will emphasize the basic terminology of the commonly used energy expressions; the goal is to enable the reader to comprehend the evaluation of computational results. For the details of a particular method, the reader is referred, primarily, to the review literature and to graduate-level texts.

1. Quantum Mechanical Methods

All of the quantum mechanical methods considered in this section begin by making the Born–Oppenheimer approximation, which assumes that the motion of the electrons is fast compared with the motion of the nuclei, or ion cores.[61] In this way the motion of the nuclei may be neglected when computing the electronic structure, in effect, separating the electronic and nuclear degrees of freedom. In this approximation the total energy of a system can be written as the sum of an electronic and a nuclear contribution, each of which depends on the positions of the atoms in the system.

All quantum mechanical methods used to compute properties of mineral surfaces rely on simplifying assumptions to the full quantum mechanical many-body problem. In the many-body problem the wavefunction depends upon the coordinates of all of the electrons in the system. This problem is often simplified by recasting it as an effective single-body problem, in which the total wavefunction for an n-electron system is expressed as a product of n-wavefunctions, $\psi_n(r)$ (one for each electron in the system).[62] This leads to a system of n one-electron Schrödinger equations,

$$\hat{H}\psi_n(r) = E_n\psi_n(r) \qquad (4)$$

where E_n are the one-electron eigenvalues of $\psi_n(r)$. Each of these n-wavefunctions depends only upon the coordinates of a single electron (hence the name *one-electron wavefunctions*). Although it is possible, in a mathematical sense, to exactly rewrite the many-body physics in terms of a one-electron-like Hamiltonian, there is no formalism for achieving this. Instead, it is necessary to make physical approximations, and it is in the nature of these approximations to the effective single-particle potential (resulting from the one-electron wavefunctions) that the various quantum mechanical methods differ.

These methods have many features in common, as well. For example, the one-particle wavefunctions typically are expressed as a linear combination of some set of basis functions.[63] The idea is to approximate the spatial character of the one-electron wavefunctions with as few functions as possible to minimize the computational expense. The proper choice of basis set is one of the critical aspects of practical computations. Many different sets of basis functions have been used. Considerable effort has gone into investigations on the construction of basis sets, and the literature available on this subject is enormous.[63]

Quantum mechanical computations can be all-electron (i.e., computations which take explicit account of every electron in the system) or valence electron only. The valence-electron-only calculations assume that the valence-shell electrons dominate the interesting chemical and physical processes in the system. A further approximation is then made to the single-particle potential as these valence electrons are then considered to move in an effective, or pseudo-, potential that includes the interactions of the nuclei and core electrons. The pseudopotential can be empirical, semiempirical, or *ab initio*, depending upon the way in which it is derived. More-detailed discussions on the construction and use of pseudopotentials are available in the literature.[64-66]

As stated previously, the differences between the various quantum mechanical methods lie in the way in which the many-body electron–electron interactions are reduced to effective one-particle, or one-electron, interactions. Perhaps the most straightforward way of illustrating these differences is to partition the one-electron electronic energy into component parts as follows:

$$E_{EL}(R) = E_{KE} + E_{EL\text{-}ION}(R) + E_H + E_X + E_C \tag{5}$$

E_{KE} is the one-electron kinetic energy, $E_{el\text{-}ion}$ is the electron-ion attraction, and the remaining terms in Equation 5 arise from the electron–electron interactions. E_H represents the two-electron (because it depends upon the coordinates of two electrons) Hartree (or coulomb) interaction; E_X represents the two-electron exchange (or Fock) interaction, which arises from the indistinguishability of the electrons, which must be reflected by the wavefunctions; and E_C is the electron correlation energy. The correlation energy corrects for the neglect of the correlated nature of the electron motion in recasting the many-electron problem into an effective one-electron problem.

In the following subsections, the way in which the SCF-LCAO (self-consistent field, linear combinations of atomic orbitals), density-functional, and tight-binding methods determine each of the terms in Equation 5 will be presented, discussed, and contrasted. It is important to note that

the presentation in the following sections will be for the *ab initio* variant
of the SCF-LCAO and density-functional methods. That is, the interactions that are included by the method will be assumed to be computed
explicitly. Empirical and semiempirical variants, for which some subset
of the interactions (i.e., Hamiltonian matrix elements) in the system are
either neglected or parameterized, also exist for each method type. The
details of these variants can be found in the literature.[67-69]

a. SCF-LCAO Methods

Typically, SCF-LCAO methods simplify the computation of the electronic energy by restricting the one-electron wavefunction to a single
Slater determinant,[70] assuring that the wavefunction is antisymmetric and
describes electrons as indistinguishable particles. It also results in the
neglect of electron correlation effects (other than correlation in the sense
that the wavefunctions obey the Pauli exclusion principle).[70] This level of
calculation is called the Hartree–Fock method.

At the Hartree–Fock level, then, Equation 5 reduces to

$$E_{EL}(R) = E_{KE} + E_{EL\text{-}ION}(R) + E_H + E_X \qquad (6)$$

where all of the terms on the right-hand side of Equation 6 are now
evaluated explicitly. To do this, the one-electron wavefunctions are computed using the effective one-electron Hamiltonian operator (referred to
in this context as the Fock[71] operator) obtained from the variational principle. The important thing about the Fock operator is that the one-electron
wavefunctions, $\psi_i(r)$, appear in its definition, requiring an iterative, or
"self-consistent" solution.

To summarize, the SCF-LCAO methods neglect electron correlation
effects by limiting the wavefunction to a single Slater determinant. The
remaining energy terms of Equation 6 — the kinetic, electron-ion, coulomb, and exchange energies — are then evaluated explicitly.

b. Density Functional Methods

The density functional theory (DFT) methods differ from the SCF-
LCAO methods in that the electron density, $\rho(r)$, is used as the variable
of interest.[72,73] It has been shown[74] that the ground-state electronic energy,
for a given external (or nuclear) potential is a functional of the electron
density. This is a surprising result for a quantum mechanical system,
where wavefunctions are the variable that usually governs physical properties. The most commonly used approach is the Kohn–Sham method.[75]
In this approach, the kinetic energy is replaced by the kinetic energy of a
system with no electron–electron interactions, $T_s[\rho(r)]$, but at the same
ground-state electron density of the original system (with electron–electron interactions). In this way the (newly defined) kinetic and electron–ion
interaction energies can be computed from the one-electron eigenvalues,

e_i, of a system of noninteracting electrons moving in the new external potential, $v_s(r)$, of the noninteracting system. The electron density is computed from the associated one-electron eigenfunctions, $\psi_i(r)$, and the Hartree energy (also referred to as $J[\rho(r)]$) is computed as in the SCF-LCAO methods.

This leaves only the exchange and correlation energies to be evaluated, along with the correction needed to account for the neglect of the electron–electron interactions in the kinetic energy computation. All of these terms are collected together as an effective one-electron term, referred to as the *exchange–correlation* energy. The specification of the exchange–correlation potential requires is the most difficult aspect of applying the Kohn–Sham DFT method. Many approaches have been taken in approximating this exchange–correlation functional, ranging from simple estimates based upon a uniform-density electron gas model[75,76] to sophisticated treatments of the nonlocal nature of the interactions.[77,78]

To summarize, the DFT methods use the electron density, rather than the wavefunction, as the system variable of interest. It is possible to exactly transform the many-body problem into an effective one-electron problem. In this effective potential, the kinetic energy is computed for a system of noninteracting electrons. The kinetic energy correction (due to the fact that real electrons interact) is then lumped together with the exchange and correlation interactions into an effective, one-electron exchange–correlation potential. The practical implementation of this method, based on DFT-LDA, is different from the Hartree–Fock method described above in that exchange and correlation are both treated using a physical approximation, whereas the Hartree–Fock method treats the exchange interaction exactly, but neglects electron correlation.

c. Tight-Binding Methods

The spirit of the empirical tight-binding methods[79] is simple: none of the terms in Equation 5 is evaluated explicitly. Instead, the Schrödinger equation (Equation 4) is recast in matrix form, and the elements of the Hamiltonian matrix are treated as adjustable parameters that are derived from experimental information about electrons in the bulk material, or the results of *ab initio* calculations for the bulk system. These parameters are then assumed to be transferable for use in computing the properties of surfaces. The range of these interactions is usually assumed to be nearest neighbor or next-nearest neighbor only, and the interaction matrix elements are assumed to have some parametric dependence upon the internuclear separation d (commonly a d^{-2} dependence for sp-bonded semiconductor systems).[80,81] The assumption of transferability (from bulk to surface) for the Hamiltonian matrix elements will be valid provided that the charge density and character of the electron bonding at the surface are not significantly different from the bulk. The success of this assumption for the covalently bonded semiconductor systems and the more "ionically" bonded insulators is an *a posteriori* justification for its use.[2-4]

The most important aspect of the empirical tight-binding methods, however, is that the electron–electron interactions are never computed, but are included empirically through the parameterization of the Hamiltonian matrix elements. This has consequences in the way the total energy is evaluated, since the electronic energy can only be expressed as the sum of the one-electron eigenvalues. This sum overestimates the electronic energy by double counting the Hartree and exchange interactions.[62] Because this extra energy cannot be explicitly accounted for, it is usually lumped together with the nuclear–nuclear repulsion, and the total energy is rewritten as the sum of an electronic term, E_{bs}, the sum of the occupied one-electron eigenvalues (commonly termed the *band structure* energy) and a pair potential (i.e., a potential that depends only upon the pairwise interactions between the atoms in the system) representing the nuclear repulsion and electron double-counting terms. Many forms have been proposed for this pair potential, the details of which can be found in the literature.[82-85]

To summarize, in the empirical tight-binding methods none of the interactions in the system is computed explicitly, but is included empirically, through the parameterization of the Hamiltonian matrix and a pair potential. The principal advantages of the method are that a greater number of atoms can be treated than with DFT or SCF-LCAO techniques and the structural relaxation of a mineral surface can be performed, since forces are readily obtained.[53]

2. Classic Potential Methods

A quantum mechanical treatment of a surface involves a computation of the electronic wavefunctions or a mathematical quantity such as a Green's function, which is closely related to the wavefunctions. The necessity for the computation of the wavefunctions arises in the evaluation of the kinetic energy of the electrons. The kinetic energy cannot be computed in any obvious way from a convergent series of two- or three-body interactions that are used to construct classic models. The potential energy, on the other hand, is due to the two-body coulomb interaction and may be described with a classic potential. Since the kinetic energy depends upon the second gradient of the wavefunction, it is extremely sensitive to small changes in the positions of atoms. For this reason quantum mechanical models of materials surfaces are more accurate than classic methods, but are often computationally prohibitive because of the necessity of computing wavefunctions. A quantum mechanical calculation scales with the cube of the number of atoms, whereas classic models scale only linearly with system size.

The empirically determined classic potential models are, by far, the most widely used in the study of "ionic" insulator surface and interface structure.[48] Typically, these potentials are central force, two-body potentials composed of three parts; a coulombic interaction, a short-range interaction, and an interaction that accounts for the polarizability of the atoms. In some cases (where directional bonding is important) a three-body term describing bond angle distortions has been found to be important and added to the potential.[48]

The coulomb potential is simply the interaction between ions, and the short-range potential is typically given by the Buckingham potential which comprises a Born–Mayer repulsive and a van der Waals attractive term.[48] The empirical constants in this potential are then fit to either experimental data or the results of more sophisticated computations. Finally, the polarizability of the atoms is usually represented by a shell model,[86] in which the atom is modeled as a spherical shell of negatively charged electron density that is harmonically coupled, with a force constant k (also empirically determined), to a positively charged ionic core. The polarization occurs when a differential displacement, W, occurs between the core and the charged shell. The long-range contribution to the associated coulomb potential is problematic since the Madelung sum[49-51] is only conditionally convergent. That is, the answer you get depends upon how you truncate the summation. This can be avoided, however, by using the summation techniques of Ewald[87] and Parry.[88,89] Nevertheless, the computation of the coulombic interactions is easily the most computationally intensive process for these models.

3. Comparison of the Methods

Of all the methods described in this section the use of empirical classic potentials is the most widespread.[48] The potential is sufficiently simple so that large numbers of atoms (on the order of hundreds) may be treated. As a result, these potentials have been used to study a wide variety of complicated defects, grain boundaries, interfaces, stepped surfaces, and adsorbed surfaces.[48] Moreover, these potentials have been used in molecular dynamics simulations to investigate the time-dependent nature of these systems. The primary disadvantage of these potentials is their obvious neglect of the quantum mechanical nature of chemical bonding. Moreover, the sensitivity of the computed results to the form and parameterization of the potential makes the general applicability of these potentials questionable.

The empirical tight-binding method is based upon quantum mechanics and is also computationally efficient. The efficiency enables routine study of the dynamics of systems with on the order of 100 atoms. It is also formulated in terms of atomiclike orbital interactions, so that the description of surface relaxations and reconstruction can be understood in terms of how these interactions differ from the bulk material at the surface. The major disadvantage of this method is that it is parameterized, and hence its results must be calibrated against more rigorous methods. In addition, the investigation of new systems requires the determination of new basis/parameter sets.

The remaining methods, semiempirical and *ab initio* local-density functional and SCF-LCAO, are all self-consistent. As such, they are computationally intensive because of the need to explicitly compute the electron–electron interactions. This typically results in the treatment of systems with a limited number of atoms (on the order of tens of atoms). The advantage to these methods is that, while computationally more difficult, they provide a first-principles determination of the surface structure independent of possible parameterization biases.

III. STRUCTURE OF MINERAL SURFACES

As stated in the introduction, one of the primary factors in determining the nature of a surface relaxation or reconstruction is the surface topology, or atomic connectivity. The presentation in this section, therefore, is organized by crystal lattice type. The most commonly studied surfaces of these lattices will be examined and their structure, as determined both experimentally and theoretically, discussed in the context of the principles set forth in the introduction.

A. Zincblende

The majority of the III-V and II-VI binary minerals crystallize in the zincblende lattice. The atoms in this lattice are fourfold coordinated, with both the anions and cations sp^3 hybridized and in a tetrahedral bonding environment. At the (110) cleavage surface, illustrated in Figure 1, the atoms are threefold coordinated, bonded to two neighbors in the surface plane and one in the plane immediately below the surface (a "backbond").[2-4] This cleavage surface is autocompensated for ZnS (Principle 1 in Section I.C), having its anion-derived surface states completely filled and its cation-derived surface states completely empty. In bulk zincblende each Zn atom contributes two valence electrons to the four bonds, or $1/2$ electron per bond. Each S atom contributes six valence electrons to four bonds, or $3/2$ electron per bond. Each ZnS unit cell then has two electrons per bond ($1/2 + 3/2$) and is stable. The truncated bulk (110) cleavage surface comprises zigzag rows of ZnS, and the surface unit cell comprises one ZnS formula unit. The surface atoms are now threefold coordinated with one dangling bond. In each unit cell, therefore, there is one Zn dangling bond (with $1/2$ electron) and one S dangling bond (with $3/2$ electron). By transferring the $1/2$ electron from the Zn dangling bond to the S dangling bond, the anion dangling bond (and anion-derived surface state) is completely filled and the cation dangling bond (and cation-derived surface state) is completely empty. The surface is autocompensated, charge neutral, and stable. The surface is also insulating (Principle 3) since an energy gap exists between the anion- and cation-derived surface states.

The question to ask now is: Can the surface lower its energy by rehybridizing the dangling bond charge density? (Principle 2). The answer to this question is: Yes, it can. In the truncated bulk geometry, the surface cations have three nearest neighbors and three electron pairs, yet remain sp^3 hybridized. The surface anions also have three nearest neighbors, but have four electron pairs. The surface energy can be lowered by rehybridizing the cation dangling bond charge density to sp^2, placing the surface cations in trigonal planar conformation, and rehybridizing the anion dangling bond charge density to p^3, placing the surface anions in trigonal pyramidal conformation.[2-4]

The issue now is whether or not changing the local conformation of the surface atoms significantly distorts the near-neighbor bond lengths (Principle 4). Because each surface atom is threefold coordinated, the topology of the surface is such that the surface atoms have one unrestricted degree of freedom and, therefore, can move into these new conformations in a way which conserves near-neighbor bond lengths. To do this, each of the surface zigzag chains tilt, with the surface cations moving down toward the bulk crystal and the surface anions moving up and out of the surface plane, as shown schematically in Figure 5.[2-4]

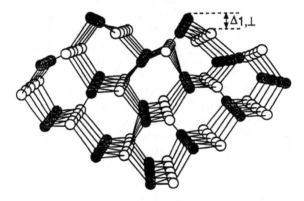

Figure 5
Illustration (side view) of the relaxed zincblende structure (110) surface. The primary structural parameter describing the surface relaxation, the perpendicular shear of the surface layer anions and cations ($\Delta_{1,\perp}$) is indicated. The relaxation in the second layer is greatly exaggerated for visibility. (From Duke, C.B.; Wang, Y.R. *J. Vac. Sci. Tech.* 1988, *B6*, 1440. With permission.)

Up to this point we have considered the class of zincblende (110) surfaces in general. The remarkable thing about these arguments, however, is that they would predict that *every* zincblende (110) surface would exhibit the *same* surface relaxation regardless of the chemical constituents which occupied the lattice, namely, a tilt of the surface zigzag chains so that the surface cations were sp^2 hybridized and the surface anions were p^3 hybridized. If the cation moved in a perfect trigonal planar conformation while precisely conserving near-neighbor bond lengths, this would correspond to a tilt of approximately 35°. It has been shown that the experimental and theoretical determinations for a variety of zincblende (110) surfaces exhibit a surface relaxation characterized by a surface chain tilt of 29 ± 3°. The actual tilt of the chains is smaller because the cation does not move into a perfect trigonal planar conformation and because near-neighbor bond lengths are not rigorously conserved. This is a remarkable result because it implies that the connectivity of the surface is the dominant factor determining surface relaxations and that the chemical nature of the constituent atoms in the lattice is of secondary importance.[2-4] The zincblende (110) surface, as a result, has been said to exhibit a "universal" relaxation, characteristic of its topology.[2-4] This concept has recently been extended to the entire surface potential energy function, including not only a common potential minimum (and therefore a common surface structure), but also a common potential curvature (and therefore a common vibrational, or phonon, structure).[90]

The structure of the ZnS (110) surface has been investigated using LEED. Dynamic LEED multiple scattering calculations found that a bond length conserving rotation of Zn and S in the top layer of 26° provided the best fit to the LEED data.[5] This experimental result is in quantitative agreement with the surface topology predicted from the simple argument outlined above. One of the more exciting, fundamental aspects of this work is the demonstration that the concept of a "universal" surface relaxation, characteristic of all zincblende (110) surfaces, extends to "ionic" materials such as ZnS.[5] The concept of a characteristic surface relaxation — that is, a relaxation that is not dependent upon the chemical nature of the atoms in the lattice, but simply depends upon the connectivity of the lattice, and the bulk lattice constant — first emerged from the study of the zincblende compound semiconductors.[5] From experimental surface structural analyses (via LEED intensity analysis), it was observed that the surface structural parameters describing the relaxation of all of the III-V and II-VI zincblende (110) surfaces were the same, within a scaling constant that depended only upon the bulk lattice constant of the material.[2-4] This implied that the topology of the surface was the primary factor governing the surface relaxation and that the chemical nature of the lattice atoms was of secondary importance. This is a remarkable result considering the vastly different small molecule coordination chemistry of these species. Even more remarkable is the fact that these concepts, developed for materials traditionally considered "covalent," can be successfully applied to ionic substances. Figure 6 shows the results of experimental and theoretical studies for a variety of zincblende (110) surfaces, illustrating this concept of a universal relaxation that depends primarily upon the bulk lattice constant of the material. As we shall see in the following section, this concept of each surface having a relaxation characteristic of its topology also extends to the wurtzite faces.

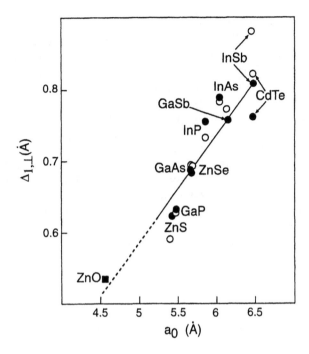

Figure 6
Illustration of the scaling of the principal surface structural parameter, $\Delta_{1,\perp}$, describing the zincblende structure (110) surface relaxation as a function of the bulk lattice constant for values of $\Delta_{1,\perp}$ determined by tight-binding total energy computations (filled circles) and LEED intensity analyses (open circles). (From Skinner, A.J.; LaFemina, J.P. *Phys. Rev. B.* 1992, 45, 3557. With permission.)

B. Wurtzite

The most studied mineral that occurs in the wurtzite structure is zincite (ZnO), and the discussion of wurtzite surfaces is centered around this material. The surface chemistry of ZnO has been studied for many years because of its catalytic importance.[91-94] Geologically, high-pressure ZnO rocksalt is a postspinel phase, important to understanding Earth's lower mantle.[95] It is regarded as the prototypical ionic semiconductor, and there have been many fundamental studies of the physics and chemistry of ZnO surfaces.[6] In fact, the cleavage faces of ZnO are, along with the Mg (001) surface, the best-characterized oxide surfaces to date.

The $(10\bar{1}0)$ and $(11\bar{2}0)$ wurtzite surfaces undergo their own relaxations characteristic of their surface topology; involving bond-length conserving motions of the surface atoms; and driven by a surface-state lowering mechanism similar to that of the zincblende (110) surface.[1-4,96] We shall consider the wurtzite structure ZnO $(10\bar{1}0)$ and $(11\bar{2}0)$ cleavage surfaces as well as the (0001) polar surfaces.

1. Wurtzite (10$\bar{1}$0) Surface

The truncated bulk wurtzite (10$\bar{1}$0) cleavage surface consists of rows of surface anion-cation dimers (Figure 7). Each surface atom is threefold coordinated, bonding to one atom in the surface layer and two atoms in the layer immediately beneath the surface layer. The surface is autocompensated, Principle 1 in Section I.C, and, moreover, the topology of this surface allows a bond-length conserving motion of the surface atoms, shown schematically in Figure 7 and pictorially in Figure 8. The definition of the independent surface structural parameters describing the relaxation is given in Figure 7.[97] The values of these parameters, determined by tight-binding, total-energy computation,[97] are listed in Table 1, along with the values determined by LEED intensity analysis.[98,99] Clearly, the agreement is quantitative. Hartree–Fock[100] and density functional theory calculations,[101] by comparison, predict a much smaller tilt angle for the surface dimer, 2.31° and 3.59°, respectively, compared with the experimental result of 11.5 ± 5°.

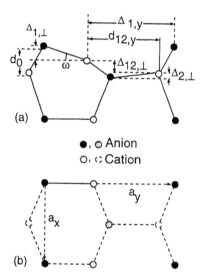

(a)

●, ○ Anion
○, ○ Cation

(b)

Figure 7
Definition of the surface structural parameters defining the relaxation of the wurtzite structure (10$\bar{1}$0) surface, (a) side and (b) top views. (From Duke, C.B.; Lessor, D.L.; Horsky, D.N.; Brandes, G.; Canter, K.F.; Lippel, P.H.; Mills, A.P.; Paton, A; Wang, Y.R. *J. Vac. Sci. Tech.* 1989, A7, 2030. With permission.)

The driving force for this relaxation, as mentioned previously, is similar to that for the zincblende (110) surface, namely, the energy lowering resulting from the redistribution of the dangling bond charge density. This similarity is understood by considering the topology of the surface: threefold coordinated atoms with the ability to undergo bond-length conserving motions.

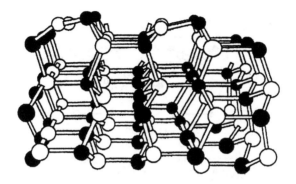

Figure 8
Perspective illustration (side view) of the relaxed wurtzite structure ($10\bar{1}0$) surface. (From Duke, C.B.; Wang, Y.R. *J. Vac. Sci. Tech.* 1988, *B6*, 1440. With permission.)

TABLE 1.

Comparison of the Structural Parameters for the ($10\bar{1}0$) Cleavage Surface of ZnO as Determined by Tight-Binding Total-Energy (TBTE) Computations and LEED Intensity Analyses

	$\Delta_{1,\perp}$	$D_{2,\perp}$	$d_{12,\perp}$	$\Delta_{1,y}$	$d_{12,y}$	d_o	Ref.
ZnO (TBTE)	0.57	0.06	0.53	3.37	2.97	0.94	97
(LEED)	0.4 ± 0.2	0.00	0.54	3.24	2.61	0.94	99

[a] The structural parameters are defined in Figure 7. Distance units are angstroms.
From LaFemina, J.P. *CRC Crit. Rev. Surf. Chem.* 1994, *3*, 297. With permission.

2. Wurtzite ($11\bar{2}0$) Surface

The wurtzite ($11\bar{2}0$) cleavage surface, shown in Figures 9 and 10, has four atoms per unit cell. The surface comprises anion–cation chains, with each surface atom threefold coordinated, bonding to two atoms in the surface layer and one in the layer directly beneath the surface. This surface also contains a glide-plane symmetry that is preserved (as observed by missing spots in the LEED pattern) in the (1 × 1) relaxation.[97,99,102,103] Early LEED analysis of ZnO ($11\bar{2}0$) concluded that the surface was unrelaxed.[103] Tight-binding total-energy computations[97] and subsequent LEED and LEPD analyses on other II-VI ($11\bar{2}0$) surfaces[104,105] have found that the surface relaxes in such a way as to preserve the glide-plane symmetry, yet still undergo large atomic motions (see Figure 10). The surface is autocompensated (Principle 1 in Section I.C), and because of the topology this surface allows a bond-length conserving motion of the surface atoms.

The mechanism for this surface relaxation is, again, identical to that for the zincblende (110) and wurtzite ($10\bar{1}0$) surfaces. The large atomic motion relaxation is driven by the energy stabilization associated with the redistribution of the dangling bond charge density, Principle 2 in Section

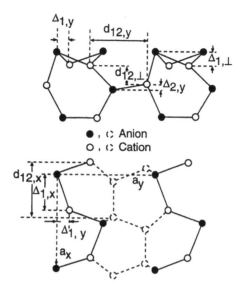

Figure 9

Definition of the surface structural-parameters defining the wurtzite structure (11$\bar{2}$0) surface relaxation, (a) side and (b) top views. (From Duke, C.B.; Lessor, D.L.; Horsky, D.N.; Brandes, G.; Canter, K.F.; Lippel, P.H.; Mills, A.P.; Paton, A; Wang, Y.R. *J. Vac. Sci. Tech.* 1989, *A7*, 2030. With permission.)

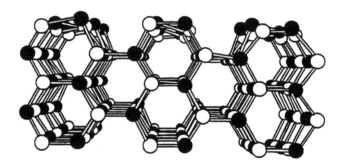

Figure 10

Perspective illustration (side view) of the relaxed wurtzite structure (11$\bar{2}$0) surface. (From Duke, C.B.; Wang, Y.R. *J. Vac. Sci. Tech.* 1988, *B6*, 1440. With permission.)

I.C. The relaxation is allowed, because the topology of the surface permits bond-length conserving motions to occur, Principle 4 in Section I.C, thereby avoiding any energy cost associated with the distortion of local bonding environments.

3. Wurtzite (0001) and (000$\bar{1}$) Surfaces

The two polar surfaces of ZnO, shown in Figure 11, comprise double layers of tetrahedrally coordinated O and Zn atoms with the surface normal parallel to the c-axis of the hexagonal cell. The Zn-terminated surface is denoted (0001) and the O-terminated surface (000$\bar{1}$), and these surfaces are known to exhibit different chemical reactivities.[106] Since both of these surfaces are polar, they are not autocompensated in their truncated bulk conformation. As explained in Section I, because of this the polar surfaces are expected to reconstruct and satisfy the autocompensation principle. Determination of surface structure using LEED for different surface preparations has led to conflicting results.[107] A polished and annealed surface was found to undergo a surface reconstruction with the (1 × 1) LEED pattern that transformed into a stable ($\sqrt{3}$ × $\sqrt{3}$)R30° when annealed above 873°C.[107] Other workers, using XPD, found no such reconstruction after annealing to similar temperatures.[108] STM results confirm that polished and annealed terraces of ZnO (0001) and (000$\bar{1}$) are atomically flat in a direction perpendicular to the surface, but lateral atomic resolution was not achieved in the experiments.[109] From a theoretical perspective, because of the large driving force for reconstruction at the nonautocompensated polar surfaces, it is surprising that there is not clearer experimental evidence for reconstruction. The wurtzite (0001) surface is similar to the zincblende (111) surface, which was discussed in Section I. The Ga-terminated (111) surface of zincblende GaAs autocompensates by defecting, with one Ga in four leaving the surface.[18] The As-terminated ($\bar{1}\bar{1}\bar{1}$) surface forms an adatom structure with As trimers autocompensating the surface.[18] Sufficient energy is derived from the cleavage process to overcome any kinetic barriers (Principle 5 in Section I.C) to the reconstruction on the GaAs polar surfaces. The ZnO polar surface may be metastable upon cleavage because of a large activation barrier to reconstruction (Principle 5) or may be stabilized by a more complicated interaction. No explanation of this phenomena has been presented in the literature.

The chemical bonding in ZnO is more ionic compared with the relatively covalent bonding in GaAs. The electrons in the ionic model of a material are associated with particular ions that make up the lattice, rather than with bonds between neighboring lattice sites. If we assume that the metal electrons are completely transferred to the anion (classic ionic bonding), then the concept of dangling bonds and dangling bond charge density is not appropriate. Instead, the surface anions retain the transferred electrons. While it may be useful to understand the role of ionicity for explanation of many physical or chemical phenomena, it is not necessary to distinguish between covalent and ionic solids when considering the autocompensation principle. All that is required to determine whether or

WURTZITE (0001)

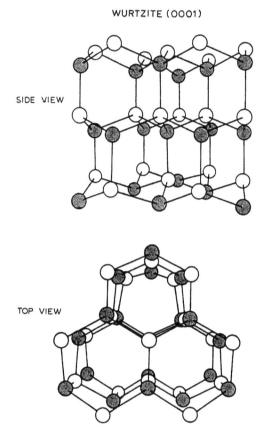

Figure 11
Top and side view schematic illustration of the truncated-bulk wurtzite structure polar (0001) surface. The surface atoms can either be cations — cation-terminated, usually designated (0001) — or anions — anion-terminated, usually designated (000$\bar{1}$). (From Duke, C.B. *Surface Properties of Electronic Materials*. D.A. King, D.P. Woodruff, Eds. Elsevier: Amsterdam, 1988; 69. With permission.)

not a surface is autocompensated is that the number of electrons per bond is the same as the bulk material. Theoretical work at the (111) polar surface of MgO found that the energy of this surface was much larger than at the nonpolar surfaces, as described in the next section.[110] Each atom has three dangling bonds at the (111) rocksalt surface, compared with the single dangling bond per atom in the wurtzite structure.

The symmetry of the polar surface dictates that any relaxation of the truncated bulk surface atoms should occur perpendicular to the plane of the surface. A LEED analysis has suggested that Zn atoms at the (0001) surface relax into the surface by 0.2 to 0.3 Å,[107] whereas angle-scanned

XPD of the polar (0001) surface strongly disagrees with this result and favors bulk termination.[108] LEED data for the (000$\bar{1}$) surface suggest termination similar to the bulk.[107]

C. Rocksalt

Discussions of the surface of rocksalt structure minerals are dominated by the (001) surface (shown in Figure 12), whose surface energy is typically significantly lower than the (110) or (111) surfaces. The rocksalt (001) and (110) surfaces contain anions and cations in equal proportions and are autocompensated. The (111) termination of the rocksalt structure gives rise to a polar surface which is not autocompensated. The cation termination is referred to as (111) and the anion termination ($\bar{1}\bar{1}\bar{1}$). Since the (111) faces are not autocompensated, the surface energy will be large according to Principle 1 in Section I.C. In the case of MgO, the (111) surfaces reconstruct to form (001) facets when the sample is annealed.[111] The (001) and (110) surfaces have one and two dangling bonds per surface atom, respectively. The single dangling bond makes the surface energy of the (001) surface substantially lower than that of the (110) and makes the (001) the preferential cleavage surface.[6]

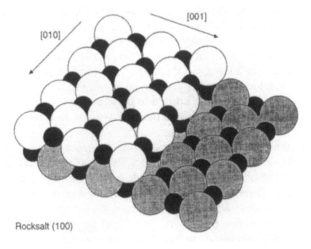

Rocksalt (100)

Figure 12
Illustration of the ideal rocksalt (001) surface with a {001} step to another (001) terrace. Large circles are anions, small circles are cations. The degree of shading indicates the depth of the atoms below the surface plane. (From Henrich, V.E. *Rep. Prog. Phys.* 1985, *48*, 1481. With permission.)

The alkaline earth oxides, alkali halides, transition-metal oxides, and sulfides crystallize in the rocksalt structure. The bonding in rocksalt structure crystals is generally ionic in nature. A significant component of the binding energy of an ionic compound is derived from the Madelung energy, which for an array of point ions in the rocksalt structure is lower than the Madelung energy for the tetrahedrally bonded networks, such as zincblende and wurtzite. We can analyze the stability of rocksalt structure surfaces with the principles outlined in Section I.C. The approach is slightly different from that often taken in the literature, where the formal ionic charges of an ionic compound are used, rather than the number of electrons in bulk bonds. As described in the previous section, the advantage of the approach presented here is that it is not necessary to know whether the bonds are ionic or covalent in nature. In the divalent metal oxides, the O atom contributes one electron to each bond, and the metal atoms, such as Mg or Ca, contribute $1/3$ electron to each bond; making a total of $4/3$ electron per bond in the bulk material. It is essential to appreciate that this does *not* conflict with the idea that each electronic level is occupied with two electrons. If we require that the surface anion dangling bond contain the same number of electrons as the bulk material, the series of ionic materials can be understood in the same conceptual framework that has been developed for covalent materials (two electrons per bond). This is clearly illustrated by the (001) surface, where each surface metal atom has a surplus of $1/3$ electron and each oxygen atom also has a surplus of one electron. To satisfy Principle 3 in Section I.C, the $4/3$ electron surplus occupies the anion-derived orbital, as in the bulk material. By comparison, the same conclusion could be reached using the ionic model in an ionic material such as MgO. The formal charges of Mg and O are 2+ and 2–, respectively, and so Mg donates $1/3$ electron, while O accepts $1/3$ electron per bond. At the (001) surface the extra $1/3$ electron on the Mg ion is accepted by a surface O to complete the sp^3 shell. While this analysis may be useful in materials that are known to be ionic, it may be less useful in more complicated materials, where mixed covalent and ionic bonding is present.

The symmetry of the (001) surface requires surface relaxations to occur perpendicular to the plane of the surface. This relaxation must be a small percentage of the bond length, by Principle 4, but a rumpling of the surface is likely due to different polarizabilities of the anion and cation. The energy cost associated with local strains quickly offsets any energy gain from the rehybridization of the surface charge density.

Experimental studies of rocksalt crystals have shown that surface relaxation of the (001) surfaces of MgO,[112-121] CaO,[122] NiO,[123,124] MnO,[125] EuO,[126,127] and CoO[128] are typically less than 5% of the bulk lattice constant and that surface rumpling is very small. A great number of papers and review articles have dealt with the surface structure of rocksalt alkali halides and oxides, and so we will not replicate these discussions in this review.[5,6] Without exception, the weight of the experimental evidence suggests that the relaxation and rumpling are very small; typically 1 to 2% of the bulk lattice constant. That is not to say that the spread of experimental and theoretical results is not large. Indeed, in the case of experiments on MgO, reported values of rumpling vary between 1 and 8% and relaxation values are between –15 and 3%.[6] (See Table 2.) The range of theoretical values is even larger.[5,6] More-recent LEED[43] and XPD[44,45] experiments, however, have settled on the smaller values and are in agreement with the set of principles from Section I.C.

TABLE 2.

Comparison of the Structural Parameters for the Relaxed MgO (001) Surface

R	C	Method (year)	Ref.
+3	–2	TBTE[a] (1991)	194
+1	0	*ab initio* HF[b] (1986)	195
+5 to +9	0	Shell Model (1978)	196
+2 to +3	–2 to 0	Shell Model (1979)	197
+11	+1	Shell Model (1985)	198
+3	–0.5	Shell Model (1985)	199
0 to +5	–3 to 0	LEED[c] (1976)	113
0 to +5	–3 to 0	LEED (1979)	114
+2 ± 2	0 ± 1	LEED (1982)	115
0	0 ± 2	LEED (1983)	116
0 to +5	0 ± 1	LEED (1991)	43
0	0 to +3	RHEED[d] (1985)	118
+8 ± 1	—	He Diffraction (1982)	119
+0.5 ± 1	–15 ± 3	ICISS[e] (1988)	200
0 to +5	0	XPD[f] (1994)	44
0	0	XPD (1994)	45

Note: R is the rumple, or differential displacement of the surface anion and cation in the direction perpendicular to the (001) surface. A positive R indicates the surface anion is displaced outward from the surface while the surface cation is displaced inward toward the surface. C is the contraction (if negative) or expansion (if positive) of the first interlayer spacing. Units are percent of the ideal interlayer spacing (2.11 Å).

[a] Tight-binding total-energy computation.
[b] *ab initio* Hartree–Fock computation.
[c] Low-energy electron diffraction intensity analysis.
[d] Reflection high-energy electron-diffraction.
[e] Impact-collision ion-scattering spectroscopy.
[f] X-ray photoelectron diffraction.

The current focus of research with rocksalt minerals leans toward the growth of high quality (100) surfaces, where the number of defect sites is minimized. For materials such as MgO this is surprisingly difficult to achieve. Bulk oxide crystals that are subject to sputter/anneal cycles in the process of surface preparation are often substantially defected because of preferential sputtering that is not readily remedied by subsequent annealing. It is well known that defects participate in many kinds of surface chemical reactions on mineral surfaces;[6] however, the presence of multiple defects on a surface precludes the possibility of determining the effect(s) that a given defect may have on chemistry. It is the understanding of the defect chemistry that drives research to fabricate higher quality mineral surfaces. Molecular beam epitaxy is being used to grow MgO on numerous substrates for this purpose.[129]

The atomic and electronic structure of the galena (PbS) (100) surface has been studied using STM and electron tunneling spectroscopy (ETS). PbS is a 0.29-eV bandgap semiconductor, and sufficient current can be passed through the material to enable the STM experiment. STM images of a freshly cleaved (100) surface show only one kind of atom under negative bias. A negative bias corresponds to electrons flowing from the sample to the STM tip. By Principle 3 in Section I.C, this site should correspond to the sulfur atoms. The positive bias image shows two types of sites, distinguished by different intensities, in the rocksalt structure. In the positive bias mode, electrons are flowing from the tip to the (empty) conduction band of the sample, and so the Pb atom also appears in the image. Surprisingly, it has been shown that the higher intensity peak in this image corresponds to the same atom as the peak that is observed under a negative bias. Further work is required to fully interpret this result.

D. Rutile Surfaces

The tetragonal rutile crystal structure consists of octahedrally coordinated cations and threefold-coordinated anions. The most widely studied rutile oxides are rutile, TiO_2, and cassiterite, SnO_2: materials with many technological applications including catalysis[130] and chemical sensor applications.[131] Consequently, the discussion in this section will focus on these two materials. Many other minerals also have the rutile structure, including CoF_2, FeF_2, MgF_2, MnF_2, and CrO_2, but almost no detailed structural studies have been undertaken for these materials.

1. Rutile (110) Surfaces

The most stable, and best-characterized, rutile surface is the (110) surface. Unfortunately, the rutile crystals, unlike the rocksalt materials, do not cleave, but fracture.[6] The resulting surface stoichiometry and structure is dependent upon the processing conditions, making detailed characterization of the properties of defect-free single-crystal surfaces more difficult. LEED and Auger studies of the SnO_2 (110) surface indicate that, depending on the annealing temperature, the surface O/Sn ratio can vary from 0.49 to 0.76 and the surface can display p(1 × 1), c(1 × 1), p(4 × 2), p(4 × 1), and amorphous structures.[132-134] New methods for preparing stoichiometric SnO_2 (110) surfaces have now been established, facilitating quantitative surface structural measurements.[135-137] The (110) surface of TiO_2, on the other hand, forms stoichiometric p(1 × 1) structures under a variety of experimental conditions, and much more work has been performed on these surfaces.[6,138-140] Annealing the TiO_2 (110) surface can also give rise to (2 × 1) LEED patterns.[141]

The stoichiometric (110) surface is shown in Figure 13. The surface is nonpolar: comprising a surface-bridging anion, two anions and two cations in the surface layer, and a subsurface-bridging anion. In this configuration, the surface-bridging anions are undercoordinated — twofold rather than threefold in the bulk — and one of the surface cations is fivefold rather than sixfold coordinated. There are twice as many O atoms as there are Ti atoms in the surface layer, in compliance with laws of valence where oxygen is –2 and Ti is +4. The autocompensation principle however is subtly different from just obeying laws of valence and is nicely demonstrated by this surface. The surface autocompensates because the nonbonding orbital of the twofold-coordinated oxygen atom becomes occupied with electrons from the fivefold-coordinated titanium.

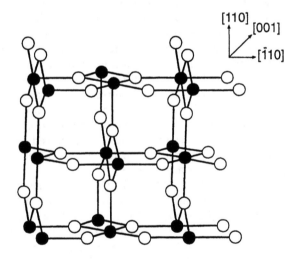

Figure 13
Illustration of the unrelaxed (truncated bulk) stoichiometric rutile (110) surface. Filled circles are metal cations, open circles are anions. (From Godin, T.J.; LaFemina, J.P. *Phys. Rev. B.* 1993, *47*, 6518. With permission.)

At this point it is useful to discuss the character of the electronic structure for the rutile materials. For the unrelaxed stoichiometric surface, the O(2p) electrons occupy both sp^2 bonding hybrids and nonbonding, lone-pair orbitals. The nonbonding, lone-pair orbitals make up the states at the top of the valence band; the bonding hybrids lie lower in energy. This is a qualitatively different situation from that found in semiconductors, where all of the electrons occupied fully hybridized bands. Because the top of the valence band is composed of these nonbonding O(2p) bands, the anion-derived surface states for the unrelaxed, stoichiometric rutile (110) surface are not expected to lie in the gap, but instead are expected to be surface resonances lying in the top of the valence band.[5] As a result, the principle governing the surface relaxations and reconstructions of these materials, Principle 2 in Section I.C, must be made more generalized. Namely, the rehybridization of the occupied nonbonding orbitals into occupied bonding hybrids can provide an energy stabilization sufficient to drive a surface relaxation or reconstruction.[5] As before, however, the atomic motion accompanying the rehybridization must be allowed by the surface topology through approximately bond-length conserving motions (Principle 4, Section I.C). If not, then the energy destabilization arising from the local strain fields as a result of the distortion of the local bonding environments will quickly offset the rehybridization energy gain and halt the relaxation.[5]

The symmetry of the (110) surface indicates that, for p(1 × 1) surfaces, any movement of the surface must be in the direction perpendicular to the surface, since there are no possible approximately bond-length conserving motions allowed for the surface atoms. Accurate total-energy calculations using the local-density approximation have been used to compute the low temperature surface structures of TiO_2.[139,142] For the (110) surface the dominant relaxations were found along the surface normal with the undercoordinated Ti and O atoms drawn into the surface. The length of bonds to subsurface atoms was found to be slightly reduced in the calculations, which modeled the surface with a slab containing six formula units. Tight-binding computations have been used to compute relaxed atomic structure of cassiterite (110). Both relaxed surfaces were found to display a small rumple of the top atomic layer and smaller counterrumples of the subsurface layers.[143]

The surface structure of the TiO_2 (110) surface has been studied using an STM in a UHV environment.[138,144-146] TiO_2 is a wide bandgap semiconductor which is made sufficiently conductive for STM studies by annealing at high temperatures.[144-146] Ti^{3+} sites are created by oxygen removal from the bulk, resulting in an n-type semiconductor. Numerous structures with periodicities larger than the bulk rutile cell were observed and were found to coexist with domain sizes of <1000 Å.[146] Domains of the stoichiometric surface described in the previous pargaraph were observed after exposure to O_2. When the surface is reduced and the surface-bridging oxygen atoms removed, the sixfold-coordinated surface cations become fourfold coordinated. The surface is no longer autocompensated in the same fashion as the bulk material, because the electrons in the cation dangling bonds have no surface anion dangling bonds to occupy. However, because of the tightly bound d electrons, first row transition elements, such as Ti, can exist in more than one valence state. For this reason there are many oxides of titanium with different stoichiometries and structures. If both the fourfold- and fivefold-coordinated surface Ti atoms were to adopt a valency of +3, the surface would be autocompensated. It seems unlikely that the fivefold-coordinated Ti would change their valency when none of their nearest neighbors has been affected by the reduction. Clearly, more work is needed to understand this surface and the similar issues that are raised by the reduced cassiterite (110) surface.[147,148]

2. Rutile (100) Surfaces

The truncated bulk (100) surface, shown in Figure 14a, is more corrugated than the (110) surface. The rectangular surface unit cell has axes along the bulk (010) and (001) directions. All metal atoms at the surface are fivefold coordinated, and the equatorial plane of the octahedra are inclined 45° with respect to the surface normal. The (100) surface is autocompensated when the bridging oxygens are present on the surface. Bond-length conserving motions of the surface atoms are possible at this surface. The dominant relaxation computed using *ab initio* LDA calculations for TiO_2 were along the [010] axis, with the surface Ti and O atoms moving in opposite directions.[142] The inward relaxation was found to be 0.7 Å for the surface Ti atoms and negligibly small for the O atoms. The inclination of the equatorial plane of the octahedra with the surface normal is reduced from 45 to 30° in this relaxed structure.

The fractured TiO_2 (100) surface exhibits a (1×1) LEED pattern,[6] but samples that have been polished and annealed were found only to exhibit (1×3), (1×5), and (1×7) patterns, with the direction of the reconstruction perpendicular to the rows of O atoms.[149] The combined use of glancing angle XRD and LEED has been used to determine atomic positions for the (1×3) surface reconstruction.[149] (See Figure 14b.) The proposed structure comprises facets of the more stable (110) surface and has been observed with an STM experiment.[150] Photoemission data suggest that the (1×3) TiO_2 surface is oxygen deficient, which is compatible with the oxygen atoms being missing from the topmost layer of the (110) facets.[149,150] The results of LDA supercell calculations[142] and empirical potential simulations[151] find that it would be energetically unfavorable for the (100) surface to form (110) facets. Further theoretical work is required to understand the underlying mechanism of this reconstruction.

3. Rutile (111) Surfaces

The truncated bulk rutile (111) surface is shown in Figure 15. As with the rutile (110) surface, two terminations are possible: one in which the bridging cations (Figure 15a) are present and one in which they are absent. Unlike the (110) surface, neither of these terminations is autocompensated, the bridged surface carrying a net negative charge because of partially filled anion dangling bonds. Recent LEED experiments on SnO_2, however, indicate a stable (111) surface with a (1×1) pattern.[152] An exhaustive analysis of all 63 possible (1×1) bulk-terminated surfaces found none of

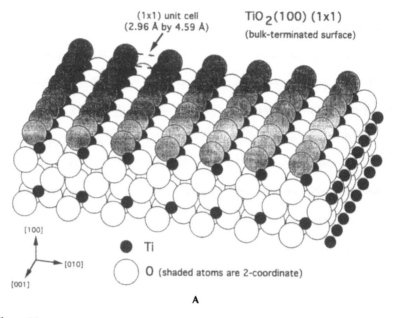

Figure 14

(A) Illustration of the ideal, truncated bulk, rutile (100) surface. (B) Illustration of the unrelaxed rutile (100) (1 × 3) surface upon which {110} microfacets have formed. The surface unit cell is denoted by the dashed lines. Small filled circles are metal cations, large open circles are anions. The shaded circles represent surface anions which are twofold coordinated. (From Henderson, M. A. *Structure and Properties of Interfaces in Ceramics, Mat. Res. Soc. Proc.* 1997, *91*, 357. With permission.)

them to be autocompensated.[153] It was concluded that none of the ideal surface terminations represented the experimentally observed surface and that to achieve autocompensation additional atoms must be added to the surface. Because the surfaces are prepared in an oxygen-rich environment, the most likely atom is oxygen. By imposing the constraints of a (1 × 1) structure, physically reasonable bond lengths, and no singly valent oxygen atoms, they found that there is only one possible structure for the observed surface. Tight-binding total-energy calculations were used to compute the equilibrium geometry shown in Figure 16. The computed relaxation, while small due to the fact that the surface does not allow approximately bond-length conserving motions of surface atoms, is sufficiently large to be detectable by a LEED intensity analysis.

E. Perovskite Surfaces

The cubic perovskite crystal structure has the general formula ABC_3 and consists of small A^{2+} cations surrounded by 12 C^{2-} anions and larger B^{4+} cations that are octahedrally coordinated by the C^{2-} anions. By far the

TiO$_2$(100) (1x3)

{110}-microfaceted surface
(unrelaxed)

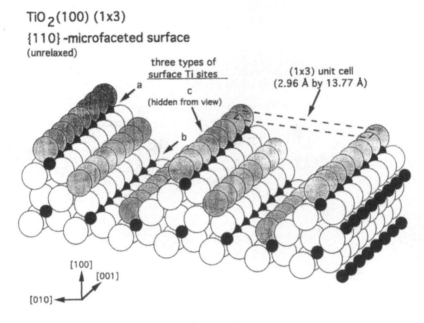

Figure 14 B

most studied are the (100) surfaces shown in Figure 17, for which there are two nonpolar autocompensated terminations: BC$_2$ (type I) and AC (type II). The AC plane has eightfold-coordinated A cations (four neighbors in the surface plane and four in the plane directly below the surface). The BC$_2$ termination consists of fivefold-coordinated B cations (four neighbors in the surface plane and one in the plane directly beneath the surface). The BC$_2$ termination consist of fivefold-coordinated B cations (four neighbors in the surface plane and one in the plane directly beneath the surface) and fourfold-coordinated anions (two B ligands in the surface plane and two A ligands in the plane directly beneath the surface).

From the topology of the two (100) surfaces it is clear that there are no motions of the surface atoms that will approximately conserve bond lengths. Consequently, we expect the surfaces to undergo small atomic motion relaxations. Moreover, the symmetry of the surfaces dictates that these motions be restricted to the direction perpendicular to the surface plane. Using classic potentials, Reiger et al.[154] computed the surface relaxation and dynamics for the (100) surfaces of KMnF$_3$ and KZnF$_3$. They found that both type I and type II surfaces rumpled in a manner analogous to that for MgO. The cations moved toward the bulk, and the anions moved out from the surface. Moreover, the rumpling was on the order of 2% of the lattice spacing, also analogous to the small atomic motion relaxation found in the rocksalt (001) surfaces.

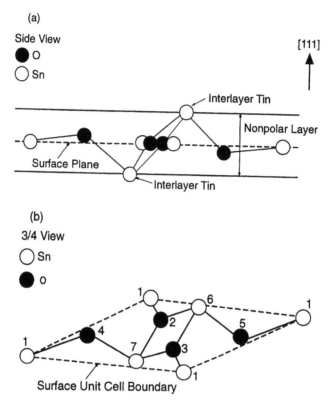

Figure 15
Schematic illustration of the ideal (truncated bulk) unrelaxed rutile (111) surface shown from two angles. Crystallographically inequivalent atoms within the surface unit cell are numbered. The primitive unit cell of an ideal surface terminated by a (111) composite layer is denoted by dashed lines. (From Godin, T.J.; LaFemina, J.P. *Surf. Sci.* 1994, *301*, 364. With permission.)

The most commonly studied surfaces of perovskite minerals are strontium and barium titanate ($SrTiO_3$ and $BaTiO_3$).[5,6] For these oxides the above description of the (100) surfaces is idealized because the materials do not cleave but fracture. Reflection high-energy electron diffraction (RHEED) studies have determined the relaxation of both TiO_2 (type I) and SrO (type II) surfaces of $SrTiO_3$.[155] These relaxations are shown in Figure 17 and are qualitatively similar to those computed by Reiger et al.[154] for $KMnF_3$ and $KZnF_3$.

Liang and Bonnell[156] have used an STM to provide atomic scale structural information on the $SrTiO_3$ (001) surface. Changes of surface morphology with annealing were observed under different annealing and oxygen exposure conditions. The resulting structures were explained in terms of the formation of reduced phases due to the different rates of sublimation for Sr, Ti, and O at the surface. The STM images reveal rowlike

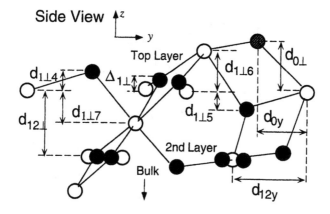

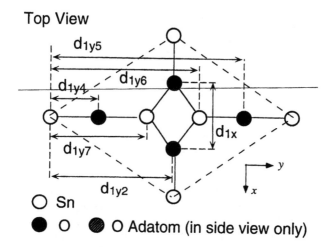

Figure 16
Schematic illustration of the autocompensated, oxygen adatom structure of the rutile (111) surface. The oxygen adatom is bonded to tin atoms 1 and 6 along the long axis of the surface unit cell (see Chapter 3, Figure 6). Also shown are a set of independent geometric parameters sufficient to specify the positions of each atom in the outermost surface layer. (From Godin, T.J.; LaFemina, J.P. *Surf. Sci.* 1994, *301*, 364. With permission.)

structures with spacings of 12 and 20 Å. The rowlike structures were found to be consistent with Sr_2TiO_4 and $Sr_3Ti_2O_7$ surface structures and compositions.

F. Corundum Surfaces

The corundum structure is similar to the rutile structure and may be derived by a simple parallel transformation or crystallographic shear. Many corundum oxides are very hard and have high melting points. Al_2O_3

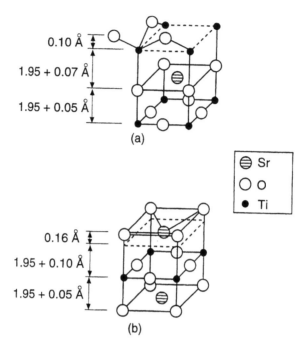

Figure 17
Ball-and-stick models of the relaxed (a) TiO₂- and (b) SrO-terminated (100) surfaces of
perovskite structure SrTiO₃. (From Hikita, T.; Hanada, T.; Kudo, M.; Kawai, M. *Surf. Sci.*
1993, *287/288*, 377. With permission.)

doped with Cr_2O_3 is the gemstone ruby, and Al_2O_3 doped with Ti_2O_3 is
sapphire. The trigonal corundum lattice, M_2O_3, comprises cations, M, in
a distorted octahedral environment and anions, usually oxygen, in a distorted tetrahedral environment. Ilmenite ABO_3, is a related structure in
which cations A and B replace the metal atoms, M, in equal proportions.
The corundum structure minerals for which some experimental and/or
theoretical surface determination has been made include corundum itself,
$\alpha\text{-}Al_2O_3$, Ti_2O_3, V_2O_3, and hematite Fe_2O_3.[6]
    A variety of $\alpha\text{-}Al_2O_3$ surfaces, each exhibiting a variety of surface
structures, have been studied experimentally. LEED studies of the (0001)
basal plane demonstrated that at low temperatures (below 1250°C) this
surface exhibits a (1×1) structure in both air and vacuum.[157-161] Annealing
to higher temperatures results in the more complex $(\sqrt{3} \times \sqrt{3})R30°$ and
$(\sqrt{31} \times \sqrt{31})R9°$ structures. The $(\bar{1}012)$ and $(11\bar{2}3)$ faces, on the other hand,
exhibit the simpler (2×1) and (4×5) structures.[162] The $(11\bar{2}0)$ surface,
studied by LEED, TEM, and a variety of reflection techniques, exhibits
both (1×2) and (1×4) reconstructions.[160,163,164]
    The question of surface termination is an important one for all of the
α-alumina surfaces, none of which is a cleavage surface. Unfortunately,

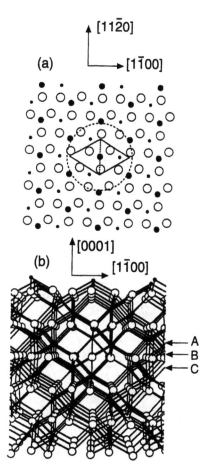

Figure 18
Illustration of the (a) top and (b) side views of the corundum (α-Al$_2$O$_3$) (0001) surface. The top view is of one surface building block comprising three atomic layers of Al–O–Al. Open circles label O atoms, and the large and small filled circles label Al atoms on the top and bottom layers. The solid line marks the surface unit cell. The three unique (0001) cleavage planes are indicated by arrows and labeled A, B, and C. (From Guo, J.; Ellis, D.E.; Lam, D.J. *Phys. Rev. B*, 1992, 45, 13647. With permission.)

no definitive experimental work has been reported, although several theoretical studies exist in the literature. Electronic structure calculations have been reported for the unreconstructed (truncated bulk) (0001) and ($1\bar{1}02$) surfaces using both semiempirical and first-principles density-functional methods (see Figures 18 and 19). In the density-functional work, the question of surface termination was explicitly considered and the cleavage energies for the three symmetry inequivalent (0001) and ($1\bar{1}02$) terminations computed.[162] These computations indicated that, for the (0001) surface, cleavage between the Al planes (labeled C in Figure 18)

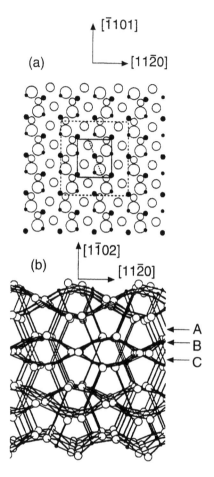

Figure 19

Illustration of the (a) top and (b) side views of the corundum (α-Al$_2$O$_3$) (1$\bar{1}$02) surface. The top view is of one surface building block comprising five atomic layers of O–Al–O–Al–O. Open circles label O atoms, filled circles label Al atoms and the smaller radii indicate greater depth from the surface. The solid line marks the surface unit cell. The three unique (1$\bar{1}$02) cleavage planes are indicated by arrows and labeled A, B, and C. (From Guo, J.; Ellis, D.E., Lam, D.J.; *Phys. Rev. B.* 1992, *45*, 13647. With permission.)

was the most stable by more than 8 eV per unit cell over the other two terminations. This (0001) surface becomes autocompensated because each surface layer aluminum atom is threefold coordinated with oxygen, compared with the sixfold coordination in the bulk. Conversely, there are three oxygen atoms just below but displaced from the surface aluminum layer that are threefold coordinated at the surface, compared with their sixfold coordination in the bulk. Autocompensation occurs by transfer of the excess electrons from the undercoordinated aluminum to electronic states derived from the undercoordinated oxygen atoms. For the ($1\bar{1}02$) surface the most stable cleavage was computed to be between O planes. Again, this cleavage was computed to be the most stable by more than 8 eV per unit cell over the competing cleavages. Of all the possible surface terminations, the ones computed to be the most stable are the ones for which the cleavage produces two charge neutral and autocompensated surfaces.

Turning to the question of the possible relaxations or reconstructions that may occur, we need to consider the topology and local chemical bonding. For the ($1\bar{1}02$) surface, the surface Al atoms are fivefold coordinated and the surface O atoms twofold coordinated. In this respect the surface topology is similar to that of the rutile (110) with no bond-length conserving motions of the surface atoms possible. The (0001) surface is more interesting because both the surface Al and O atoms are threefold coordinated. This topology allows a bond-length conserving rotation in which the Al atoms can move down toward the bulk into an approximately sp^2 hybridization, while the surface O atoms "pucker" out from the surface into a distorted pyramidal hybridization. Ab initio Hartree–Fock[165] and local density-functional computations[166] of thin slabs of the (0001) surface, in which the positions of the surface atoms perpendicular to the surface were optimized, have demonstrated that the surface Al atoms relax to position on the order of 0.4 to 0.7 Å away from their truncated bulk positions. Unfortunately, the positions of the surface O atoms were not fully optimized.

The fully optimized geometry of the α-Al$_2$O$_3$ (0001) surface has been computed with a tight-binding total-energy model.[167] The unrelaxed and relaxed geometries are shown schematically in Figure 20a and b. The surface Al atoms move down toward the surface, while the surface O atoms have buckled out from the surface. Figure 20 also contains the definitions of the independent surface structural parameters specifying the relaxation, and the computed values of these are listed in Table 3. It is important to note that the relaxed atomic positions differ from the bulk atomic positions by as much as 0.7 Å for the surface Al atoms. This is much larger than the resolution of typical LEED intensity analyses for atomic displacements normal to the surface plane (0.05 Å). The transverse displacements of the O atoms, however, are much smaller (approximately 0.2 Å). This length is comparable to the usual resolution of LEED for transverse displacements but should be measurable by ion-scattering techniques.

Qualitatively, this relaxation is similar to the relaxation that occurs in the zincblende (110) surface. The driving force for the corundum relaxation, however, is different. For the (110) surface the relaxation is driven by the energy gained by the stabilization of the occupied, anion-derived surface state, following the redistribution of the dangling bond charge density into the surface and back bonds. The electronic structure of the α-Al$_2$O$_3$ (0001) surface is different from the zincblende (110). The unrelaxed truncated bulk (0001) surface has no occupied states in the bandgap.

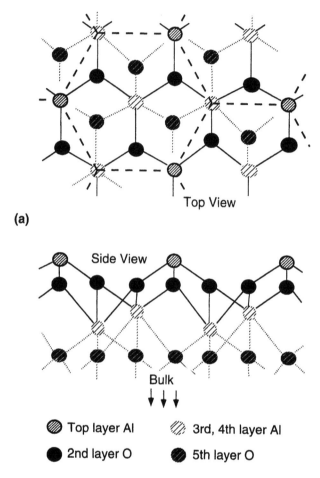

Figure 20
Schematic illustration of the top and side views of the α-A_2O_3 (a) unrelaxed and (b) relaxed (0001) surface. The boundaries of the surface unit cell are given by dashed lines. Subsurface atoms and bonds are shaded. (From Godin, T.J., LaFemina, J. P. *Phys. Rev. B.* 1994, *49*, 7691. With permission.)

As a result, the surface relaxation is driven by the rehybridization of the electron density around the undercoordinated surface O atoms.

G. Silica Surfaces

The interest in silica (SiO_2) and silicate materials is widespread. Geologically, silicate minerals (including zeolites) make up the greater part of the Earth's mantle.[168,169] Amorphous silica is of obvious importance to the microelectronics industry.[170] Crystalline SiO_2 exists in nine different allotropes: α- and β-quartz,[171,172] α- and β-cristobalite,[171,173] α- and β-tridymite,[171,174]

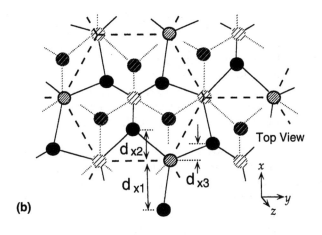

(b)

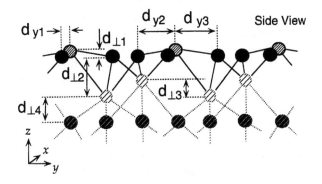

Figure 20 (continued)

TABLE 3.

Values of Geometric Parameters (Figure 20) for Relaxed and Unrelaxed Corundum (0001) Surfaces as Computed by a Tight-Binding Total-Energy Model[16]

	d_{x1}	d_{x2}	d_{x3}	d_{y1}	d_{y2}	d_{y3}	$d_{\perp1}$	$d_{\perp2}$	$d_{\perp3}$	$d_{\perp4}$
Unrelaxed	0.9	1.6	0.7	1.4	0.1	1.7	0.8	0.8	0.5	0.8
Relaxed	1.1	1.8	0.7	1.5	0.2	1.5	0.1	0.8	0.6	0.8
LDA relaxed (approximate)	—	—	—	—	—	—	0.1	0.8	0.3	1.0

Note: Units are angstroms. The values are compared to LDA calculations.[166] Hyphens denote values that were not reported in Reference 166.

From LaFemina, J.P. *CRC Crit. Rev. Surf. Chem.* 1994, 3, 297. With permission.

coesite,[175,176] keatite,[177] and stishovite.[178,179] All of the allotropes, with the exception of stishovite, consist of silicon tetrahedra and twofold-coordinated oxygen atoms. Stishovite has the rutile crystal structure comprising distorted silicon octahedra and threefold-coordinated oxygen atoms. Only α-quartz has had its surface structural properties examined in any detail theoretically and experimentally, and so the discussion in this section will focus on this allotrope. An early LEED study[180] indicated that the (10$\bar{1}$0), (0001), and (10$\bar{1}$1) surfaces all displayed a (1 × 1) pattern at room temperature. Upon heating to temperatures greater than 500°C, the surface was disrupted and the LEED patterns disappeared. In addition, surface degradation by the incident electron beam was also observed. The only other experimental information available are EELS[181] and XPS[182] measurements taken on ill-defined surfaces.

Only two computations of a surface electronic structure for α-quartz surfaces have been reported utilizing both a tight-binding and an *ab initio* Hartree–Fock potential.[237] Both computations were performed for the (10$\bar{1}$0) surfaces[183,184] shown in Figure 21. As is evident from the structural diagram of Figure 21, there are many ways a (10$\bar{1}$0) surface may be formed, two of which are shown. These two surfaces, labeled a and b, are the surfaces formed by cutting the bonds labeled 1 and 1'. Surface a has half of the surface Si atoms threefold coordinated with one dangling bond, while the remaining Si atoms have the full tetrahedral coordination, but are attached to singly coordinate O atoms. While this surface is autocompensated, the singly coordinate surface O atoms in the unrelaxed surface are in an energetically unfavorable local bonding environment and result in the appearance of occupied O 2p – derived surface states in the bandgap for the *ab initio* derived potential,[184] contrary to the EELS data.[181] The tight-binding computations showed no density of states in the gap.[183]

For the b surface of Figure 21, all of the surface Si atoms are threefold coordinated with one dangling bond. Moreover, as with the a surface, one of the surface O atoms is singly coordinated. Again, the tight-binding computations showed no density of states in the bandgap for the unrelaxed surface,[183] while the *ab initio* computations found occupied O 2p – derived states in the gap.[184] Heggie et al.[184] considered a relaxation in which the singly coordinated surface O atom forms a shortened double bond with the undercoordinated Si to which it is attached and found that a 20% reduction of the bond length was sufficient to push the occupied levels out of the gap.[184] The problem with this surface is that it is not autocompensated, since there are O dangling bonds which are partially filled, but no Si dangling bonds.

On both of these surfaces, the singly coordinated O atom is in an energetically unfavorable bonding environment and can be expected to attempt some relaxation in which to satisfy its valence.[183,184] For the b surface, this can only be accomplished (so that the surface becomes charge neutral) by doubling the size of the unit cell and forming oxygen dimers. It is unclear, however, whether these dimers would be stable with respect to desorption and the formation of gaseous O_2 molecules. For the a surface, the situation is more complex, and it has been suggested that the surface will attempt to relax by overlapping the dangling bonds of the singly coordinated O and threefold-coordinated Si atoms.[183] This would satisfy

Surface a

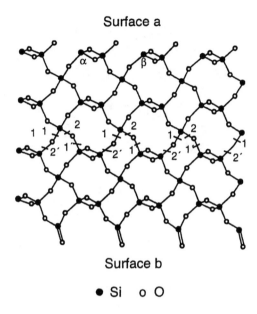

Surface b

● Si   o O

Figure 21
Illustration of the ideal (truncated bulk) a and b α-quartz (10$\bar{1}$0) surfaces. Bonds are labeled 1, 1′, 2, and 2′ for discussion in the text. (From Heggie, M.; Jones, R.; Nylén, M. *Philos. Mag. B.* 1985, *51*, 573. With permission.)

the valences of the surface atoms while leaving the surface autocompensated and charge neutral. The relaxation can occur in two ways, with the singly coordinated O bonding to the Si atoms labeled α or β (see Figure 21). Clearly, bonding to the β-Si involves significant distortions of the surface and subsurface bonding environments and, therefore, is unlikely to be energetically favorable. Bonding to the α-Si, however, results in a "bridging" Si – O – Si linkage that satisfies all of the surface atom valences, yet does not induce large local strain fields since the distortions are primarily in bond angles, rather than in bond lengths.

H. Carbonate Surfaces

The cleaved (10$\bar{1}$4) face of $CaCO_3$ is a nonpolar cleavage surface (and hence it is autocompensated) with cationic Ca^{2+} and anionic CO_3^{2-} groups connected via a rhombahedral network. There are two carbonate groups which are distinguished by different in-plane rotations. In an unrelaxed, bulk termination model of the surface, there are oxygen atoms both 0.8 Å above and 0.8 Å below the surface, as well as in the surface plane. The two oxygen atoms lying above the surface form a zigzag pattern over the (10$\bar{1}$4) surface with a lateral deviation of approximately 0.8 Å, as shown in Figure 22.

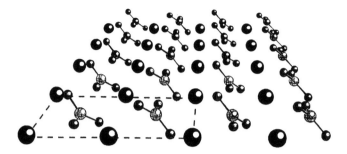

Figure 22
Perspective illustration (side view) of the truncated bulk calcite (10$\bar{1}$4) cleavage surface. The dashed lines define the surface unit cell. The black circles are Ca atoms, the medium-sized shaded circles are C atoms, and the small checkered circles are O atoms.

The (10$\bar{1}$4) surface of calcite has been studied using LEED and AFM.[22,34,185,186] Stipp and Hochella[185] used LEED to examine the surface after exposure to a variety of treatments. A dominant (1 × 1) pattern was observed by LEED that suggested the top few monolayers of the calcite surface are very similar to the bulk material. The appearance of additional and very weak reflections were conjectured to result from slightly different orientations of some CO_3 groups; however, the diffuseness of the extra reflections implied that the misorientation did not exhibit consistent long-range order. Surface structures observed with an AFM in aqueous environment were found to be consistent with this result.[186] The magnified constant force AFM image is shown in Figure 23, and the rectangular surface unit cell is marked for clarity. In addition to the (1 × 1) structure, close examination of the figure reveals features at the corner sites that are broader and brighter than the features at the center of the cell, suggesting that the surface carbonate groups are inequivalent. Different domains, where the brighter and less bright spots are interchanged, were observed in the experiment. This asymmetry was not observed in earlier AFM work on the (10$\bar{1}$4) calcite surface, where the surface was believed to be bulk terminated, and has led to the speculation that there is an atomic relaxation of the carbonate groups within the (1 × 1) cell.

IV. DISCUSSION

As discussed throughout this chapter, the study of mineral surfaces is in its infancy, and much more experimental and theoretical work is needed before a library of detailed mineral surface structures is available. There is, however, a set of simple chemical and physical principles that can be used to understand the structure of mineral surfaces and interfaces. These principles include the autocompensation of surfaces via the redistribution of dangling bond charge density, the role of microscopic stress,

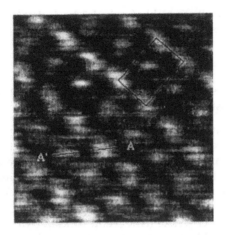

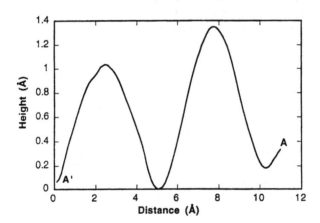

Figure 23
(a) A constant force image showing the rectangular surface unit cell and the (1×1) symmetry of the (1014) surface in basic solution. The difference in intensity between the rows containing the corners of the surface unit cell and the rows running through the interior of the surface unit cell is also evident. (b) A cross-sectional profile showing the different vertical heights between the two rows. (From Liang, Y.; Lea, A.S.; Baer, D.R.; Engelhard, M.H. *Surf. Sci.* 1994, submitted.)

and the importance of surface topology and processing conditions in determining the nature of surface relaxations and reconstructions. The utility of these principles was established in Section III, where they were used not only to explain observed mineral surface structures, but also to predict feasible classes of structures for mineral surfaces which have not yet been studied in detail experimentally.

The application of these principles to mineral surfaces also has raised an important issue: why are these concepts, originally developed from the study of covalent semiconducting materials, applicable to wide-band-gap materials traditionally thought of as ionic? And what implication does this have for the conventional view of ionic materials and their surface physics and chemistry?

A. Autocompensation and the Rehybridization of Dangling Bond Charge Density

The autocompensation principle states that surfaces are most stable when the anion-derived dangling bonds are completely filled and the cation-derived dangling bonds completely empty. In the world of covalent semiconductors, this has always been interpreted as meaning that each anion-derived dangling bond contains two electrons. In the application of this principle to minerals, however, a generalization is necessary in order to fit some perovskite (ABC_3) and rocksalt (AB) structure minerals into this conceptual framework. The problem with these materials is that, in the bulk, there is simply not a sufficient number of valence electrons to have two electrons per bond. To illustrate with NaCl: each Na atom contributes one electron to six bonds ($1/6$ electron/bond) while each Cl atom contributes seven electrons to six bonds ($7/6$ electrons/bond) for a total of $4/3$ electron per bond. The generalized autocompensation principle requires only that the surface bonds possess the same number of electrons that bulk bonds possess, thus fulfilling the general stability condition that the valence bands of a material be completely filled while the conduction bands are completely empty. After all, it is unfair to require the surface anion dangling bonds to contain two electrons when the bulk bonds contain only $4/3$ electron! In this way, the stability of perovskite and rocksalt cleavage surfaces can be understood within the same conceptual framework as all other nonmetallic materials.

This is a powerful concept, that the surfaces of all nonmetallic materials can be understood within this simple conceptual framework based upon autocompensation. A potential exception to what we have discussed in this chapter are the wurtzite ZnO (0001) polar surfaces, which appear to be stable in their truncated bulk conformation. This could be an issue of kinetics, with the truncated bulk surface metastable, or it could be due to a more complicated interaction. Clearly, this is a fertile area for more detailed experimental and theoretical studies.

Once a surface has autocompensated, which fixes the surface stoichiometry, the redistribution of the dangling bond charge density determines how the surface atoms will move. In competition with these effects is the energy cost incurred if the relaxation or reconstruction significantly distorts the local bonding environment (i.e., gives rise to large local strain fields).

The surface dangling bond charge density can be rehybridized in several ways. New bonds can be formed at the surface, either between surface atoms or between surface atoms and adsorbates. In addition, for many of these surfaces the dangling bond charge density is localized on an anion with lone-pair electrons. A primary driving force for surface relaxation or reconstruction on these surfaces comes from the energy gain associated with the rehybridization of these nonbonding electrons into bonding orbitals. Such is the case for the rutile (110), α-corundum (0001), and calcite ($10\bar{1}4$) surfaces.

B. Surface Stress and the Importance of Surface Topology

The creation of local strain fields, arising from the distortion of local bonding environment, is the prime competitor of the drive to rehybridize the surface dangling bond charge density and move the surface atoms into more favorable local bonding environments. The energy cost associated with the creation of these local strain fields inhibits the motion of the surface atoms and limits the extent to which the surface and subsurface atoms can move to lower the surface electronic energy. Consequently, the surface topology, or atomic connectivity, is an important factor in determining the details of surface relaxations and reconstructions. When the topology of the surface is such that approximately bond-length conserving motions of the surface atoms are allowed, large atomic motion relaxations are observed as with the α-corundum (0001), zincblende (110), perovskite (100), and wurtzite ($10\bar{1}0$) and ($11\bar{2}0$) surfaces. When such motions are prohibited by the surface topology, the surface undergoes small atomic motion relaxations as are observed as with the rocksalt (001), rutile (110), and calcite ($10\bar{1}4$) surfaces.

C. Ionic vs. Covalent Bonding: Implications for Surface Structure

It has long been recognized that the classification of materials and bonds as ionic or covalent, while useful at times, is neither a complete nor a definitive description.[187-189] This recognition, however, has just recently begun to be extended to surface properties.[5,6,190,191] Still, the literature is replete with references to the idea that the cleavage process for ionic crystals does not cut bonds, but merely separates ions,[192,193] or to the idea that the surface properties of ionic crystals are determined solely by classic electrostatics; hence, the nonpolar surfaces of ionic crystals are unrelaxed and unreconstructed.[13] As we have seen in this chapter, these ideas are simply not true. And while classic electrostatics has been useful for understanding many bulk properties of these materials, it is an incomplete conceptual framework for the consideration of surface properties. For example, classic electrostatics correctly predicts the small relaxation of the rocksalt (001) surface. In this case, because the topology of the surface precludes any large bond-length conserving motions of the surface atoms, electrostatic considerations are successful in predicting the surface relaxation. But so are models which contain only short-range, orbital overlap interactions. And as we have discussed in the previous section, the relaxation and surface stability can be understood within the generalized autocompensation framework.

We have also examined many surfaces for which the ideas of classic electrostatics fail completely to predict the surface structure, including the zincblende (110), wurtzite (10$\bar{1}$0) and (11$\bar{2}$0), perovskite (100), and α-corundum (0001). All of these surfaces undergo large atomic motion relaxations, allowed by the surface topology and driven by the redistribution of the dangling bond charge density at the surface. These surfaces can also be understood within the generalized autocompensation framework discussed in the previous section. Indeed, the only mineral surfaces studied which potentially do not fit within this conceptual framework are the polar ZnO (111), and this may be an issue of reconstruction kinetics.

If, in fact, this generalized autocompensation framework holds, the need to invoke a qualitatively different bonding for ionic materials vs. covalent materials is replaced by a single type of bonding which, for bulk materials, completely fills the valence band and completely empties the conduction band and, for surfaces, results in filling the surface anion dangling bonds (with the same number of electrons as in the bulk bonds) and emptying the surface cation dangling bonds.

One implication of this is that every surface would exhibit a relaxation or reconstruction characteristic to its crystalline topology and *not* dependent upon the particular atomic constituents (or ionicity) of the lattice. This is exactly what happens for zincblende (110) surfaces, the only surface for which sufficient experimental and theoretical information exists to test this idea.[2-4] As discussed in Section III, all zincblende (110) surfaces exhibit the same characteristic relaxation comprising a tilt of the surface zigzag chains by 29 ± 3°. This includes materials such as GaAs, with a Phillips ionicity of 0.31, and ZnSe and ZnO, which have Phillips ionicities of 0.62 and 0.63, respectively.

D. Areas for Future Research

A detailed microscopic understanding of the atomic structure and chemical bonding at mineral surfaces is the essential first step toward understanding the important geochemical processes which occur at these surfaces. As we have seen, this understanding is just beginning to emerge for the most simple minerals, and advancements are intimately tied to the development of new computational and experimental techniques necessary for probing more-complex systems. Two critical areas involve the preparation of well-characterized single-crystal surfaces that can be studied with the full range of surface scientific tools that have been successfully applied to metal and semiconductor surfaces, along with the development of new surface techniques capable of interrogating the surface atomic structure and chemical bonding of complex mineral systems both under ideal conditions and *in situ*.

On the theoretical front, the development of simple quantum mechanical models is also essential to achieving a firm understanding of the factors which control surface structure and chemistry. The application of the simple chemical and physical principles outlined in this chapter looks promising. There are, however, only a few specific mineral surfaces, for example, MgO (001) and zincblende (110), for which there is sufficient theoretical and experimental information available to claim that a detailed understanding exists. The extension of these concepts to encompass the full range of mineral surfaces, especially the polar surfaces, defected surfaces, and chemisorbed interfaces, represents an enormous opportunity.

Mineral systems are also fertile grounds for exploring the nature of surface preparation and sample history effects upon surface chemistry. The development of the processing–function relationships requires a thorough understanding of the surface energetics and mechanisms that control important geochemical phenomena, such as growth and precipitation/dissolution reactions. This includes detailed knowledge of the surface potential energy function, including the characterization of metastable structures and their transition states, the mechanisms of surface structural and phase transitions, and the role of surface dynamics. In fact, the enumeration of these processing–function relationships and the identification of the processing conditions necessary to realize specific surface structures and chemistries is a major frontier in understanding and controlling real-world mineral surface chemistry.

ACKNOWLEDGMENTS

The authors are indebted to Dr. Charles B. Duke for many fruitful discussions and collaborations, which have led to many of the themes discussed in this chapter, and for his constant interest and encouragement. We would also like to thank Dr. Victor Henrich for many pleasurable and informative discussions. Finally, we are grateful to Drs. Don Baer, Dave Blanchard, Tom Godin, Roger Haydock, Jim McCoy, Yong Liang, and Andrew Skinner for our continuing collaborations and for allowing the use of results prior to publication. The authors acknowledge the support of the Pacific Northwest Laboratory during the preparation of this manuscript. The Pacific Northwest Laboratory is a multiprogram national laboratory operated for the U.S. Department of Energy by Battelle Memorial Institute under Contract DE-AC06-76RLO 1830. All figures taken of other sources are used with the permission from at least one author.

REFERENCES

1. Duke, C.B. *Appl. Surf. Sci.* 1982, *11/12*, 1.
2. LaFemina, J.P. *Surf. Sci. Rep.* 1992, *16*, 133.
3. Duke, C.B. *Surface Properties of Electronic Materials*; D.A. King; D.P. Woodruff, Eds. Elsevier: Amsterdam, 1988; pp. 69-118.
4. Duke, C.B. *Reconstructions of Solid Surfaces*; K. Christman; K. Heinz, Eds. Springer: Berlin, 1990.
5. LaFemina, J.P. *Crit. Rev. Surf. Chem.* 1994, *3*, 297.
6. Henrich, V.E.; Cox, P.A. *The Surface Science of Metal Oxides.* Cambridge University Press: Cambridge, 1994.
7. Van Hove, M.A.; Weinberg, W.H.; Chan, C.-M. *Low-Energy Electron Diffraction: Experiment, Theory, and Surface Structure Determination.* Springer-Verlag: Berlin, 1986.
8. Chambers, S.A. *Adv. Phys.* 1991, *40*, 357.
9. van der Veen, J.F. *Surf. Sci. Rep.* 1985, *5*, 199.
10. Chen, C.J. *Introduction to Scanning Tunneling Microscopy.* Oxford University Press: New York, 1993.
11. Kittel, C. *Introduction to Solid State Physics*; 5th ed. John Wiley and Sons: New York, 1976.
12. Somerjai, G.A. *Chemistry in Two Dimensions: Surfaces.* Cornell University Press: Ithaca, NY, 1981.
13. Zangwill, A. *Physics at Surfaces.* Cambridge University Press: Cambridge, 1988.
14. Hoffmann, R. *Solids and Surfaces: A Chemist's View of Bonding in Extended Structures.* VCH Publishers: New York, 1988.
15. Lannoo, M.; Friedel, P. *Atomic and Electronic Structure of Surfaces.* Springer-Verlag: Berlin, 1991.
16. Harrison, W.A. *Electronic Structure and the Properties of Solids.* Dover: New York, 1980.
17. Biegelson, D.K.; Brigans, R.D.; Northrup, J.E.; Swartz, L.E. *Phys. Rev. B.* 1990, *41*, 5701.
18. Harrison, W.A. *J. Vac. Sci. Tech.* 1979, *16*, 1492.
19. Hochella, M.F., Jr.; White, A.F.; Eds. *Mineral-Water Interface Geochemistry; Rev. Mineral.* 1990, *23*.
20. Wiesendanger, R.; Guntherodt, H.-J.; Eds. *Scanning Tunneling Microscopy I & II.* Springer-Verlag: Berlin, 1993.
21. Sarid, D. *Scanning Force Microscopy.* Oxford University Press: New York, 1991.
22. Gratz, A.J.; Manne, S.; Hansma, P.K. *Science.* 1991, *251*, 1343.
23. Doyen, G.; Koetter, E.; Vigneron, J.P.; Scheffler, M. *Phys. Rev. B.* 1990, *51*, 281.
24. Tersoff, J.; Hamann, D.R. *Phys. Rev. Lett.* 1983, *50*, 1998.
25. Novak, D.; Garfunkel, E.; Gustafsson, T. *Phys. Rev. B.* 1994, *50*, 5000.
26. Eggleston, C.M.; Hochella, M.F., Jr. *Am. Mineral.* 1992, *77*, 221.
27. Eggleston, C.M.; Hochella, M.F., Jr. *Am. Mineral.* 1992, *77*, 911.
28. Binnig, G.; Gerber, C.; Stoll, E.; Albrecht, T.R.; Quate, C.F. *Europhys. Lett.* 1987, *3*, 1281.
29. Meyer, G.; Aner, N.M. *Appl. Phys. Lett.* 1990, *56*, 2100.
30. Meyer, E.; Heinzelmann, H.; Rudin, H.; Guntherodt, H.-J. *Z. Phys.* 1990, *79*, 3.
31. Heinzelmann, H.; Meyer, E.; Brodbeck, D.; Overney, G.; Guntherodt, H.-J. *Z. Phys. B*, 1992, *88*, 321.
32. Meyer, E.; Guntherodt, H.-J.; Haefke, H.; Gerth, G.; Krohn, M. *Europhys. Lett.* 1991, *15*, 319.
33. Rachlin, A.L.; Henderson, G.S.; Goh, M.C. *Am. Mineral.* 1992, *77*, 904.
34. Ohnesorge, F.; Binnig, G. *Science.* 1993, *260*, 451.
35. Overney, G. *Scanning Tunneling Microscopy III*; R. Wiesendanger; H.-J. Guntherodt, Eds. Springer-Verlag: Berlin, 1993; chapter 10.
36. Tomanek, D. *Scanning Tunneling Microscopy III*; R. Wiesendanger; H.-J. Guntherodt, Eds. Springer-Verlag: Berlin, 1993; chapter 11.
37. Wiesendanger, R.; Guntherodt, H.-J.; Eds. *Scanning Tunneling Microscopy III.* Springer-Verlag: Berlin, 1993.

38. Hillner, P.E.; Manne, S.E.; Gratz, A.J.; Hansma, P.K. *Ultramicroscopy.* 1992, *42–44*, 1387.
39. Kahn, A. *Surf. Sci. Rep.* 1983, *3*, 193.
40. Schrödinger, E. *Ann. Phys.* 1926, *79*, 361, 489.
41. Schrödinger, E. *Ann. Phys.* 1926, *80*, 437.
42. Schrödinger, E. *Ann. Phys.* 1926, *81*, 109.
43. Blanchard, D.L.; Lessor, D.L.; LaFemina, J.P.; Baer, D.R.; Ford, W.K.; Guo, T. *J. Vac. Sci. Tech.* 1991, *A9*, 1814.
44. Chambers, S.A.; Tran, T.T. *Surf. Sci.,* 1994, *314*, L867; *Appl. Surf. Sci.* 1994, *81*, 161.
45. Varma, S.; Chen, X.; Zhang, J.; Davoli, I.; Saladin, D.K.; Tonner, B.P. *Surf. Sci.* 1994, *314*, 145.
46. Hehre, W.J.; Radom, L.; Schleyer, P.V.R.; Pople, J.A. *Ab initio Molecular Orbital Theory,* John Wiley & Sons: New York, 1986.
47. Hoffmann, R. *J. Chem. Phys.* 1963, *49*, 36.
48. Colbourn, E.A. *Surf. Sci. Rep.* 1992, *15*, 281.
49. Madelung, E. *Gött. Nach.* 1909, 100.
50. Madelung, E. *Gött. Nach.* 1910, 43.
51. Madelung, E. *Phys. Z.* 1910, *11*, 898.
52. Wulff, G. *Z. Kristallog.* 1901, *34*, 449.
53. Feynman, R. *Phys. Rev.* 1939, *56*, 340.
54. Hirabayashi, K. *J. Phys. Chem. Jpn.* 1969, *27*, 1475.
55. Joannopoulos, J.D.; Cohen, M.L. *Phys. Rev. B.* 1974, *10*, 5075.
56. Lieske, N.P. *J. Phys. Chem. Solids.* 1984, *45*, 821.
57. Applebaum, J.R.; Hamann, D.R. *Rev. Mod. Phys.* 1976, *48*, 479.
58. Yndurain, F.; Falicov, L. *J. Phys. C.* 1975, *8*, 1571.
59. Yndurain, F.; Falicov, L. *Solid State Commun.* 1975, *17*, 855.
60. Tsukada, M.; Adachi, H.; Satoko, C. *Prog. Surf. Sci.* 1983, *14*, 113.
61. Born, M.; Oppenheimer, J.R. *Ann. Phys.* 1927, *84*, 457.
62. Lowe, J.P. *Quantum Chemistry.* Academic Press: New York, 1978; chapter 9.
63. Feller, D.; Davidson, E.R. *Reviews in Computational Chemistry.* VCH Publishers: New York, 1990; pp. 1–45.
64. Pickett, W.E. *Comp. Phys. Rep.* 1989, *9*, 117.
65. Szasz, L. *Pseudopotential Theory of Atoms and Molecules.* John Wiley & Sons: New York, 1985.
66. Zunger, A. *J. Vac. Sci. Tech.* 1979, *16*, 1337.
67. Stewart, J.J.P. *Reviews in Computational Chemistry.* VCH Publishers: New York, 1990; pp. 45–83
68. Pople, J.A.; Beveridge, D.L. *Approximate Molecular Orbital Theory.* McGraw- Hill: New York, 1970.
69. Anderson, O.K.; Klose, W.; Nohl, H. *Phys. Rev. B.* 1978, *17*, 1209.
70. Slater, J.C. *Phys. Rev.* 1929, *34*, 1293.
71. Fock, V.A. *Fundamentals of Quantum Mechanics;* 3rd ed. Mir Publishers: Moscow, 1978.
72. Jones, R.O.; Gunnarsson, O. *Rev. Mod. Phys.* 1989, *61*, 689.
73. Parr, R.G.; Yang, W. *Density Functional Theory of Atoms and Molecules.* Oxford: New York, 1989.
74. Hohenberg, P.; Kohn, W. *Phys. Rev. B.* 1964, *136*, 864.
75. Kohn, W.; Sham, L.J. *Phys. Rev.* 1965, *140*, A1133.
76. Ceperley, D.M.; Alder, B.J. *Phys. Rev. Lett.* 1980, *45*, 566.
77. Salahub, D.R.; Zerner, M.C. *The Challenge of d and f Electrons;* D.R. Salahub; M.C. Zerner, Eds. American Chemical Society: Washington, D.C., 1989; pp. 1–16.
78. Becke, A.D. *J. Chem. Phys.* 1993, *98*, 5648.
79. Slater, J.C.; Koster, G.F. *Phys. Rev.* 1954, *94*, 1498.
80. Harrison, W.A. *Surf. Sci.* 1976, *55*, 1.
81. Harrison, W.A. *Phys. Rev. B.* 1981, *24*, 5835.
82. Chadi, D.J. *Phys. Rev. Lett.* 1978, *41*, 1062.
83. Chadi, D.J. *Phys. Rev. B.* 1979, *19*, 2074.

84. Chadi, D.J. *Vacuum.* 1983, *33*, 613.
85. Chadi, D.J. *Phys. Rev. B.* 1984, *29*, 785.
86. Dick, G.B.; Overhauser, A.W. *Phys. Rev.* 1958, *112*, 90.
87. Ewald, P.P. *Ann. Phys.* 1921, *64*, 253.
88. Parry, D.E. *Surf. Sci.* 1975, *49*, 433.
89. Parry, D.E. *Surf. Sci.* 1976, *54*, 83.
90. Godin, T.J.; LaFemina, J.P.; Duke, C.B. *J. Vac. Sci. Tech. A*, 1992, *10*, 2059.
91. Göpel, W. *Ber. Bunsenges. Phys. Chem.* 1978, *82*, 744.
92. Anderson, A.B.; Nichols, J.A. *J. Am. Chem. Soc.* 1986, *108*, 1385 and references therein.
93. Frost, J.C. *Nature.* 1988, *334*, 577.
94. Au, C.T.; Hirsch, W.; Hirschwald, W. *Surf. Sci.* 1989, *221*, 113.
95. Liu, L. *High-Pressure Research: Applications in Geophysics*; M.H. Manghnani; S. Alimoto, Eds. Academic Press: New York, 1977; pp. 245–253.
96. Skinner, A.J.; LaFemina, J.P. *Phys. Rev. B.* 1992, *45*, 3557.
97. Duke, C.B.; Wang, Y.R. *J. Vac. Sci. Tech.* 1988, *B6*, 1440.
98. Duke, C.B.; Lubinsky, A.R.; Chang, S.C.; Lee, B.W.; Mark, P. *Phys. Rev. B.* 1977, *15*, 4865.
99. Duke, C.B.; Meyer, R.J.; Paton, A.; Mark, P. *Phys. Rev. B.* 1978, *18*, 4225.
100. Jaffe, J.E.; Harrison, N.M.; Hess, A.C. *Phys. Rev. B.* 1994, *49*, 11153.
101. Schroer, P.; Kruger, P.; Pollman, J. *Phys. Rev. B.* 1994, *49*, 17092.
102. Duke, C.B. *Crit. Rev. Solid State Sci.* 1978, *8*, 69.
103. Lubinsky, A.R.; Duke, C.B.; Chang, S.C.; Lee, B.W.; Mark, P. *J. Vac. Sci. Tech.* 1976, *13*, 189.
104. Kahn, A.; Duke, C.B.; Wang, W.R. *Phys. Rev. B.* 1992, *44*, 5606.
105. Duke, C.B.; Lessor, D.L.; Horsky, T.N.; Brandes, G.; Canter, K.F.; Lippel, P.H.; Mills, A.P., Jr.; Paton, A.; Wang, Y.R. *J. Vac. Sci. Tech.* 1989, *A7*, 2030.
106. Gay, R.R.; Nodine, M.H.; Henrich, V.E.; Seiger, H.J.; Solomon, E.I. *J. Am. Chem. Soc.* 1980, *102*, 6752.
107. Chang, S.C.; Mark, P. *Surf. Sci.* 1974, *46*, 293.
108. Sambi, M.; Granozzi, G.; Rizzi, G.A.; Casarin, M.; Tondello, E. *Surf. Sci.* 1994, *319*, 149.
109. Rohrer, G.S.; Bonnell, D.A. *Surf. Sci. Lett.* 1991, *247*, L195.
110. Gibson, A.; Haydock, R.; LaFemina, J.P. *Vac. Sci. Tech. A.* 1992, *10*, 2361.
111. Henrich, V.E. *Surf. Sci.* 1976, *57*, 385.
112. Legg, K.O.; Prutton, M.; Kinniburgh, C. *J. Phys. C: Solid State Phys.* 1974, *7*, 4236.
113. Kinniburgh, C. *J. Phys. C: Solid State Phys.* 1975, *8*, 2382.
114. Prutton, M.; Walker, J.A.; Welton-Cook, M.R.; Felton, R.C.; Ramsey, J.A. *Surf. Sci.* 1979, *89*, 95.
115. Welton-Cook, M.R.; Berndt, W. *J. Phys. C: Solid State Phys.* 1982, *15*, 5961.
116. Urano, T.; Kanaji, T.; Kaburagi, M. *Surf. Sci.* 1983, *134*, 109.
117. Ichimiya, A.; Takeuchi, Y. *Surf. Sci.* 1983, *128*, 343.
118. Maksym, P. *Surf. Sci.* 1985, *149*, 157.
119. Rieder, K.H. *Surf. Sci.* 1982, *118*, 57.
120. Cui, J.; Jung, D.R.; Frankl, D.R. *Phys. Rev. B.* 1990, *42*, 9701.
121. Jung, D.R.; Cui, J.; Frankl, D.R. *J. Vac. Sci. Tech. A.* 1991, *9*, 1589.
122. Prutton, M.; Walker, J.A.; Welton-Cook, M.R.; Felton, R.C.; Ramsey, J.A. *J. Phys. C: Solid State Phys.* 1979, *12*, 5271.
123. Kinniburgh, C.G.; Walker, J.A. *Surf. Sci.* 1977, *63*, 274.
124. Welton-Cook, M.R.; Prutton, M. *J. Phys. C: Solid State Phys.* 1980, *13*, 3993.
125. Lad, R.J.; Henrich, V.E. *Phys. Rev. B.* 1988, *38*, 10860.
126. Bas, E.B.; Baninger, U.; Mulethaler, H. *Jpn. J. Appl. Phys. Suppl. 2, Pt. 2, Proc. 2nd Intern. Conf. on Solid Surfaces,* 1974, 671.
127. Felton, R.C.; Prutton, M.; Matthew, J.A.D. *Surf. Sci.* 1979, *79*, 117.
128. Felton, R.C.; Prutton, M.; Tear, S.P.; Welton-Cook, M.R. *Surf. Sci.* 1979, *88*, 474.
129. Chambers, S.A.; Tran, T.T.; Hileman, T.A. *J. Mat. Res.* 1994, *9*, 2944.
130. Linsebigler, A.L.; Lu, G.; Yates, J.T., Jr. *Chem. Rev.* 1995, *95*, 735.
131. Janata, J. *Principles of Chemical Sensors.* Plenum Press: New York, 1989.

132. de Frésart, E.; Darville, J.; Gilles, J.M. *Solid State Commun.* 1980, *37*, 13.
133. de Frésart, E.; Darville, J.; Gilles, J.M. *Appl. Surf. Sci.* 1982, *11/12*, 259.
134. de Frésart, E.; Darville, J.; Gilles, J.M. *Surf. Sci.* 1983, *126*, 518.
135. Cox, D.F.; Fryberger, T.B.; Semancik, S. *Phys. Rev. B.* 1988, *38*, 2072.
136. Cox, D.F.; Fryberger, T.B.; Semancik, S. *Surf. Sci.* 1989, *224*, 121.
137. Cox, D.F.; Fryberger, T.B. *Surf. Sci. Lett.* 1990, *227*, L105.
138. Onishi, H.; Iwasawa, Y. *Surf. Sci. Lett.* 1994, *313*, L783.
139. Vogtenhuber, D.; Podloucky, R.; Neckel, A.; Steinemann, S.G.; Freeman, A.J. *Phys. Rev. B.* 1994, *49*, 2099.
140. Tait, R.H.; Kasowski, R.V. *Phys. Rev. B.* 1979, *20*, 5178.
141. Sadehgi, H.R.; Henrich, V.E. *Appl. Surf. Sci.* 1984, *19*, 330.
142. Ramamoorthy, M.; Vanderbilt, D.; King-Smith, R.D. *Phys. Rev. B.* 1994, *49*, 16271.
143. Godin, T.J.; LaFemina, J.P. *Phys. Rev. B.* 1993, *47*, 6518.
144. Rohrer, G.S.; Henrich, V.E., Bonnell, D.A. *Science.* 1990, *250*, 1239.
145. Rohrer, G.S.; Henrich, V.E., Bonnell, D.A. *Mat. Res. Soc. Symp. Proc.* 1991, *209*, 611.
146. Sander, M.; Engel, T. *Surf. Sci. Lett.* 1994, *302*, L263.
147. Cox, P.A.; Egdell, R.G.; Harding, C.; Patterson, W.R.; Tavener, P.J. *Surf. Sci.* 1982, *123*, 179.
148. Themlin, J.M.; Sporken, R.; Darville, J.; Caudano, R.; Gilles, J.M.; Johnson, R.L. *Phys. Rev. B.* 1990, *42*, 11914.
149. Chung, Y.W.; Lo, W.J.; Somorjai, G.A. *Surf. Sci.* 1977, *64*, 588.
150. Murray, P.W.; Leibsle, F.M.; Fischer, H.J.; Flipse, C.F.J.; Muryn, C.A.; Thornton, G. *Phys. Rev. B.* 1992, *46*, 12877.
151. Oliver, P.M.; Parker, S.C.; Purton, J.; Bullett, D.W. *Surf. Sci.* 1994, *307*, 1200.
152. Blanchard, D.L., Jr.; Baer, D.R. *J. Vac. Sci. Tech. A.* 1992, *10*, 2237.
153. Godin, T.J.; LaFemina, J.P. *Surf. Sci.* 1993, *301*, 364.
154. Reiger, R.; Prade, J.; Schröder, U.; deWette, F.W.; Kress, W. *Phys. Rev. B.* 1989, *39*, 7938.
155. Hikita, T.; Hanada, T.; Kudo, M.; Kawai, M. *Surf. Sci.* 1993, *287/288*, 377.
156. Liang, Y.; Bonnell, D.A. *Surf. Sci. Lett.* 1993, *285*, L510.
157. Charig, J.M. *Appl. Phys. Lett.* 1967, *10*, 139.
158. Chang, C.C. *J. Appl. Phys.* 1968, *39*, 5570.
159. French, T.M.; Somorjai, G.A. *J. Phys. Chem.* 1970, *74*, 2489.
160. Susnitzky, D.W.; Carter, C.B. *J. Am. Ceram. Soc.* 1986, *69*, C-217.
161. Baik, S.; Fowler, D.E.; Balkely, J.M.; Raj, R. *J. Am. Ceram. Soc.* 1985, *68*, 281.
162. Guo, J.; Ellis, D.E.; Lam, D.J. *Phys. Rev. B.* 1992, *45*, 13647.
163. Hsu T.; Kim, Y. *Surf. Sci.* 1991, *243*, L63.
164. Yao, N.; Wang, Z.L.; Cowley, J.M. *Surf. Sci.* 1989, *208*, 533.
165. Causà, M.; Dovesi, R.; Pisani, C.; Roetti, C. *Surf. Sci.* 1989, *215*, 259.
166. Manassidis, I.; De Vita, A.; Gillan, M.J. *Surf. Sci. Lett.* 1993, *285*, L517.
167. Godin, T.J.; LaFemina, J.P. *Phys. Rev. B.* 1994, *49*, 7691.
168. Liu, L.-G.; Bassett, W.A. *Elements, Oxides, and Silicates: High Pressure Phases with Implications for the Earth's Interior.* Oxford University Press: New York, 1986.
169. Liebau, F. *Structural Chemistry of Silicates: Structure, Bonding, and Classification.* Springer-Verlag: Berlin, 1985.
170. Pantelides, S.T.; Lucovsky, G.; Eds. SiO_2 and Its Interfaces. Materials Research Society: Pittsburgh, 1988.
171. Wyckoff, R.W.G. *Crystal Structures;* Vol. I. John Wiley & Sons: New York, 1963.
172. Wright, A.F.; Lehmann, M.S. *J. Solid State Chem.* 1981, *36*, 371.
173. O'Keefe, M.; Hyde, B.G. *Acta Crystallogr. B.* 1976, *32*, 2923.
174. Leadbetter, A.J.; Wright, A.F. *Philos. Mag.* 1976, *33*, 105.
175. Smyth, J.R.; Smith, J.V.; Artioli, G.; Kvick, Å. *J. Phys. Chem.* 1987, *91*, 988.
176. Geisinger, K.L.; Spackman, M.A.; Gibbs, G.V. *J. Phys. Chem.* 1987, *91*, 3237.
177. Shropshire, J.; Keat, P.P.; Vaughn, P.A. *Z. Krist.* 1959, *112*, 409.
178. Stishov, S.M.; Popova, S.V. *Geokhimiya.* 1961, *10*, 837.
179. Hill, R.J.; Newton, M.D.; Gibbs, G.V. *J. Solid State Chem.* 1983, *47*, 185.

180. Jánossy, I.; Menyhand, M. *Surf. Sci.* 1971, *25*, 647.
181. Bermudez, V.M.; Ritz, V.H. *Phys. Rev. B.* 1979, *20*, 3446.
182. Görlich, E.; Haber, J.; Stoch, A.; Stoch, J. *J. Solid State Chem.* 1980, *33*, 121.
183. Nylén, M. *Phys. Stat. Sol. b.* 1984, *122*, 301.
184. Heggie, M.; Jones, R.; Nylén, M. *Philos. Mag. B.* 1985, *51*, 573.
185. Stipp, S.L.; Hochella, M.F., Jr. *Geochim. Cosmochim. Acta.* 1991, *55*, 1723.
186. Liang, Y.; Lea, A.S.; Baer, D.R.; Engelhard, M.H. *Surf. Sci.* 1994. Submitted.
187. O'Keeffe, M. *Structure and Bonding in Crystals*; Vol. 1. M. O'Keeffe; A. Navrotsky, Eds. Academic Press: New York, 1981; p. 299.
188. Shull, H. *J. Appl. Phys.* 1962, *33*, 290.
189. Slater, J.C. *Quantum Theory of Molecules and Solids*; Vol. 2. McGraw-Hill: New York, 1965.
190. Goniakowski, J.; Noguera, C. *Surf. Sci.* 1994, *319*, 68.
191. Goniakowski, J.; Noguera, C. *Surf. Sci.* 1994, *319*, 81.
192. Munnix, S.; Schmeits, M. *Phys. Rev. B.* 1983, *28*, 7342.
193. Munnix, S.; Schmeits, M. *Phys. Rev. B.* 1984, *30*, 2202.
194. LaFemina, J.P.; Duke, C.B. *J. Vac. Sci. Tech.* 1991, *A9*, 1847.
195. Causà, M.; Dovesi, R.; Pisani, C.; Roetti, C. *Surf. Sci.* 1986, *175*, 551.
196. Welton-Cook, M.R.; Prutton, M. *Surf. Sci.* 1978, *74*, 276.
197. Martin, A.J.; Bilz, H. *Phys. Rev. B.* 1979, *19*, 6593.
198. Lewis, G.V.; Catlow, C.R.A. *J. Phys. C.* 1985, *18*, 1149.
199. Schröder, U.; deWette, F.W.; Kress, W. *Proc. 2nd Int. Conf. Phonon Physics.* World Scientific: Philadelphia, 1986; p. 653.
200. Nakamatsu, H.; Sudo, A.; Kawai, S. *Surf. Sci.* 1988, *194*, 265.

Chapter 2

ELECTRONIC STRUCTURE OF MINERAL SURFACES

Victor E. Henrich

CONTENTS

0-8493-8351-X/96/$0.00+$.50
© 1996 by CRC Press, Inc.

I. INTRODUCTION

A knowledge of the electronic structure of mineral surfaces is a prerequsite to understanding many of their properties. Surface electronic structure is determined not only by the bulk electronic properties of the material but also by surface geometry, defects, adsorbates, etc. The purpose of this chapter is to address electronic structure at the most fundamental level: on the atomic scale for atomically clean surfaces. Only in this way can experimental results be compared with theoretical calculations in order to obtain as full an understanding of the surface as possible. Experimentally, this means carefully controlled surface science measurements on well-characterized single-crystal surfaces.

In practice, this goal limits us to a fairly small subgroup of minerals. Most of the surface science research performed to date has been motivated by other than geologic concerns. Thus, the choice of materials studied does not necessarily mirror the geologist's primary interests. The only class of materials other than metals and semiconductors that has been investigated in any detail is metal oxides. However, many of the geologically important oxides have yet to be studied. For example, there have been no measurements on any aluminosilicate, and even SiO_2 has been examined only cursorily. On the other hand, some of the metal oxides that have been studied in depth are not minerals. Lower oxides of the transition metals, such as Ti_2O_3 and V_6O_{13}, do not occur in nature, since in an atmospheric or aqueous environment they would oxidize to higher oxides of the metal cation (i.e., TiO_2 or V_2O_5). Also, only a few ternary metal oxides have been studied. Another important class of minerals — the metal sulfides — has not yet been studied with regard to their surface electronic structure. This chapter will thus be restricted to a discussion of oxides.

However, even with the limited range of materials that have been studied to date, it is possible to get a good general picture of the electronic structure of oxide minerals. Measurements and calculations on both transition-metal and non-transition-metal oxides cover the entire spectrum of types of surface electronic structure; the surface properties of most oxide minerals will be extensions of these concepts. We will concentrate on oxide minerals whenever possible, but consideration of nonmineral compounds is also included since it is important for a full understanding of electronic properties, particularly of defects.

Although this chapter is devoted to the electronic structure of mineral surfaces, one of the primary geologic concerns is the interaction of minerals with the environment. Many of the surface science studies of metal oxides that have been performed have examined the chemisorption of molecules onto well-characterized surfaces. Particularly important in determining chemisorption behavior is the defect structure of the surface and the changes in electronic structure that defects produce. In ionic compounds such as metal oxides, point defects (usually O vacancies) drastically alter the surface electronic structure. That electronic structure is in turn altered by chemisorption. We will thus include some general aspects of chemisorption on metal-oxide surfaces here, along with a few examples for specific systems.

There are two basically different ways in which atoms and molecules interact with metal-oxide surfaces. The predominant type of interaction is *donor/acceptor*, or *acid/base*, which involves the overlap of filled electronic orbitals on donor (Lewis base) ions with empty ones on acceptor (Lewis acid) ions. Surface cations are acidic, while surface O^{2-} ions are basic. Although orbitals overlap in the bonding, there is no change in the net charge (i.e., valence state) of the adsorbate or of the substrate ions. Arguably the most important donor/acceptor reaction on metal-oxide surfaces is the dissociation of water to form adsorbed hydroxyl ions:

$$H_2O + O^{2-}_{lattice} \rightarrow OH^-_{lattice} + OH^-_{ads} \qquad (1)$$

The other type of chemisorption on metal oxides involves *reduction/oxidation*, or *redox*, reactions. Two categories of redox reaction are possible. One involves the actual transfer of electronic charge between adsorbate and surface, with a change in the oxidation state of each. A simple example would be

$$Cl_2 + 2e^-_{\text{lattice}} \rightarrow 2Cl^-_{\text{ads}} \qquad (2)$$

Such reactions are rare on non-transition-metal oxides since too much energy is required to change the valence state of the substrate ions. However, they are common on transition-metal oxides, where the energy difference between two different valence states on the substrate cations is small.

The other type of redox reaction involves transfer of lattice O ions to or from the substrate. An example is

$$CO + O^{2-} \rightarrow CO_2 + 2e^- \qquad (3)$$

Oxygen-exchange redox reactions are the reason that metal oxides can be effective selective oxidation catalysts for hydrocarbons.

Because of the limited space available in this chapter, it is not possible to consider all of the nuances of the electronic structure of, or chemisorption on, oxide surfaces. Readers interested in greater detail in any of the areas covered in this chapter are referred to Reference 1.

II. MODELS OF ELECTRONIC STRUCTURE

Conceptual approaches to the electronic structure of solids can be divided into two classes: *localized* and *itinerant*.[2] In localized models, one begins with an atomic description in which all electrons occupy atomic orbitals on specific atoms. The interaction between atoms in a solid is then treated as a perturbation of the atomic orbitals via electron–electron interactions between different atoms. Itinerant descriptions, on the other hand, consider the electrons as waves extending throughout an infinite periodic lattice. It is intuitively reasonable that localized models should be appropriate for insulating, highly ionic compounds, while itinerant models should best describe metals. For the most part, that is the case. But there exists a wide variety of materials that do not fall neatly into either of those categories. Unfortunately, most metal oxides are of that type. It is, therefore, not obvious how to describe the electronic structure of many metal oxides. The localized and itinerant models have such different origins that there is no easy way to formulate a description that contains the features of both. This is currently a very active area in solid state physics, and metal oxides are playing a crucial role as test cases for the various theoretical approaches. So it will not be possible for us to give complete descriptions of the electronic structure of many of the materials that are of interest here. However, the fundamental properties of the electronic structure of metal oxides can be understood by considering the range of models that has been used and the strengths and limitations of each.

Throughout this chapter we will consider two classes of metal oxide. In *non-transition-metal oxides*, such as MgO, Al_2O_3, ZnO, and SnO_2, the cation orbitals that are involved in crystal bonding have s and p symmetry. With the exception of Sn, which can exist either as Sn^{2+} or Sn^{4+}, those cations occur in only one valence state in metal oxides. It requires such a large amount of energy to add an electron to, or remove an electron from, a non-transition-metal cation that such a process does not occur under normal circumstances. In *transition-metal oxides*, on the other hand, the cation bonding orbitals have d symmetry. Those cations each have two or more valence states that lie close in energy, so that it is relatively easy to change the valence of a cation, for example, in the vicinity of a defect. The electronic structure of non-transition-metal oxides is significantly simpler than that of transition-metal oxides, and their properties are correspondingly less varied. We will therefore consider the electronic structure of each in separate sections below.

A. Energy Bands

At one extreme of the spectrum of approaches to electronic structure is the fully delocalized, or *energy band*, concept. In this picture, the electrons that are involved in crystal bonding are represented by fully extended states, spread throughout the entire crystal, whose properties depend primarily upon the translational symmetry of the crystal lattice. (Tightly bound core electrons are still considered as atomic orbitals, but they play no direct role in conduction or chemical properties.) The individual atoms are important in that they provide the spatially varying potential in which the electrons move, but none of the outer electrons is considered to be associated with any particular atom. The two parameters that characterize each electronic state are its energy, E, and its wavevector, **k**, or (equivalently) crystal momentum, $\hbar\mathbf{k}$. (When magnetic effects are important, as they are in magnetite and bunsenite, for example, the spin of the electron must also be considered.) The properties of electrons are entirely specified by the *dispersion relation*, E(**k**). Plots of E(**k**), usually in a few high symmetry directions in k-space, are referred to as *energy-band diagrams*.

An example of such a band diagram is given in Figure 1 for the mineral periclase, MgO.[3] The ordinate gives the energy of the allowed electronic states, and the abscissa is the electron wavevector **k**. All of the information on electronic states is contained in values of **k** that lie in the first Brillouin zone, and the letter designations along the abscissa refer to high symmetry points on the zone boundary (except Γ, which specifies the zone center). Energy bands are defined as those regions of energy in which allowed electronic states exist, whether or not the states are occupied. The highest-lying band of states that is occupied in the ground state of MgO thus extends from 0 to –6 eV. This is referred to as the *valence band*, and comparison with localized models of electronic structure (to be

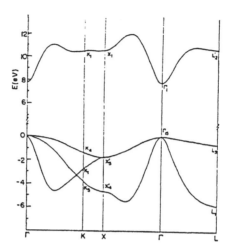

Figure 1

Bulk band structure of MgO calculated by the LCAO (tight-binding) method. Only the upper part of the valence band and the lower part of the conduction band (O 2p and Mg 3s contributions, respectively) are shown. (From Lee, V.C.; Wong, H.S. *J. Phys. Soc. Jpn.* 1978, *45*, 895. With permission.)

discussed below) shows that the electrons in this band are derived primarily from O 2p orbitals. The lowest-lying empty band extends upward in energy from 7.6 eV and is referred to as the *conduction band*. In a localized picture its electrons would correspond predominantly to Mg 3s and 3p orbitals. The energy difference between the top of the valence band and the bottom of the conduction band is referred to as the *bandgap* of the material. The Fermi energy, E_F, which is the line of demarcation between filled (below E_F) and empty (above E_F) states, lies somewhere in the bulk bandgap.

In reality, the valence band contains a significant component of cation character, and the conduction band contains some anion character. This is referred to as *hybridization* or *covalency* and will be discussed in Section II.B below. The O 2p orbitals that comprise the upper part of the valence band (i.e., the smaller binding energy region) are those with very little bonding character (i.e., those O 2p orbitals that do not interact appreciably with electrons on the Mg atoms). The states near the bottom of the valence band are the O 2p orbitals that directly participate in bonding to the Mg atoms. Thus, the upper and lower parts of metal-oxide valence bands are sometimes referred to as *nonbonding* and *bonding*, respectively.

The energy-band description of electronic structure is most applicable to simple metals, such as Na or Al, and to elemental semiconductors such as Si. Its validity for metal oxides is questionable. It can be shown that for closed-shell nonmetallic solids, such as MgO and Al_2O_3, the band and localized pictures are formally equivalent descriptions of the *ground state* of the material. However, this is not true for solids having partially filled bands or for *excited states* of any solid. Since any measurement perturbs the system that it is measuring, the band structure approach cannot be used directly to interpret most experimental results. Nonetheless, many metal oxides do have itinerant electrons, and the concept of bands arises, as we shall see below, in localized electronic structure pictures as well.

B. Localized Electron Approaches

The other extreme in conceptualizing electronic structure is to consider the electrons to be localized on particular atoms. The simplest localized approach is the *ionic model*, in which the solid is considered to be composed of positive cations and negative anions that interact only weakly with each other. For metal oxides, such a model is most applicable to the pre-transition-metal oxides MgO, CaO, SrO, and Al_2O_3. The translational symmetry of the crystal lattice is no longer important, and it is the *local ligand environment* of ions that determines their electronic structure. For MgO, in which the Mg ions are very nearly Mg^{2+} and the O ions are O^{2-}, the valence and conduction levels derived from the ionic model are shown in Figure 2.[24] Column (a) gives the energies for various changes in oxidation state on the free ions. Energies are referred to the vacuum level, which is defined as the state in which an electron has been removed from an atom or ion and placed at rest an infinite distance away. (Practically, "infinite" need be no farther than hundreds of angstroms from the atom or ion.) Although oxygen anions are almost always O^{2-} in metal oxides, the free ion O^{2-}/O^- level is positive with respect to the vacuum level by about 9 eV, and an isolated O^{2-} ion is unstable. In order to stabilize O^{2-}, it is necessary to put the ions in a lattice where they are surrounded by positive cations. They are then subject to the *Madelung potential*, due to the other ions, which is sufficient to bring the O^{2-}/O^- level below the vacuum level. The Mg^+/Mg^{2+} level is the negative of the second ionization energy of Mg. In ionic oxides the Madelung potential stabilizes electron orbitals on anions and destabilizes them on cations.

In the ionic model for MgO, which is shown in column (b) in Figure 2, the occupied orbitals having the highest energies (i.e., those that are least tightly bound) are the 2p orbitals on the O^{2-} ions, and the lowest unoccupied orbitals are the Mg 3s. Molecular orbital terminology is often used to describe these orbitals: HOMO refers to the highest occupied molecular orbital, and LUMO refers to the lowest unoccupied molecular orbital. The energy difference between the HOMO and LUMO is equivalent to the bandgap in the energy-band model. When only the Madelung potential is included, the ionic model gives bandgaps that are too large. In MgO, the ionic value is about 24 eV, while the experimentally measured value is only 7.8 eV. The origin of this discrepancy lies in having neglected *polarization* and *orbital overlap*. Electrostatic polarization refers to the deformation of the electron distribution on an atom or ion that takes place

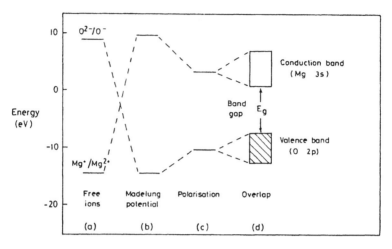

Figure 2
Derivation of the valence- and conduction-band energies in MgO using the ionic model. Cation and anion energy levels are shown (a) for free ions; (b), (c), and (d) including Madelung potentials, relaxation, and bandwidths, respectively.

in the presence of an electric field, such as that generated by its ligands. Polarization tends to lower the energy of the LUMO and raise the energy of the HOMO, thus reducing the bandgap, as shown in column (c). Orbital overlap, which is sometimes referred to as *bandwidth*, must also be included because, in a solid, the outer electron wavefunctions on adjacent atoms or ions must partially occupy the same region of space. The equilibrium interatomic spacing in an ionic solid is determined by a balance between the attractive coulomb force between ions having opposite charges, and a repulsive force, of purely quantum mechanical origin, that the Pauli exclusion principle requires when one tries to put too many electrons into the same small volume. Thus, orbital overlap is necessary in order to keep a solid from collapsing. It is not possible to determine the size of orbital overlap effects from the ionic model alone; it must be determined either from band-structure calculations or from experiment. In most metal oxides the O 2p bandwidth is around 6 eV, as shown in column (d). Bandwidths for the empty cation states are usually larger than that, since the electron wavefunctions are more extended spatially. As in the energy-band picture, atomic core levels, such as O 1s and 2s and Mg 1s, 2s, and 2p, need not be considered explicitly since they have high binding energies and are very localized spatially.

It is possible to consider electrons as localized on particular ions without using a fully ionic model. As the overlap of electronic wavefunctions on adjacent ions increases, it is useful to think in terms of orbital *hybridization*, in which the electronic states on a particular ion are composed of both cation and anion wavefunctions. (Obviously, hybridization *must* occur in any solid, so the point at which it is useful to treat it explicitly is merely one of degree.) The HOMO orbitals can then be thought of as bonding — and the LUMO orbitals as antibonding — combinations of cation and anion wavefunctions. For the post-transition-metal oxides ZnO and SnO_2, which are believed to be significantly less ionic than the pre-transition-metal oxides, this is probably a better way of considering their electronic structure. In this description the bonding orbitals will be concentrated on the O^{2-} ions and the antibonding orbitals on the positive cations, which is consistent with the purely ionic picture.

One of the most useful calculational approaches for describing localized electron materials is to consider a small *cluster* of ions in, or at the surface of, the solid as a molecule, and then use molecular-orbital methods to determine its allowed electronic states. Those states correspond to electrons that are localized *on* the cluster, but delocalized *within* the cluster. It is extremely important, however, to take proper account of the presence of the rest of the solid by some suitable *embedding* technique; the accuracy of cluster calculations depends strongly on the type of embedding potential used. For example, neglecting the polarization of the ions outside the cluster can shift the calculated energy levels by several electron volts. It is not surprising, therefore, that cluster calculations may not give good values for the bandgap.

The results of a cluster calculation of the bulk electronic structure of MgO are shown in Figure 3.[5,6] The "molecule" chosen for explicit calculation was an octahedral $[MgO_6]^{10-}$ cluster. The molecular-orbital energy levels were then calculated by a discrete variational $X\alpha$ (DV-$X\alpha$) method. The empty orbitals labeled $6a_{1g}$ and $6t_{1u}$ correspond closely to the Mg 3s and 3p levels, respectively, although they are slightly hybridized with O 2p. The primarily O 2p orbitals are those lying between -5 and -8 eV; there is, of course, some Mg 3s and 3p hybridization in those levels.

The cluster results show clearly the transition from localized to delocalized behavior. As the cluster size increases, the number of molecular orbitals in any energy range increases, ultimately merging to form bands. Since metal oxides have both localized and delocalized components to their electronic structures, it is important to keep the full range of models in mind and not force a particular material into a specific model; the truth will in general lie somewhere in between.

C. Models of Transition-Metal Oxides

The discussion of the conceptual approaches to electronic structure given above applies to all types of metal oxide. However, in transition-metal oxides, whose bonding orbitals consist of d electrons, additional complications occur that necessitate more sophisticated approaches and concepts. One of the most important properties of the d electrons in transition-metal oxides is the possibility of *variable oxidation states*. Mn, for example, can occur as ions having valences of 2+, 3+, 4+, 6+, or 7+. One

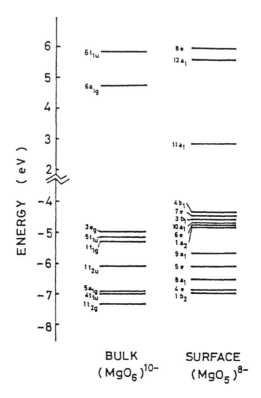

Figure 3
Orbital energies calculated by the DV-Xα method for MgO clusters, representing the bulk (left) and the (100) surface (right). (From Satoko, C.; Tsukada, M.; Adachi, H. *J. Phys. Soc. Jpn.* 1978, *45*, 1333. With permission.)

of the consequences of this is that it is much easier to produce defects whose electronic structure is significantly different from that of the stoichiometric compound. For example, in rutile TiO_2 the Ti ions are normally Ti^{4+}. (The valence of oxygen in stoichiometric metal oxides is always assumed to be formally O^{2-}.) If an O^{2-} ion is removed from either the bulk or the surface, it is necessary to have two additional electrons associated with the defect site in order to maintain *local charge neutrality*.[1] (Coulomb forces are both very strong and very long range, so having a net charge on a crystal is not energetically favorable.) As we shall see below, it costs very little energy to place an electron on a Ti^{4+} ion, changing it (formally) to Ti^{3+}, and that is how the extra electrons are accommodated at defect sites. But the electronic levels of Ti^{3+} are very different from those of Ti^{4+}, particularly at the surface, and the result is the creation of occupied electronic states in the bulk bandgap of TiO_2.

The variety that results from multiple cation valence in the 3d-transition-metal oxides is shown in Table 1, where the stoichiometric compounds

TABLE 1.

d-Electron Configuration vs. Crystal Structure for 3d-Transition-Metal Oxides

	Bixbyite	Rutile	Corundum	Rocksalt	Spinel	Other
$3d^0$	Sc_2O_3	TiO_2	—	—	—	TiO_2 (anatase and brookite) V_2O_5 (orthorhombic) CrO_3 (orthorhombic)
$3d^1$	—	VO_2 (T ≥ 340 K)	Ti_2O_3	—	—	—
$3d^2$	—	CrO_2	V_2O_3	TiO_x (0.6 < x ≤ 1.28)	—	—
$3d^3$	—	β-MnO_2	Cr_2O_3	VO_x (0.8 ≤ x ≤ 1.3)	—	—
$3d^4$	Mn_2O_3	—	—	—	Mn_3O_4	—
$3d^5$	—	—	α-Fe_2O_3	MnO	Fe_3O_4	—
$3d^6$	—	—	—	FeO	Co_3O_4	—
$3d^7$	—	—	—	CoO	—	—
$3d^8$	—	—	—	NiO	—	—
$3d^9$	—	—	—	—	—	CuO (monoclinic)
$3d^{10}$	—	—	—	—	—	Cu_2O (cubic) ZnO (wurtzite)

are classified according to their cation d-electron configuration and their crystal structure.[1] Two features are apparent from this table. Since only a few different crystal structures are represented, there exist large isostructural families of oxides, all having the same geometric structure, but different cation d-electron configurations. By using isostructural oxides, it has been possible to separate geometric from electronic effects in the study of the properties of these materials. The other feature of the 3d-transition-metal oxides that is apparent from Table 1 is the variety of stable oxidation states that occur with many of the transition metals. The most obvious example in the table is vanadium. The highest valence state of V is 5+, and V_2O_5 is referred to as the *maximal valency oxide* of vanadium. The oxides having valences of 4+, 3+, and 2+, all of which are stable in bulk phases, are called *lower oxides*. The situation is even more complex than it appears from Table 1, however. For the V and Ti oxide systems, there are additional stable bulk phases, referred to as *Magnéli phases*, which have stoichiometries between those of the oxides shown in the table.[4] Those crystal structures contain crystallographic shear planes that are associated with periodic arrays of bulk oxygen vacancies. The cations on the shear planes have a lower valence state than those in the unsheared regions of the structure.

A different type of nonstoichiometry can also exist in many transition-metal oxides. As indicated in Table 1, TiO_x and VO_x can have values of x that cover a rather large range of composition; a similar effect occurs for Fe_xO.[7] These nonstoichiometric oxide structures are associated with defect formation in the bulk, where the defect density can be as large as 15% in TiO_x and VO_x. The ease of defect formation in such compounds can have an effect on their surface properties, since the surface stoichiometry may well be different from that of the bulk.

Energy-band descriptions of the electronic structure of transition-metal oxides can be formulated in much the same way as they are for the simpler non-transition-metal oxides. The O 2p orbitals will always be filled and will be the major component of the valence band. However, unlike non-transition-metal oxides, the occupation of the cation d states can range from d^0 to d^{10}. For insulating maximal valency d^0 oxides such as TiO_2 and V_2O_5, the band structure will look very much like that for MgO in Figure 1. There will be a bandgap separating the primarily O 2p valence band from the primarily Ti or V 3d conduction band, and the Fermi level will lie somewhere within the bandgap. However, for oxides having a d^n configuration, the situation becomes more complicated. The d-electron orbitals are significantly more spatially localized than are s or p orbitals. (The 4f orbitals in the rare earth elements are even more highly localized, and f-electron wavefunctions on adjacent atoms hardly overlap at all.) This means that a purely itinerant band picture, in which all valence and conduction electron states extend throughout the entire crystal, may not be as accurate a description of the system. That has, in fact, proved to be the case. Account must be taken of the localization of the d electrons on specific ions, so conceptual approaches that begin with an atomic picture are often more useful for describing these oxides.

In describing the excited states of transition-metal compounds, one must take account of the coulomb repulsion between the d electrons on a particular atom. A good example of this occurs in bunsenite, NiO. The electronic ground state of NiO consists, in a purely ionic picture, of the configuration $Ni^{2+}O^{2-}$. To generate an excited electronic state of this oxide, such as for describing electrical conduction or optical reflectivity, one needs to consider removing a d electron from one cation and placing it on an adjacent cation. The ionic description of this process is

$$Ni^{2+} + Ni^{2+} \rightarrow Ni^{1+} + Ni^{3+} \tag{4}$$

This process requires a net input of energy, the value of which is referred to as the *Hubbard U*.[4] The quantity U results primarily from the d–d coulomb repulsion between electrons when they are on the same ion. For ions in free space (i.e., where their electron wavefunctions do not significantly overlap), U can be as large as 20 eV. In the solid, *screening* of the excess electronic charge by other electrons and *polarization* of adjacent ions in response to the excess charge (which was mentioned previously in the discussion of cluster calculations) tend to greatly reduce that value. For the high-Z transition-metal compounds, including NiO, values of U are in the range of 5–7 eV. For the low-Z transition-metal compounds, such as those of Ti and V, U is closer to 3 eV.

Even in a localized picture of electronic structure in solids, account must be taken of the effects of the overlap of the electron wavefunctions on adjacent atoms. This results in a broadening of the atomic orbitals into bands, the width of which increases with increasing wavefunction overlap. This *bandwidth*, W, which tends to delocalize electrons, competes with U in determining the electronic structure of transition-metal oxides. The full Hubbard theory is quite complicated, but the basic idea of how the relative magnitudes of U and W determine the electronic structure of transition-metal compounds can be easily understood. When U > W, electron repulsion is too strong to permit the electrons to delocalize, and the material will be an insulator. However, if W > U, orbital overlap is so strong that it can overcome the coulomb repulsion, and the material will be a metal. In NiO, U > W, and thus stoichiometric NiO is insulating. In NiO and other high-Z transition-metal oxides, spin effects are also very important, giving rise to *magnetic insulators*. This important class of compounds will be discussed more fully in Section II.D below.

In compounds whose cations have different valence states in the stoichiometric material, electronic conduction can often occur much more easily than in the example of NiO above. In the case of magnetite, Fe_3O_4, some of the Fe ions are Fe^{3+}, while others are Fe^{2+}. Conduction would then correspond to

$$Fe^{2+} + Fe^{3+} \rightarrow Fe^{3+} + Fe^{2+} \tag{5}$$

Magnetite is indeed far more highly conducting than NiO for this reason. However, full metallic conductivity is not achieved because of additional effects such as polarization; further discussion of those nuances is beyond the scope of this chapter.

Another effect that is of crucial importance in determining the electronic structure of transition-metal oxides is *crystal-field* or *ligand-field splitting*. In the case of MgO discussed above, the LUMO, or bottom of the conduction band, consisted of spherical, nondegenerate s orbitals. In transition-metal oxides, however, the conduction orbitals are d electrons, which are fivefold degenerate and have spatially highly directional wavefunctions. The electric fields generated by the neighboring ions in the solid lift this degeneracy, with the manner in which the d orbitals are then split in energy depending upon the d orbital involved and the geometric arrangement of the surrounding ligands. By far the most common coordination geometry of the cations in transition-metal oxides is *octahedral*, in which each cation has six O^{2-} nearest-neighbor ligands. The O^{2-} ion arrangement is an undistorted octahedron in the rocksalt structure, and the group-theoretical terminology used to describe electron wavefunctions in that O_h symmetry is often used even if the O^{2-} octahedron is distorted. That distortion is small in the rutile structure, and somewhat larger in corundum. ZnO is the only common metal oxide that has a purely *tetrahedral* coordination, in which four O^{2-} ligands surround each cation. In minerals containing Si, the Si atoms are usually surrounded by a tetrahedron of ligands because of sp^3 hybridization. The spinel (and inverse spinel) crystal structures, such as that of magnetite, have both octahedral and tetrahedral cation sites.

The orientations of the five cation d-electron orbitals with respect to the six O^{2-} ligands in an octahedral site are shown in Figure 4.[4] The orbitals fall into two groups: three of t_{2g} symmetry and two of e_g symmetry. The e_g orbitals have their lobes of maximum charge density pointing directly toward the ligands, while t_{2g} orbitals are directed in between. The coulomb field of the ligands, along with orbital overlap and bonding interactions, split the d orbitals so that the energy of the t_{2g} orbitals is lowered relative to that of the e_g orbitals. The magnitude of this *crystal-field splitting*, Δ, is about 1 to 2 eV in the 3d-transition-metal oxides. The e_g orbitals can form σ bonds with the appropriate O 2p orbitals, and the t_{2g} levels can form π bonds; these bonding orbitals constitute the cation d-electron admixture, or *covalency*, in the primarily O 2p valence band. The corresponding antibonding orbitals form what are referred to as the *d levels*, or the conduction band in d^0 oxides.

In compounds where the transition-metal cations have a tetrahedral ligand coordination, the different symmetry changes the splitting of the d-electron levels. The t_{2g} orbitals lie at higher energy than the e_g, and the magnitude of the crystal field splitting is reduced. In lower symmetry environments, the e_g and t_{2g} levels themselves will be split as well.

D. Magnetic Insulators and Configuration Interactions

An important class of transition-metal oxides consists of those having partially filled d levels that are also insulators when stoichiometric. Examples are FeO, MnO, CoO, and NiO, all of which have the rocksalt structure and are antiferromagnetic. These materials posed a serious problem in the early days of energy-band theory, since materials having only filled and empty energy bands should be insulators, while those having partially filled bands, as would be the case for the d^n configuration with $0 < n < 10$, should be metals. Part of the explanation lay in considering the spin of the electrons and the way in which it modifies electron–electron interactions. These interactions result from the purely quantum mechanical *exchange* terms in the Hamiltonian of the system, which couple the spins of different electrons. In the rocksalt monoxides, exchange energy terms are comparable to crystal field splittings, so the ordering in energy of the various d orbitals can be quite different from that described above. The magnitude of the effect can be seen in Figure 5, which shows how the various perturbations affect the 3d orbitals in NiO and MnO.[8] The case of MnO, which has a half-filled, high-spin d^5 configuration as a result of a very large exchange energy, could be partially accounted for within an energy-band picture by assuming an appropriate antiferromagnetic

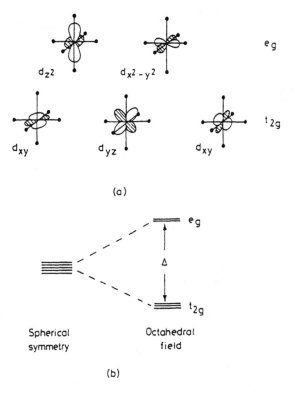

Figure 4
The crystal-field splitting of d orbitals in octahedral coordination: (a) orientation of d orbitals with respect to surrounding oxygens; (b) the resulting orbital energies.

order for the spins on adjacent cations. Antiferromagnetic order doubles the size of the real-space unit cell, thereby halving the dimensions of the Brillouin zone in reciprocal (or momentum) space. This creates a gap in the middle of the 3d band, so that the five majority-spin electrons just fill one band, and the five empty minority-spin levels constitute a separate band. However, the calculated magnitude of the bandgap was much smaller than the experimentally observed value. In addition, such arguments would not work for the other rocksalt monoxides which did not have an exactly half-filled d band. While a great deal of progress has been made in the past few years in describing magnetic insulators by band pictures, the situation is still far from clear, and controversies persist.

The Hubbard approach to the treatment of electron repulsion mentioned earlier was an attempt to explain the electronic structure of magnetic insulators. While band theory was unable to explain the insulating behavior of NiO and CoO, even when their antiferromagnetic order was included, the Hubbard model, which ignores the interaction between electrons on different atoms but explicitly treats electron–electron interactions

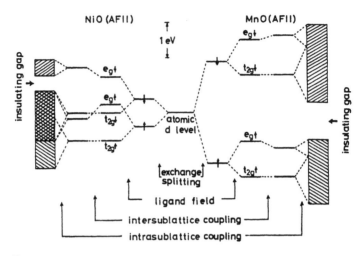

Figure 5
Schematic diagram showing how the energy bands of d states are constructed in antiferro-magnetic NiO and MnO. (From Terakura, K.; Oguchi, T.; Williams, A.R.; Kübler, J.; *Phys. Rev. B.* 1984, *30*, 4734. With permission.)

on the same atom (i.e., electron *correlation*), could account for their insu-lating behavior as long as U > W. Correlation splits the d-electron levels into two bands, referred to as the upper and lower Hubbard bands. In the example of NiO given above, the ground-state configuration [Ni^{2+} + Ni^{2+}] constitutes the lower band, while the upper, normally empty band corresponds to [Ni^{1+} + Ni^{3+}]. Exchange interactions cause the spins in the lower Hubbard band to align parallel, thus giving a net magnetic moment on each cation. Although more recent electronic structure calculations have gone far beyond the original Hubbard model, it is still used as the basis for many treatments of the properties of magnetic insulators.

Localized cluster methods are perhaps the most useful for describing the electronic structure of magnetic insulators. However, the simple ionic, or ligand field, picture is not adequate. Simple ligand field models assume that the ground state of a solid contains only one cation electron config-uration, say d^n, and that the O anions are O^{2-}. So the ground-state wave-function would be

$$\psi_g = a\,|\,d^n\rangle \tag{6}$$

Applying this model to the interpretation of valence-band photoemission experiments, which will be described in Section III.B below, is found to give a poor description of the spectra. A much better description is given by the *configuration–interaction model*,[9,10] which assumes that both the ground and excited states of the system are composed of a mixture of several different electron configurations. For the ground state, this is writ-ten as

$$\psi_g = a \left| d^n \right> + b \left| d^{n+1}\underline{L} \right> + c \left| d^{n+2}\underline{L}^2 \right> \tag{7}$$

where $\underline{L}$ denotes a hole on an O ligand. The last two terms on the right thus describe configurations in which electrons are transferred from the O^{2-} ligands to the cation; this is referred to as *ground-state covalency* and describes departures from a purely ionic picture. This model has been applied very successfully to the interpretation of photoemission spectra of magnetic insulators.

One of the successes of the configuration–interaction model has been in predicting the nature of the bandgap in transition-metal oxides.[11] In the Hubbard picture, exciting an electron from the HOMO (valence band) to the LUMO (conduction band) involves only a change in the configuration of the cation d electrons; the O 2p electrons do not play any direct role in bandgap transitions. Materials for which this picture is appropriate are called *Mott–Hubbard insulators*. For many years NiO was considered to be a prototypical Mott–Hubbard insulator. However, analysis of valence-band photoemission spectra from NiO using the configuration–interaction model showed that, for the ground state $\left| d^8 \right>$, the dominant configuration describing the upper edge of the valence band after emission of the photoelectron was $\left| d^8\underline{L} \right>$. In other words, exciting an electron from the highest-lying valence-band states (which are the initial states involved in bandgap optical transitions) could best be described as excitation of an O 2p electron. Such materials are now referred to as *charge-transfer insulators*, and the term Mott–Hubbard insulator is reserved for materials, like Ti_2O_3 and V_2O_3, where the states on both sides of the bandgap have primarily cation d-electron character. However, it is important to remember that for either charge-transfer or Mott–Hubbard insulators, the existence of upper and lower Hubbard bands is crucial. One can think of charge-transfer insulators as materials in which the lower Hubbard band lies below the O 2p band. At present, this area of oxide electronic structure is in flux, and more research is necessary before classifications can be considered to be definitive.

As we shall see in discussing measurement techniques below, one of the most important features of the configuration–interaction model is its ability to treat the large perturbations of electronic structure that occur when an electron is removed from an atom or ion. Band-theory approaches to the electronic structure of magnetic insulators are completely incapable of doing this.

III. MEASUREMENT TECHNIQUES

The surfaces of mineral specimens are in general extremely complex. Given the current status of the determination of the electronic structure of metal oxides, it would not be possible to meaningfully characterize most mineral samples. Experimental efforts to date have thus concentrated on well-characterized oxide surfaces — using single crystals whenever possible — so that the results can be interpreted by comparison with various theories. In this section we will discuss the methods that are employed to produce well-characterized oxide surfaces and the surface-sensitive techniques that are used to study them. For the reader who is interested in further details on any of the experimental techniques, there are several excellent books.[12-14]

A. Surface Preparation and Characterization

The goal of single-crystal studies of mineral surfaces is to understand their electronic structure in detail on the atomic scale. Experimental measurements are of much greater value if the surface is sufficiently well characterized that the results can be meaningfully compared with theoretical calculations. As has been mentioned above, the complexity of the excited states of metal oxides, and in many cases the lack of adequate knowledge of their bulk electronic structure, often requires the comparison of experimental spectra with calculations for even initial interpretation. Therefore, it is extremely important that surfaces be as well characterized structurally as possible.

For elemental materials, the preparation of atomically clean, well-ordered surfaces is a relatively easy task. Contamination can be removed from single-crystal surfaces in an ultrahigh vacuum (UHV) surface analysis system by ion bombardment (the use of inert gas ions such as Ar^+ or He^+ is implied unless stated otherwise), which is the atomic equivalent of sandblasting. Annealing in UHV then generally produces well-ordered surfaces, although surface reconstructions, such as the infamous (7×7) on Si (111), may be present. For any *compound*, however, the situation is more complicated. Ion bombardment can always be used to remove contamination, but it usually also removes one component of the compound more readily than it does others. This *preferential sputtering* creates clean but nonstoichiometric surfaces. It is no mean feat to find a subsequent annealing procedure that will restore both the geometric order and the stoichiometry to the surface, and in most cases it is not possible to do so.

The most likely procedure for producing at least stoichiometric surfaces is *cleavage* or *fracture* in UHV. Cleavage refers to easy breakage along specific bulk crystal planes, while fracture occurs in a variety of directions in samples that have no low-energy faces. (It is important to distinguish cleavage from *parting*; the latter refers to the fracture of mineral specimens along planes of structural weakness and generally occurs in twinned samples or ones that have been subjected to certain kinds of pressure. Cleavage, on the other hand, if it occurs, does so repeatedly in all specimens.) The low-energy surfaces that are produced by cleavage in UHV are not necessarily related to the naturally occurring faces of a crystal (i.e., its *habit.*). The latter are faces of low energy *in the ambient in which the crystal was formed*. For example, MgO crystals cleave only along the (100) plane, while periclase mineral samples often occur as octahedral crystals having predominantly (111) faces; the same situation occurs for bunsenite, NiO. The rocksalt (111) plane is polar, however, and in vacuum it is inherently unstable (see Figure 7 below).

Most of the surface science research that has produced meaningful electronic structure information on metal oxides has been performed on either cleaved or fractured samples. Some oxide surfaces can also be prepared by ion bombardment cleaning followed by annealing [for example, MgO (100), TiO_2 (110), Ti_2O_3 ($10\bar{1}2$) with difficulty], but they are in the minority. And measurements taken on those surfaces must at least initially be compared with ones from cleaved or fractured surfaces in order to establish their validity. In some cases, such as NiO (100), annealing of a stoichiometric UHV-cleaved surface destroys the stoichiometry of the surface.[1] Thus, most of the experimental results to which we refer below will be for cleaved or fractured single-crystal surfaces.

The experimental methods for characterizing the atomic scale geometric structure of surfaces are imperfect. The most frequently used technique is low-energy electron diffraction (LEED). It requires long-range order on the surface in order to produce diffraction patterns, and only those areas of the surface that exhibit long-range order contribute to the patterns. LEED is thus not very sensitive to point defects or random step distributions, and "good" LEED patterns can be produced in cases where the majority of the surface does not possess long-range order. However, in a number of careful experiments LEED has given detailed information on the relaxation and reconstruction of metal-oxide surfaces.

Local probing of the geometric structure of solid surfaces is possible by using the more recent techniques of scanning tunneling microscopy (STM) and atomic force microscopy (AFM). On metals and semiconductors these techniques have demonstrated the ability to image on the atomic scale, although such resolution is just beginning to be achieved on metal oxides. Since STM and AFM are local probes, they also have the potential to image individual defects. While point defects on metal oxides remain just beyond reach, steps have been imaged on several oxides.[1] In the near future STM and AFM will give us a much better understanding of the atomic arrangement on mineral surfaces.

B. Photoelectron Spectroscopy

The most powerful spectroscopic technique for determining the electronic structure of surfaces in the vicinity of the Fermi and vacuum levels, which is the region of interest for the interaction of a surface with the environment, is photoelectron spectroscopy. Monochromatic photons of energy $\hbar\omega$ incident on a sample in vacuum are absorbed partially by the excitation of electrons from filled states below E_F to normally empty allowed states above E_F, with the energy of the electrons increasing by $\hbar\omega$. If the final state of the electron lies above the vacuum level of the sample, there is a chance that the electron can be emitted from the surface into vacuum, where its kinetic energy is analyzed with an electron spectrometer. The initial-state energy of the electron can then be determined by subtracting $\hbar\omega$ from its final-state energy. Depending upon the energy of the photon, the process is usually classified as either ultraviolet or X-ray photoelectron spectroscopy (UPS or XPS, respectively), although the availability of the entire range of photon energies through the use of synchrotron radiation has resulted in other names being used. Two of the most common laboratory-based photon sources are He gas-discharge atomic lines: He I at 21.22 eV and He II at 40.84 eV.

Although photons are absorbed to great depths into the sample, photoemission is surface sensitive to the extent that the inelastic mean-free-path of the excited electrons is short. For electrons moving in solids with kinetic energies from a few tens to a few hundreds of eV, the mean-free-path can be as short as one or two monolayers. Thus, the main contribution to the intensity in UPS spectra comes from only the first few atomic layers. However, there is always some contribution from the bulk of the sample as well. The relative weighting of the bulk and surface contributions depends upon electron kinetic energy, angle of electron emission, etc.; these variables are often used to separate the bulk and surface components, although the procedure is not always straightforward.

Since negatively charged electrons are emitted from the sample during the photoemission process, while the incident photons are neutral, electrons must be able to move from the (usually grounded) sample support through the sample to its surface in order to neutralize the positive surface charge left after the emission of an electron. It requires only a small amount of bulk conductivity in the sample for this to occur, but many interesting minerals are extremely good insulators: MgO, Al_2O_3, many silicates, etc. There can thus be a serious problem in performing photoemission measurements on such materials. It is possible in principle to stabilize the surface potential by flooding the surface with a high flux of low-energy (a few eV) electrons, which are attracted to the positive surface charge and neutralize it. However, in practice this procedure does not always work well, and, as a result, very few highly insulating materials have had their electronic structure determined in any detail by photoemission. This is one of the primary reasons that we know as little about the electronic structure of mineral surfaces as we do.

The simple picture of the photoemission process given above works fairly well for emission from the valence band of nearly-free-electron metals and some semiconductors. But it is inadequate to explain the photoemission spectra of either deeper-lying core levels in *any* compound or even valence-band photoemission from ionic materials. In core-level emission, the electrons remaining on the atom or ion move spatially so as to screen the positive charge created by the core hole. There are associated shifts in energy of the electrons, and those shifts are mirrored in the kinetic energy of the photoemitted electrons. This gives rise to both shifts in the apparent binding energy of the core level from which the electron is emitted and additional peaks, often referred to as *satellites*, which correspond to different final-state screening processes. Thus, care must be taken in interpreting the features in any core-level photoemission spectrum, and in some cases the spectrum must be compared with theoretical calculations in order to determine the origin of various features.

The extent of the failings of the simple model in describing valence-band photoemission from ionic compounds has only been fully realized within the past few years, and interpretation of both UPS and XPS spectra for those materials is still a very active area of research. The origin of the problem, which is particularly severe with transition-metal oxides, has been alluded to in Sections II.C and II.D. For transition-metal oxides, the energy-band model upon which the simple interpretation of photoemission is based may not be applicable even to the ground-state electronic structure; it certainly does not apply to excited states of the system. In the discussion of the configuration–interaction model in Section II.D, Equation 7 for the ground state of the system already includes transfer of charge from the ligands to the cations (i.e., ground-state covalency) explicitly. After photoemission of a d electron from an n–electron ground state, the final state of the system contains only (n − 1) electrons. The wavefunction of the final state would then be written as

$$\psi_f = \alpha \left| d^{n-1} \right> + \beta \left| d^n \underline{L} \right> + \gamma \left| d^{n+1} \underline{L}^2 \right> \tag{8}$$

where the final-state screening of the valence-band hole is embedded in the coefficients β and γ. Full configuration–interaction calculations, often utilizing data from both core-level and valence-band photoemission, may thus be necessary in order to interpret experimental spectra.

There are three basic ways in which photoemission spectra can be taken, and each gives a different type of information about the electronic structure.[15] The methods can be illustrated most simply by using an energy-band picture of the electronic structure of the sample. Using a fixed incident-photon energy, a spectrum can be obtained by scanning the kinetic energy at which electrons are detected; this simply involves scanning the bandpass energy of the electron spectrometer. Such a spectrum, which is the type taken unless otherwise specified, is referred to as an *energy distribution curve*, or EDC. In terms of the density-of-states of electrons in the bands, each point in such a spectrum consists of a convolution of the filled and empty density-of-states corresponding to the initial and final electron states. (The final states are normally empty levels lying above the vacuum level.) If a tunable photon source is available, however, the joint density-of-states convolution can be partially eliminated. In *constant initial state* (CIS) spectroscopy, the electron analyzing energy is increased as the photon energy is increased, so that the photoelectrons observed always originate from the same initial state. The initial density-of-states is thus removed from the spectrum, although changes in excitation probabilities with changing final-state symmetry will still be present, so the spectra do not simply reflect the final density-of-states. In *constant final state* (CFS) spectroscopy, the photoelectron analyzing energy is held constant as the photon energy is increased, so the resulting spectra mirror the initial density-of-states; of course, the same caveat concerning transition probabilities applies here also. While the latter two types of measurement have been widely applied to metals and semiconductors, only a few ionic materials or metal oxides have yet been studied in this way.

Our discussion of photoemission so far has considered only the energy of the electron in its initial and final states. In an energy-band picture, however, the state of an electron is also characterized by the three components of its crystal momentum, $\hbar\mathbf{k}$. Since the momentum of even far-ultraviolet photons is extremely small on the scale of a Brillouin zone (i.e., on the scale of the crystal momentum possessed by electrons in solids), the momentum of an electron would not be changed significantly in the photoemission process. Such a transition is referred to as a *direct transition*. However, the coupling of electrons to the lattice in solids is very strong, so *phonons* can also play a role in the photoemission process.[2] The energy of even optical phonons in solids is small compared with the energy changes involved in photoemission, but phonons do possess a large amount of momentum. Thus, if a phonon is created or annihilated in the photoemission process, the momentum of the electron in the final state will be very different from that in the initial state. Such transitions are referred to as *indirect transitions*. Both types of transition occur in photoemission, and it is usually difficult, if not impossible, to separate them. In order to determine the momentum of a photoemitted electron, it is necessary to measure not only its final-state energy but also its *direction* of emission. Such measurements are referred to as *angle-resolved photoemission spectroscopy* (ARPES, ARUPS, etc.). The component of the momentum of an electron parallel to the surface is conserved as the electron is emitted from the solid, but, because of the lack of three-dimensional periodicity normal to the surface, its normal component is not. However, if one makes various assumptions about the potentials seen by the electron, its initial-state momentum in the direction normal to the surface can also be determined. Thus, in an energy-band picture, it is in principle possible to determine the full dispersion relation, $E(\mathbf{k})$, of the electrons in the ground state. In metal oxides the utility of this approach is questionable. The ionic nature of the bonding results in very strong electron–phonon coupling, and, as has been mentioned several times before, the band picture is of marginal utility at best in describing the electronic structure of metal oxides. In spite of this, however, a few band-mapping experiments have been conducted on metal oxides.

In addition to the CIS and CFS photoemission spectra described above, another important type of experiment can be performed with the use of tunable radiation. If the photon energy is swept through the optical absorption threshold for excitation of an electron from a core level to the lowest-lying empty orbital on the same atom or ion, large resonance effects are seen in the intensity of the emission from the outer valence orbitals. The effect has been studied primarily in transition-metal oxides. The origin of this *resonant photoemission* is as follows for a 3d-transition-metal oxide having a ground state configuration of $3d^n$ in the photon energy range around the 3p $\rightarrow$ 3d optical absorption threshold.[16] The process described above for photoemitting a d electron from one of the occupied 3d orbitals, which is referred to as the *direct process*, can be written as

$$3p^6 3d^n + \hbar\omega \rightarrow 3p^6 3d^{n-1} + e^- \tag{9}$$

This process occurs for any photon energy high enough that the final state of the excited electron lies above the vacuum level of the sample and can thus escape. However, when $\hbar\omega$ is greater than the 3p $\rightarrow$ 3d threshold, a competing two-step process can also occur. In the first step, a 3p electron is excited to a normally empty 3d orbital

$$3p^6 3d^n + \hbar\omega \rightarrow \{3p^5 3d^{n+1}\}^* \tag{10}$$

leaving the ion in an excited state. This is a very short-lived state, and the excited ion decays via a super-Coster–Kronig decay

$$\{3p^5 3d^{n+1}\}^* \rightarrow 3p^6 3d^{n-1} + e^- \tag{11}$$

The final state in Equation 11 is exactly the same as the final state in Equation 9, so the two emission processes interfere. This interference gives rise to changes in the intensity of the 3d photoemission that are similar to the Fano lineshapes derived theoretically to explain such processes. Resonant photoemission is being used increasingly to identify the origin of various features in photoemission spectra, since a resonance at a particular energy is ion or atom specific. Examples of its use will be given below for several materials.

UPS is also a very powerful technique for the study of adsorption on surfaces. Photoemission spectra from a surface that contains less than about one monolayer of an atomic or molecular adsorbate contain substrate electronic structure features, as well as electrons emitted from the molecular orbitals of the adsorbate. If one assumes that the interaction between the adsorbate and substrate is small, then, to first order, the UPS spectrum from a surface containing adsorbates can be considered as a sum of emission from the unperturbed substrate and the adsorbate molecular orbitals. If a spectrum for the clean surface is then subtracted from the spectrum from an adsorbate-covered surface, the resulting *difference spectrum* will be simply the emission from the adsorbate. There is, of course, always a finite adsorbate/substrate interaction, and proper account of this must be taken in interpreting difference spectra. But the technique is a powerful one for determining what species are adsorbed on solid surfaces.

C. Electron-Energy-Loss Spectroscopy

One of the other useful techniques for determining the surface electronic structure of materials is electron-energy-loss spectroscopy (EELS). A monoenergetic beam of electrons is incident on the surface of the sample, and, while some of the electrons will be reflected at the surface, others will enter the sample into allowed, but normally empty, electronic states. By interacting with the other electrons in the sample, those electrons will lose various amounts of energy as they create excitations in the sample. One important type of excitation in ionic compounds consists of phonons, but we will not consider that loss channel in this chapter as it is not directly relevant to the determination of electronic structure. Another loss channel consists of an incident electron exciting an electron from its ground state to some normally empty level above E_F. The incident electron can then be scattered back out of the sample, and the amount of energy that it lost in the excitation process can be measured. The resulting loss spectrum contains peaks that correspond to the energies of the various excitation processes in the sample. The short mean-free-path of the electrons makes EELS even more surface sensitive than photoemission. The features in an energy-loss spectrum that originate from electron excitation (or *electron-hole pair creation*) consist of a convolution of the filled and empty densities-of-states in the sample. Unlike photoemission, however, a particular peak in an EELS spectrum only characterizes the *difference* in energy between the initial and final state of an electron, without determining the absolute energy of either state. The electronic structure information is thus much more convoluted in EELS than in photoemission, and the technique is correspondingly less useful. However, important electronic structure information on metal oxides has been obtained by using EELS, some of which will be discussed below.

EELS has one practical advantage over photoemission on insulating samples. Since both the incident and emitted particles are electrons, it is often possible to find conditions under which the surface potential of the sample can be stabilized. Thus, EELS spectra can sometimes be obtained on samples where it is not possible to use photoemission.

If EELS spectra can be taken with sufficiently high energy resolution, they are capable of determining the vibrational spectra of surface phonons or the molecular vibrations of adsorbates. The technique is then referred to as high-resolution electron-energy-loss spectroscopy (HREELS) and is an important complement to UPS difference spectroscopy in studying adsorption on surfaces. Primary-electron energies of only a few electron-volts are used, making the technique extremely surface sensitive. Energy resolutions of 1 to 5 meV are readily achievable and are sufficient to observe most important intramolecular and molecule/substrate vibrational frequencies. For ionic materials such as metal oxides, however, low-energy electrons couple so strongly to substrate surface phonon vibrations that those loss peaks can be more than an order of magnitude larger than the loss features due to adsorbed molecules. Unfortunately, many of the interesting molecular vibrational frequencies occur in the same energy-loss region as do lattice modes. There are, however, several methods of at least partially removing some of the substrate phonon loss peaks so that adsorbate vibrational modes can be observed.[1]

D. Other Techniques

A potentially useful technique for determining the spectrum of normally empty electronic states in solids is inverse photoelectron spectroscopy (IPS). In IPS a beam of monoenergetic electrons is incident on the sample, as in EELS, and again we consider only those electrons that enter the sample. (When incident-electron energies of several keV are used, the process is referred to as Bremstrahlung isochromat spectroscopy, or BIS.) However, instead of measuring the spectrum of electrons that are reemitted from the sample, the photons that are emitted as the incident electrons drop into lower-lying empty orbitals are detected. Then, either the incident-electron energy can be swept as a fixed emitted photon energy is monitored, or the electron-beam energy can be fixed and the spectrum of emitted photons recorded. This process is formally the time reversal of photoemission. The filled electronic states in the sample do not participate in IPS, so this technique gives information complementary to that of photoemission. Like photoemission, however, IPS is susceptible to surface charging on insulating samples. That, coupled with the difficulty of detecting the emitted photons, has kept IPS from being applied very widely to metal oxides.

The interpretation of IPS spectra of ionic materials is also as complicated as that for photoemission. If the initial state of the system (before the incident electron has entered the sample) contains n electrons, then the final state has (n + 1) electrons. Proper interpretation of the spectra thus involves calculating the energies, etc. for both the n- and (n + 1)-electron states. To date this has only been done for IPS on a few metal oxides. Also, for the electron and photon energies usually used, IPS and BIS are relatively bulk-sensitive techniques, so little data have been obtained on surface electronic structure.

The scanning tunneling microscope (STM) can be used in a mode referred to as scanning tunneling spectroscopy (STS) that is capable of measuring both filled and empty densities-of-states within a few electron-volts of the Fermi level. The current that flows between the tip and the sample in an STM depends not only upon the magnitude and polarity of the voltage applied between them, but also on their densities-of-states near E_F. If the tip density-of-states is known, it is then possible to deconvolute the density-of-states of the sample from a spectrum of tunneling current as a function of tip-to-sample voltage as the tip remains fixed in space over the sample. Although the range of energy over which the electronic structure can be measured is only a few eV, STS offers the exciting possibility of determining the density-of-states at particular atomic sites on a surface. This could be a powerful method of determining the electronic structure at particular defect sites on surfaces, although the use of STS at that level of sophistication on metal oxides has yet to be realized. STS is, however, an exciting tool for future research.

IV. NON-TRANSITION-METAL OXIDES

The non-transition-metal oxides are those whose lowest-lying empty cation orbitals are of s or p symmetry. Those whose well-characterized single-crystal surfaces have been studied are Al_2O_3, MgO, CaO, SrO, BaO, SnO_2, and ZnO. All of these are insulators when stoichiometric, having bandgaps of 9.5, 7.8, 6.9, 5.3, 4.4, 3.6, and 3.4 eV, respectively. The electronic properties of the alkaline earth oxides and Al_2O_3 are similar and will be discussed together; the narrower gap oxides ZnO and SnO_2 possess somewhat different properties and will be discussed later. While Cu_2O has a filled 3d band similar to that in ZnO, it will be discussed under transition-metal oxides since its Cu 3d orbitals directly overlap the O 2p valence band and the Cu ions can exist in more than one valence state.

A. Alkaline Earth Oxides and Al_2O_3

An important property of these oxides is that their cations can only have a single valence state: 2+ for the alkaline earth cations, and 3+ for Al. This has important ramifications for the electronic structure of surface defects. For transition-metal oxides, whose cations can change valence state at defect sites, we shall see below that the electronic structure at surface defects can be drastically different than that of stoichiometric surfaces. For the non-transition-metal oxides, however, the surface stoichiometry tends to be close to that of the bulk even when the surface is highly disordered or stepped, and the changes in electronic structure that occur when surface defects are created are small.

1. MgO

Periclase and the other alkaline earth oxides have the rocksalt crystal structure. By far the most stable surface plane for these compounds is (100), since its energy is much lower than that of any other crystal face;

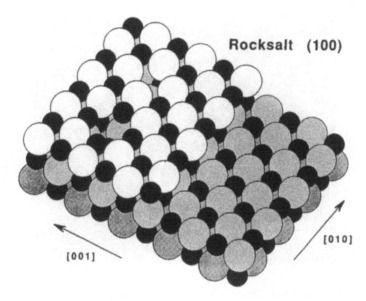

Figure 6
Model of the rocksalt (100) surface. Large circles are O anions, small circles are metal cations. A [010] step to another (100) terrace is shown, as are both missing anion and missing cation point defects.

all of the alkaline earth oxides cleave along (100). Figure 6 shows an idealized model of the (100) rocksalt surface, assuming it to be an ideal termination of the bulk structure. (A surface whose atoms are in the same positions that they would have in the bulk is said to be *unrelaxed*.) The large spheres are the O^{2-} ions, with their shading increasing with depth below the surface plane; the smaller black spheres are the metal cations. A step from one (100) terrace to another is also shown, as are cation and anion point defects; these will be discussed later. The stoichiometric (100) plane contains equal numbers of cations and anions. Since they have equal and opposite ionic charge, the plane has no net charge; such a surface is referred to as *nonpolar*. In contrast, the (111) planes in the rocksalt structure contain all one type of ion, with alternating planes of cations and anions in the [111] direction. Thus, the (111) surface would have a net charge, and the crystal structure would have a net dipole moment normal to the surface. Such surfaces are referred to as *polar*, and they are inherently unstable. It is not even possible to calculate the surface energy for a polar surface, since some terms in the calculation diverge. Thus, it is believed that any polar surface will restructure, facet, etc. to reduce the magnitude of the surface dipole. Figure 7 shows a scanning electron micrograph of an MgO (111) surface that was annealed to about 1400 K in UHV; it has faceted entirely to (100) planes.[17]

Because of the lack of periodicity in the direction normal to the surface, no real surface is expected to be truly unrelaxed. A number of

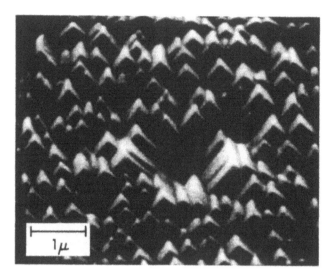

Figure 7
Scanning electron micrograph of MgO (111) annealed at about 1400 K for 1 min. (From Henrich, V.E. *Surf. Sci.* 1976, *57*, 385. With permission.)

experiments, which will not be considered here, have been performed to determine what relaxation or reconstruction takes place on MgO (100). The consensus is that the surface is very nearly a truncation of the bulk crystal structure, with no more than a 2.5% decrease in the average position of the outermost layer of atoms relative to the bulk, and less than about 2% rumpling of the top plane, with the Mg^{2+} ions moving inward and the O^{2-} ions moving outward relative to the unrelaxed surface.

The bulk electronic structure of MgO has already been considered in some detail in Section II above. It is one of the most ionic of metal oxides, and its ground state can be well described by a purely ionic approximation: $Mg^{2+}(3s^0)$ $O^{2-}(2p^6)$. MgO is highly insulating, and it cannot be made conducting even when heavily doped. The reduced ligand coordination at the surface gives the possibility of having a smaller bandgap on the surface than in the bulk. While there is no irrefutable experimental verification, EELS measurements indicate that the bandgap on the (100) surface is reduced from the bulk value by 1.5 to 2 eV.[18-21] Experiments using diffuse reflectance and total-electron reflection have also seen transitions at energies lower than that of the bulk bandgap.[22,23]

The highly insulating nature of MgO makes UPS measurements of its surface electronic structure difficult, but some measurements have been reported on (100) single-crystal samples.[24-26] Figure 8 shows angle-resolved UPS spectra for annealed MgO (100) taken at both He I and He II photon energies. Only in the He II spectra is the full 6 eV width and the two-peaked structure of the bonding and nonbonding O 2p orbitals in the valence band seen; in the He I spectra only the nonbonding region of the

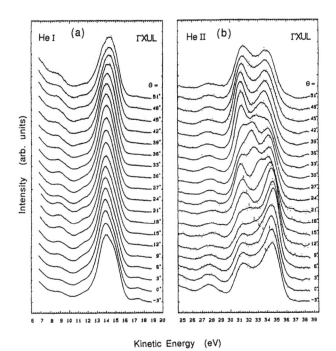

Kinetic Energy (eV)

Figure 8
Angle-resolved (a) He I and (b) He II UPS spectra for MgO (100) in the ΓXUL emission plane, normalized to the integrated valence-band intensity. (From Tjeng, L.H.; Vos, A.R.; Sawatzky, G.A. *Surf. Sci.* 1990, *235*, 269. With permission.)

band is visible. The reason for this is not yet fully understood. By analyzing the spectra as a function of emission angle (i.e., wavevector **k**), the authors separated the surface energy bands from those of the bulk, and their resultant surface band dispersion is shown in Figure 9. The shaded regions are the projection of the bulk bands onto the (100) surface, and the squares give the dispersion of the surface levels. Note that in all but one region the surface levels overlap the bulk bands; such states are referred to as *surface resonances* since they couple strongly to the bulk bands, and electrons in them have short lifetimes. To date, no independent verification of this surface band structure has been provided by other experimental techniques.

Two important surface electronic structure parameters are the *work function*, which is the distance in energy from the Fermi level to the vacuum level, and the *electron affinity*, which is the distance from the bottom of the conduction band to the vacuum level. The electron affinities of MgO and CaO have been the subject of speculation for several years.[27] The highly insulating nature of these oxides has precluded their direct measurement by electron emission, a technique commonly used for semiconductors. X-ray-scattering measurements conclude that the electron

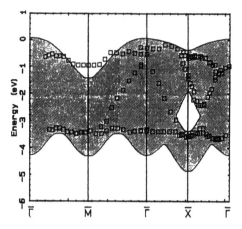

Figure 9

Experimentally determined surface band structure of MgO (100) (squares) superimposed onto the projection of the bulk energy bands (shaded region). (From Tjeng, L.H.; Vos, A.R.; Sawatzky, G.A. *Surf. Sci.* 1990, *235*, 269. With permission.)

affinities of MgO and CaO are 1 and 0.7 eV, respectively. However, theoretical calculations of impurity stability in those oxides suggest that both electron affinities may be *negative*, i.e., that the vacuum levels lie below the bottom of the conduction bands. The suggested values are –1.3 and –0.9 eV for MgO and CaO, respectively. While the disagreement between these two results has not yet been resolved, it is interesting that the secondary-electron yield (i.e., the number of electrons emitted from the surface for each electron incident on the surface from the outside) is larger for MgO than for any other material. This observation is consistent with a negative electron affinity, since then the secondary electrons that had been excited into the conduction band by collisions with an incident electron could literally "fall out" of the material at the surface (i.e., there would be no surface potential barrier for electrons approaching the surface to overcome).

Several theoretical calculations of the surface electronic structure of MgO (100) have been performed;[3,6,28-31] most of those were for an unrelaxed surface. Some band-structure and cluster calculations were discussed in Section II above. Both types predict a narrowing of the bandgap at the surface, in qualitative agreement with the EELS results mentioned above. However, since those are only calculations of the ground state, they are not sufficient to interpret experiments that strongly perturb the electronic structure; no such calculations have been performed to date for the alkaline earth oxides. The EELS results agree quantitatively better with the cluster calculations than with band models.

Important for crystal growth, dissolution and chemisorption is the electronic structure in the vicinity of steps on mineral surfaces. Figure 6 shows a [010] step between two (100) terraces in the rocksalt structure.

That such steps exist on rocksalt oxides, sometimes in large numbers, is well known.[1] What is not well known is how steps affect the electronic structure of alkaline-earth oxides. Decoration experiments, in which relatively inert atoms such as Au are evaporated in submonolayer amounts onto a surface, show clearly that steps on ionic solids are preferential sites for adsorption. But there is no direct evidence of the changes that steps produce in surface electronic structure. Indirect evidence (i.e., that MgO surfaces prepared in different ways, and hence presumably having different step densities, exhibit similar spectra) suggests that the changes are small. Theoretical calculations have been performed of the geometric relaxation that should occur at steps on MgO (100),[32] but not of the changes in electronic structure. The charge distribution and the polarizability of the O ions at step and kink sites have also been considered by Hartree–Fock cluster calculations, and, although the polarizability depends fairly strongly upon the geometric site, there is little change in charge distributions. Specifically, the O ions at steps are only slightly perturbed from their bulk O^{2-} configuration.

As has been alluded to previously, point defects on transition-metal oxides produce large changes in surface stoichiometry and electronic structure. This is not true for MgO. The MgO (100) surface can be bombarded with inert gas ions of several keV without changing its stoichiometry more than a few percent. This is presumably due to the inability of either the O^{2-} or Mg^{2+} ions to exist in any other valence state. Although ion bombardment does etch away the sample, the removal of one type of ion must destabilize the bonding of adjacent ions of the other type so that preferential sputtering does not occur. Such an interaction between the surface ions would result in only small deviations from bulk stoichiometry regardless of surface treatment.

The angle-resolved UPS measurements in Reference 26 were unable to detect any significant change in the valence-band density-of-states with ion bombardment. However, EELS measurements do exhibit one major change when defects are present on the surface.[18,20,33,34] Figure 10 shows first-derivative EELS spectra for UHV-cleaved MgO (100) taken at incident-electron energies of 100 and 2000 eV;[34] the former is very surface sensitive, while the latter more heavily weights bulk loss processes. Two features increase dramatically in the surface-sensitive spectrum. The feature at about 6 eV loss energy corresponds to the narrowing of the bandgap at the surface and is present in EELS spectra from all types of MgO surfaces. The feature at about 2 eV (higher resolution measurements place the loss at 2.3 eV[20]), however, can be eliminated from the spectra by annealing in UHV or by exposing the surface to O_2. It can be created by ion bombardment of the surface. The exact nature of this defect state, which presumably is in the empty density-of-states, is not understood. It has been attributed to both a surface O vacancy (i.e., a surface F-center)[18,33] and to a surface cation vacancy (i.e., a surface V-center),[20,34] but additional work will be required to uniquely determine its origin. Theoretical cluster

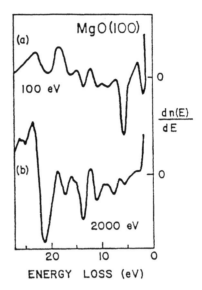

Figure 10
First-derivative ELS spectra for UHV-cleaved MgO (100) for incident-electron energies of
(a) 100 eV (surface sensitive) and (b) 2000 eV (bulk sensitive). (From Henrich, V.E.; Kurtz,
R.L. *J. Vac. Sci. Tech.* 1981, *18,* 416. With permission.)

calculations have only been performed for surface O vacancies on MgO
(100);[6] they do not clarify the situation.

The simplicity of the electronic structure of MgO (100) surfaces and
the fact that it is little perturbed from that of the bulk oxide may be largely
responsible for the inertness of MgO toward molecular adsorption. No
molecule has been shown unequivocally to adsorb onto stoichiometric
MgO (100) surfaces at room temperature. (Of course, molecules can always be physisorbed or condensed on surfaces at sufficiently low temperatures, but that process is of little interest in geochemistry.) Defect sites
have been shown to promote adsorption of some molecules, including
H_2O, CH_3OH, and $HCOOCH_3$, but even in those cases the interaction is
weak.[1] H_2O adsorption is believed to be dissociative, with the creation of
adsorbed OH^- ions via Equation 1 above. Part of this interaction may take
place at step sites, but that has not been clearly demonstrated. This is a
major difference between the behavior of non-transition-metal oxides,
where the cations can neither donate nor accept electrons, and the cations
in transition-metal oxides that can change their valence state upon adsorption.

2. CaO, SrO, and BaO

Only a few studies have been performed on well-characterized surfaces of these alkaline-earth oxides. The results obtained are in general

similar to those for MgO. The surface bandgap narrowing that was found in MgO has also been observed in CaO and SrO. EELS measurements for CaO and SrO exhibited loss features at 4.97 and 4.3 eV, respectively, which are less than the bulk bandgaps of 6.9 and 5.3 eV.[35,36] No loss features at energies less than the bulk bandgap were found in EELS measurements on single-crystal BaO, however. Low-energy loss peaks were observed on defective surfaces of all three oxides, having energies of 1.2, 0.9, and 0.6 eV for CaO, SrO, and BaO, respectively. They have been attributed to surface cation vacancies, but the same caveats discussed for point defects on MgO apply here.

3. Al_2O_3

The most stable polymorph of alumina is corundum, α-Al_2O_3. (Unless noted otherwise, the designation Al_2O_3 will imply the corundum form here.) Corundum is not only an important mineral, but it is vital to the fields of ceramics, catalysis, microelectronics, etc. It is surprising, therefore, that little work has been done to determine the electronic structure of well-characterized Al_2O_3 surfaces. The primary reason is that, like MgO, it is an excellent insulator that cannot be made conducting by doping. Unlike MgO, Al_2O_3 does not cleave. Several different surfaces have comparable surface energies, and experiments have been conducted on (0001) and ($10\bar{1}2$).

In spite of surface charging problems, two UPS[37,38] and several XPS[37-40] measurements have been performed on Al_2O_3 (0001). Figure 11 presents both UPS and XPS spectra of the O 2p valence band from one of those studies, in which the surface potential was stabilized by use of an electron flood gun. The width of the valence band is about 8 eV, slightly larger than that measured for MgO, but 2 eV smaller than the theoretically predicted value.[41-43] A similar two-peaked structure was also observed, corresponding to bonding and nonbonding orbitals.

EELS measurements indicate a narrowing of the bandgap at the surface of Al_2O_3,[40] presumably because normally empty surface states are pulled down out of the Al 3s,p conduction band. Empty surface states were observed on Al_2O_3 (0001) at 1.0 and 4.0 eV below the bottom of the bulk conduction band. The Al_2O_3 (0001) surface exhibits a series of reconstructions as a function of surface treatment, most of which correlate to the degree of reduction of the surface.[1] EELS measurements have shown that the surface bandgaps are different for the different reconstructions.[39] Inert-gas ion bombardment disorders the Al_2O_3 (0001) surface, but does not reduce it, similar to the behavior of MgO.

Several calculations of the surface electronic structure of Al_2O_3 have been reported. Figure 12 shows the energy-band structure of Al_2O_3 (0001) calculated by a self-consistent extended-Hückel tight-binding slab method;[43] both the surface bands and the projection of the bulk bands onto the

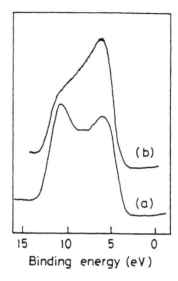

Figure 11
(a) $Al_{K\alpha}$ XPS and (b) He II UPS spectra for polished and annealed Al_2O_3 (0001). A high-energy electron flood gun was used to stabilize the surface potential. (Redrawn from Ohuchi, F.S.; Kohyama, M. *J. Am. Ceram. Soc.* 1991, *74*, 1163.)

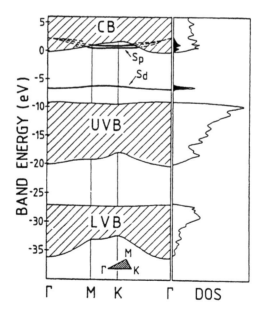

Figure 12
Band structure and density-of-states calculated for the Al_2O_3 (0001) surface. CB, UVB, and LVB are projections of the bulk conduction, upper-valence and lower-valence bands, respectively, and S_p and S_d are predicted surface states. (From Ciraci, S.; Batra, I.P. *Phys. Rev. B.* 1983, *28*, 982. With permission.)

(0001) surface are shown. The total density-of-states is shown on the right side of the figure. The Al–O bonding on the surface was calculated to be less ionic than in the bulk. The band labeled S_d, which lies about 3 eV above the top of the O 2p valence band, is composed of the Al 3s and $3p_z$ surface "dangling bonds." The other surface state, S_p, lies at the same energy as the bulk conduction band. The effect of possible surface relaxation on the electronic structure of Al_2O_3 (0001) has been addressed by using a Hartree–Fock slab method.[41,42] Relaxation of the outermost Al layer was found to be as large as 0.4 Å, although there are no experimental measurements with which to compare. This relaxation partially restores the ionic nature of the surface Al–O bonds. A surface state was also found in the bandgap of the relaxed surface, but at an energy higher than that of the calculations mentioned above.

There have been no experimental studies of the nature of defects on Al_2O_3 surfaces. Point defects have been considered theoretically, however, using extended-Hückel tight-binding methods.[43] Surface O vacancies give rise to states at energies of 1.3, 2.7, and 8.1 eV below the bottom of the conduction band. The upper two of those comprise primarily Al 3s,p orbitals redistributed around the O vacancy, while the lowest state is predominantly O 2p. Surface Al vacancies do not give rise to any bandgap surface states in this calculation.

Largely because of the difficulty of doing electron spectroscopy on highly insulating surfaces, there have been no reports of experimental studies of the adsorption of any molecule onto well-characterized alumina surfaces. However, from the similarity between the electronic structure of Al_2O_3 and MgO, it can be safely assumed that stoichiometric corundum surfaces are quite inert. Surface defects most likely lead to only weak adsorption in some instances and do not induce any adsorption in others.

B. Post-Transition-Metal Oxides: ZnO and SnO_2

The only two post-transition-metal oxides that have been studied in single-crystal form are ZnO and SnO_2. ZnO is one of the few binary metal oxides whose cations are tetrahedrally coordinated with O^{2-} ions. While many of the electronic properties of these oxides are similar to those of the pre-transition-metal oxides discussed above, their bandgaps are significantly smaller. This results in less ionic character to their bonding. There are also filled d bands close to the O 2p band, although for most purposes they can be considered as shallow core levels that do not participate in bonding. A major difference in electronic structure occurs for SnO_2: the Sn ion has two stable valence states, 2+ and 4+. This results in very different surface defect properties and easy reduction of the surface. In that respect SnO_2 is similar to the transition-metal oxides that will be discussed in Section V below.

1. ZnO

Zincite has important applications as a gas sensor, which has motivated a great many studies of its surface properties. ZnO can be made
slightly conducting by the creation of either O vacancies or Zn interstitials
(the defect chemistry of ZnO is quite complex), so surface charging is not
a problem in electron spectroscopic measurements. Figure 13 shows UPS
spectra for the four major low-index faces of ZnO: the polar (0001)-Zn
and (000$\bar{1}$)-O faces and the nonpolar (or *prism*) (10$\bar{1}$0) and (11$\bar{2}$0) faces.[44]
The valence band can be seen to extend from about 3 to 8 eV, with the
emission from 3 to 5 eV corresponding to the nonbonding O 2p orbitals
and that from 5 to 8 eV to the O 2p–Zn 4s bonding states. The Zn 3d band
lies at about 10 eV binding energy. Although the basic electronic structure
of the four surfaces is similar, the detailed shape of the valence-band
emission is quite different for each face.

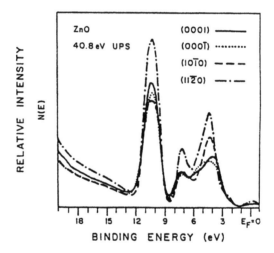

Figure 13
Angle-integrated He II UPS spectra for the nonpolar (10$\bar{1}$0) and (11$\bar{2}$0) and the polar (0001)-
Zn and (000$\bar{1}$)-O faces of ZnO. (From Gay, R.R.; Nodine, M.H.; Henrich, V.E.; Zeiger, H.J.;
Solomon, E.I. *J. Am. Chem. Soc.* 1980, *102*, 6752. With permission.)

Several EELS measurements have been performed on various ZnO
crystal faces, and the results are often contradictory.[45-51] The overall characteristics of the spectra from various groups are consistent with those
expected based upon the dielectric constant of ZnO, but there are sufficient
differences in the details of the various spectra that it has not been possible
to determine anything about the empty surface states on ZnO.

There have been a large number of theoretical calculations of the electronic structure of ZnO surfaces.[6,52-57] The consensus is that there are either
no surface states lying in the bulk bandgap, or those that do are positioned
very close to the conduction and valence band edges; only DV-Xα cluster

calculations predict that defect surface states will exist in the bandgap. Although most of the calculations assumed unrelaxed surfaces, one calculation considered the effects of relaxation. Allowing the ions on the (10$\bar{1}$0) surface to relax to the positions determined by LEED measurements did not affect some of the resonances, while others shifted a significant amount.

The presence of point defects on ZnO surfaces is crucial to their performance as gas sensors. As for most other metal oxides, the dominant type of point defect produced by either particle (including photon) bombardment or high temperature annealing is O vacancies. There are changes in the valence-band density-of-states upon creation of surface defects, and they have been studied by a variety of techniques.[58-60] However, as with other non-transition-metal oxides, O-vacancy defects do not produce new filled electronic states in the bulk bandgap. The removal of O ions from the surface requires that electronic charge move into the region of the defect in order to maintain local charge neutrality, but the charge is distributed in a delocalized accumulation layer at the surface, which results in an attendant increase in surface electrical conductivity. This accumulation layer is strongly affected by adsorbed atoms or molecules, and it is the resultant modulation of surface conductivity that gives ZnO its gas-sensing properties.

A great number of surface science studies have been performed on the interaction of all types of molecules with both polar and nonpolar ZnO surfaces. Many of the results are in conflict, and the safest interpretation of the large body of work is that the chemistry of ZnO surfaces is extremely complex. Surface Zn sites are acidic, and O^{2-} ions are basic, so donor/acceptor interactions can easily occur. But O atom transfer to and from the ZnO surface is also possible, which, when combined with the ill-understood defect chemistry of ZnO, leads to a plethora of possible adsorption mechanisms. We will not even attempt a summary of the results here; the interested reader is referred to other sources.[1]

Early adsorption experiments on single-crystal ZnO surfaces, however, provide excellent examples of the use of photoemission difference spectroscopy in identifying adsorbed species and determining the nature of molecule/surface interactions. Figure 14a presents UPS spectra for the clean and benzene-covered ZnO (10$\bar{1}$0) surface, and the difference between the two spectra (after scaling the clean surface spectrum by 0.5 to account for attenuation of the substrate photoelectrons by the adsorbed C_6H_6) is shown in Figure 14b.[61] Comparison of this difference spectrum, which to first order represents adsorbate molecular orbital emission, with the gas-phase photoemission spectrum of benzene (Figure 14c) gives a great deal of information about the adsorption process. That the two spectra exhibit the same set of peaks shows that C_6H_6 adsorbs molecularly, and the small shift to higher binding energy of the three lowest binding-energy molecular orbitals shows that they are the ones primarily involved

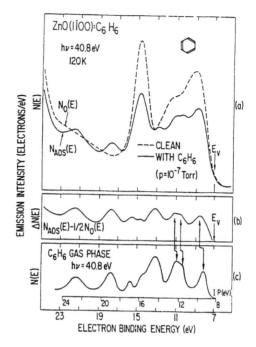

Figure 14
(a) He II UPS spectra for ZnO (10$\bar{1}$0) at 120 K, clean (dashed curve) and in the presence of a 10^{-7} Torr C_6H_6 ambient; (b) difference spectrum, as described in the text; and (c) gas-phase C_6H_6 UPS spectrum. (From Rubloff, G.W.; Lüth, H.; Grobman, W.D. *Chem. Phys. Lett.* 1976, 39, 493. With permission.)

in the bonding. Such comparisons have been very important in elucidating adsorption mechanisms on many metal-oxide surfaces.

2. SnO$_2$

The oxide cassiterite, SnO$_2$, is different from all other non-transition-metal oxides in one important respect: its Sn cations can exist in either a 2+ or a 4+ valence state. The energy difference between those valence states is fairly small in SnO$_2$,[62] which manifests itself in the ease of creation of surface O-vacancy point defects. Like ZnO, and for the same reasons, SnO$_2$ is important in gas-sensing applications. It can also be doped to produce optically transparent but electrically conducting coatings.

Cassiterite has the tetragonal rutile crystal structure. The thermodynamically most stable surface, and the one that has received the most attention both experimentally and theoretically, is (110).[63-69] Figure 15 shows an idealized model of that surface, assuming that it is simply a truncation of the bulk structure. Also shown in the figure are two O-vacancy point defects, one in a row of "bridging" O ions, and one in the main surface plane. Figure 16 shows valence-band UPS spectra for both

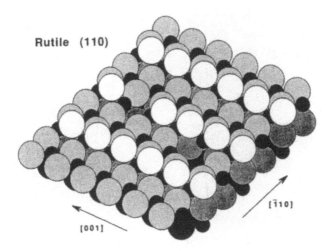

Figure 15
Model of the rutile (110) surface. Two types of O-vacancy defect are shown.

stoichiometric and reduced SnO_2 (110) surfaces. The valence band in SnO_2 has a very different shape than that in most other metal oxides. The nonbonding O 2p orbitals are almost dispersionless, giving rise to a sharp peak at the upper edge of the band. All Sn ions are 4+ on the stoichiometric surface; as O^{2-} ions are removed, enough Sn ions become 2+ to compensate for the loss of negative charge on the O ions.

Theoretical calculations have been performed of the electronic structure of several of the low-index faces of SnO_2: (110), (100), and (001).[71-74] Two types of (110) surface have been considered: the "stoichiometric" surface, which contains all of its bridging O ions and has all Sn^{4+} ions in the surface plane; and the "compact" plane, from which all of the bridging O ions have been removed, and thus all surface Sn ions are 2+. There are differences in the surface-state structure for the two surfaces, but in all cases the surface states overlap the bulk bands, and no new electronic states appear in the bulk bandgap. This is in basic agreement with the results on all other non-transition-metal oxides.

There have been a great many experimental studies of the properties of defects on SnO_2 surfaces. Much of the effort has been directed at removal of the bridging O ions from the (110) surface. The bridging O ions can be easily removed in a variety of ways, and a compact surface can be produced that gives (1 × 1) LEED patterns similar to those of the stoichiometric surface. The electrons on the Sn^{2+} ions on the compact surface are found to be localized and do not contribute to the electrical conductivity of the surface. It is not until after all of the bridging O ions have been removed that further surface reduction increases the conductivity.[75] This is apparent in Figure 16b, since the increased density-of-states produced by removing bridging O ions is localized near the valence-band

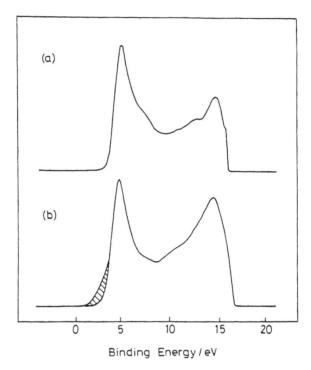

Binding Energy / eV

Figure 16
He I UPS spectra for (a) nearly perfect and (b) defective SnO_2 (110). The zero of binding energy is taken at the Fermi level. The surface-defect-induced feature is indicated by hatching. (From Egdell, R.G.; Eriksen, S.; Flavell, W.R. *Solid State Commun.* 1986, *60*, 835. With kind permission from Elsevier Science Ltd, The Boulevard, Langford Lane, Kidlington OX5 1GB, UK.)

maximum and not at the Fermi level.[70] When further reduction does increase the surface conductivity, UPS spectra show that occupied states exist in the bulk bandgap.

Most surface science studies of molecular adsorption on SnO_2 surfaces have been concerned with the changes in surface electrical conductivity that can be used in gas-sensing applications.[1] Since bridging O ions can be easily removed from the SnO_2 (110) surface with the attendant creation of Sn^{2+} ions, both stoichiometric and compact surfaces have been studied. O_2 does not adsorb on the stoichiometric surface, although it does interact with reduced surfaces. It is an electron acceptor, thus decreasing the surface conductivity. The opposite behavior is found for H_2, which probably dissociates to give adsorbed H^+ ions and an attendant increase in surface conductivity. Water also acts as an electron donor on SnO_2 (110), increasing the surface conductivity. However, since large surface conductivity changes can often be produced by a relatively small density of adsorbates, it has not been easy to quantify adsorption on SnO_2 surfaces.

V. TRANSITION-METAL OXIDES

The primary distinguishing characteristic of the transition-metal oxides is that their cations can exist in more than one valence state. This gives rise to a much more varied electronic structure, particularly with respect to defects; this will be discussed in Section V.A below. It also gives the transition-metal oxides a greater variety of chemisorption and catalytic properties. The fact that their cation valence electrons are of d rather than s,p symmetry is also important in the types of chemical reaction that can occur on transition-metal-oxide surfaces.

A. Surface Defects and Charge Neutrality

We have already referred to the properties of point defects and deviations from surface stoichiometry in transition-metal oxides in Section II.C above. Let us now consider two specific examples for illustration. The most thoroughly studied transition-metal oxide is rutile, TiO_2. As for cassiterite, its most stable surface is (110). Consider the removal of a single bridging O^{2-} ion in Figure 15. In the ionic approximation, all Ti ions on the stoichiometric surface are 4+, with a $3d^0$ electronic configuration. When an O^{2-} ion is removed, two electrons must remain at the defect site in order to maintain local charge neutrality. The lowest-lying electronic states that those electrons can occupy are the Ti 3d orbitals. Since the energy difference between Ti^{4+} and Ti^{3+} is small, partial population of the 3d orbitals does occur. This give rise to totally new electronic states that do not exist either on terraces or at steps on the stoichiometric surface. This is illustrated in Figure 17, which shows UPS spectra for stoichiometric and reduced TiO_2 (110) surfaces. In addition to the appearance of defect surface states in the bulk bandgap (which never occurred for non-transition-metal oxides), the structure in the O 2p valence band is smeared out as a result of the surface disorder, and the valence band moves almost 1 eV to higher binding energy as a result of population of some of the Ti 3d orbitals.

The electronic structure changes that occur at surface O-vacancy defects on oxides whose stoichiometric surfaces have partially occupied cation d orbitals are generally less dramatic than that shown in Figure 17, but the general idea is the same. Figure 18 shows UPS spectra for both stoichiometric and reduced NiO (100) surfaces. Bunsenite is an insulator in which the ground-state electronic configuration of the Ni cations is Ni^{2+} $3d^8$. When surface O^{2-} ions are removed, the dominant change in the valence-band UPS spectra is the appearance of a shoulder on the upper (low binding energy) edge of the band. This emission corresponds to electrons localized at defect sites on the surface. They partially populate the lowest-lying empty Ni 3d orbitals on adjacent cations, so those ions can be thought of roughly as Ni^+ $3d^9$. The appearance of such shoulders

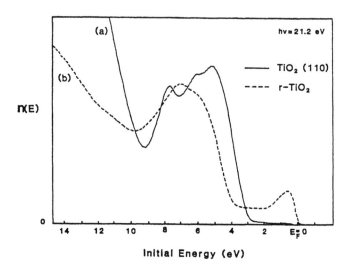

Figure 17
UPS spectra of valence-band region from polished and annealed surfaces of (a) stoichiometric and (b) reduced TiO$_2$ (110). (From Sadeghi, H.R.; Henrich, V.E. *J. Catal.* 1988, *109*, 1. With permission.)

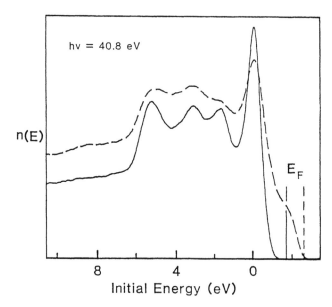

Figure 18
He II UPS spectra of UHV-cleaved NiO (100) (solid curve) and the steady-state 500 eV Ar$^+$-ion-bombarded surface (dashed curve). (From McKay, J.M.; Henrich, V.E. *Phys. Rev. B.* 1985, *32*, 6764. With permission.)

in UPS spectra is more usual on d^n oxides than are totally new emission bands such as that on rutile.

B. d^0 Binary and Ternary Oxides

Transition-metal oxides whose d orbitals are empty (d^0) behave in some ways like non-transition-metal oxides. They are insulators when stoichiometric, having bandgaps of 3 to 4 eV. The valence band is nominally composed of O 2p orbitals, although there is, of course, hybridization with the d orbitals of the metal cations. The more complex nature of the d-electron wavefunctions results in more structure in the valence band than for s,p oxides. Unlike most non-transition-metal oxides, however, these compounds can be fairly easily reduced, either in the bulk or on the surface.

1. TiO_2

The most thoroughly studied d^0 transition-metal oxide is rutile, TiO_2 (see Reference 1 for references to UPS and EELS work). The ground state of d^0 oxides can be described by an energy-band picture; for excited states, however, that model is only an approximation. Figure 19 shows the results of a calculation of the total density-of-states of the valence and conduction bands of TiO_2, including the contribution made by both the O 2p and Ti 3d orbitals.[78] While the valence band is composed primarily of O 2p wavefunctions, there is a significant contribution of the Ti 3d electrons in the higher binding energy, or bonding, region of the band. The splitting of the Ti 3d levels into t_{2g} and e_g components can also be seen. TiO_2 can be reduced by heating in a reducing atmosphere such as UHV. The bulk O ions become somewhat mobile at temperatures of several hundred Kelvin, and electronic carrier concentrations of 10^{18} cm^{-3} or greater can be achieved in this way. It is therefore easy to perform the full range of surface spectroscopic experiments on TiO_2 without any problems due to surface charging.

Since the (110) face has the lowest surface energy, that surface is the most easily prepared and, hence, the most extensively studied. TiO_2 does not cleave well, so most experimental studies have prepared surfaces by ion bombardment and annealing.[1] The UPS spectra in Figure 17 were taken on such a surface; spectra for UHV fractured (110) surfaces are similar. The Fermi level is pinned at the bottom of the conduction band since the sample had been bulk reduced. The O 2p valence band is about 6 eV wide, which is typical of most transition-metal oxides. (The emission intensity below about 10 eV is the background of inelastically scattered electrons.) The bandgap measured on this surface is the same as that of the bulk; this is also true for the (100) and (110) surfaces. No intrinsic surface states are found in the bulk bandgap for any of those faces. EELS

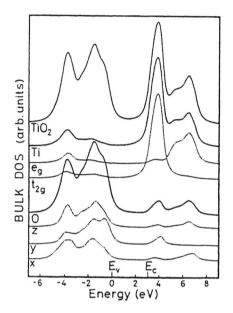

Figure 19

Calculated partial and total densities-of-states (DOS) for TiO_2. The upper-most curve shows the total density-of-states for valence and conduction bands, and the lower curves present the breakdown into different atomic-orbital contributions. E_V is the valence-band maximum and E_C the conduction-band minimum. (From Munnix, S.; Schmeits, M. *Phys. Rev. B.* 1984, *30*, 2202. With permission.)

measurements on TiO_2 surfaces also do not give any evidence for the existence of empty intrinsic surface states in the bandgap.

Several different theoretical approaches have been used to calculate the surface electronic structure of TiO_2. Early band models illustrated some of the difficulties in treating transition-metal oxides. A linear combination of atomic orbitals (LCAO) calculation predicted that there would be a band of intrinsic surface states, split off from the Ti 3d band, in the bulk bandgap for the (110) surface.[79] When those states were not observed experimentally, the theory was modified to include cation d-electron coulomb repulsion on the surface.[80] One-electron calculations using the linear combination of muffin tin orbitals (LCMTO) method also predicted a significant narrowing of the bandgap at the surface.[81] Early cluster calculations also predicted a narrower bandgap at the surface of TiO_2, although the origin of the narrowing was different than that in the band calculations.[2,82,83] More recent calculations using the scattering-theoretical LCAO method have been more successful in predicting the full bulk bandgap on the surface of all major low-index faces of TiO_2.[74,78,84] However, when surface point defects are treated by that approach, results in disagreement with experiment are obtained.

The nature of point defects on rutile surfaces has been discussed above. Predominantly O-vacancy defects are generated by electron or ion bombardment, intense photon irradiation, or annealing to high temperature, where a measurable density of point defects is required by thermodynamics, and then quenching to room temperature. That the bandgap defect states are primarily of Ti 3d character has been verified by resonant photoemission measurements.[85] The additional charge on the surface Ti ions can also be seen in XPS spectra of the Ti 2p core levels.

The detailed spectrum of bandgap defect states depends strongly upon their density. When the surface O-vacancy defect density is low enough that most point defects do not interact, the defect state lies entirely within the bandgap. As the density of defects is increased, the defect state grows in amplitude and its centroid moves away from the valence band. By the time significant near-neighbor defect interactions occur, the defect peak extends to the Fermi level, thus creating a conducting layer on the surface. That is the situation shown for the reduced surface in Figure 17.

As adjacent surface defects begin to interact on TiO_2 surfaces, a new feature appears at about 1 eV loss energy in EELS spectra. Although the transition is characteristic of interacting surface defects, its origin is not clear. The complications mentioned in Section II.C above concerning excited states in transition-metal oxides raise the possibility that it may involve an interionic transition on two adjacent Ti cations. A simple interpretation in terms of a band picture is almost certainly not correct.

IPS measurements have also been performed on both stoichiometric and reduced TiO_2 (110) surfaces.[86] Upon the creation of O-vacancy defects, a new empty electronic state appears about 3 eV above E_F. Its location agrees with theoretical calculations of the empty state density on reduced surfaces.

Resonant photoemission measurements have been performed on both stoichiometric and reduced TiO_2 (110) surfaces.[85] The intensity of the bandgap emission exhibits a strong resonant enhancement at the Ti 3p → 3d optical absorption threshold, clearly identifying the defects as of predominantly Ti 3d character. However, for both reduced and stoichiometric surfaces, the valence band also exhibits strong resonant effects. Through detailed studies of the resonant behavior of the various components of the valence band, it has been determined that both the Ti 3d and Ti 4s,p orbitals hybridize with the O 2p orbitals in that band. The changes in resonant behavior that occur upon defect creation suggest that, while there is more overall covalent mixing between O and Ti ions on the reduced surface, the hybridization between the Ti 3d and the nonbonding O 2p orbitals is reduced on the defective surface.

Several theoretical methods have been applied to the study of defects on TiO_2 surfaces. All of the methods predict that bandgap surface states will appear when the surface is reduced, but there are major differences in what type of surface defect structure is necessary to produce them.

Cluster calculations predict that even isolated O-vacancy defects will produce bandgap states consisting primarily of Ti 3d wavefunctions;[83,84,87-89] the location of the defect state within the gap is close to that measured experimentally. However, the scattering-theoretical LCAO method that correctly predicted the absence of intrinsic bandgap states on stoichiometric TiO_2 surfaces does not predict defect surface states in agreement with experiment.[90,91] While all experiments agree that even a small density of isolated O vacancies produces occupied bandgap surface states, the LCAO calculations do not predict any occupied bandgap surface states until all of the bridging O ions have been removed, and then O ions in the second atomic plane have also been removed. That result is in clear contradiction to experiment.

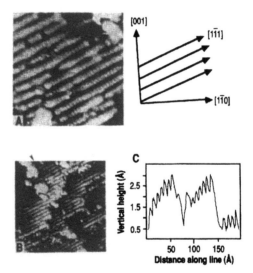

Figure 20
(A) An STM image showing a 109 Å by 109 Å area of the (110) surface of heavily reduced TiO_2 (110). (B) A 246 Å by 246 Å area that includes several steps. (C) Profile taken along the line indicated by the arrows in (B), showing the height of the corrugations and the steps. (From Rohrer, G.S.; Henrich, V.E.; Bonnell, D.A. *Science*. 1990, 250, 1239. With permission.)

When TiO_2 is heavily bulk reduced, a new type of surface defect structure can be produced. Figure 20 shows STM images of the (110) surface of TiO_2 samples that had been very heavily reduced by annealing in UHV at high temperatures for long periods of time.[92] While the row structure observed looks superficially like the rows of bridging O ions on that surface, both the orientation and distances are incorrect. (Atomic resolution was not achieved in this study.) The right-hand side of the figure shows the crystallographic orientation of the surface in the STM image. The rows of O ions lie along the [001] direction, while the observed structure is oriented along [1̄11]. This is just the direction along which bulk (121) Magnéli planes would intersect the (110) surface if the bulk of the crystal contained an ordered array of such crystallographic shear planes. While some scanning tunneling spectra were taken of this surface structure, no definitive conclusions could be drawn about its electronic structure.[93,94]

There have been a great many studies of the adsorption of a variety of molecules on TiO_2.[1] With the possible exception of H_2O discussed below, the stoichiometric surface is inert at room temperature, and all chemisorption activity occurs at defects. The most active defect sites, not surprisingly, are O vacancies. Figure 21 shows the changes that occur in UPS spectra of reduced TiO_2 (110) as it is exposed to O_2. This is an excellent example of a redox interaction, with the O_2 molecules initially dissociating (the difference spectrum in Figure 21b is characteristic of adsorbed O^{2-} ions) and removing charge from the Ti 3d orbitals of cations adjacent to the defects. For O_2 exposures higher than 20 to 100 L, the adsorbate also reduces the surface, but the nature of the adsorbed moiety is uncertain.

One of the strongest redox reactions that occurs on reduced TiO_2 surfaces is with SO_2. Adsorbed SO_2 dissociates at O-vacancy defects with essentially unity sticking coefficient, depopulating the Ti 3d defect states more rapidly than does O_2. Adsorption ceases as soon as the surface is fully reoxidized. This interaction will be discussed in more detail below when considering adsorption on Ti_2O_3.

Because TiO_2 was the first semiconducting oxide electrode used in the photocatalytic decomposition of water by photoelectrolysis,[95] the interaction of H_2O with various types of TiO_2 surfaces has been studied by many groups. The results point out the difficulties inherent in preparing well-characterized single-crystal metal oxide surfaces. Studies of "stoichiometric" TiO_2 surfaces have reached conclusions ranging from there being no interaction with H_2O to dissociative chemisorption. The variability in the results most likely arises from the range of defects present on the surfaces. The majority of studies on reduced surfaces show that H_2O *does* dissociate at defect sites, producing adsorbed hydroxyl ions.

2. $SrTiO_3$, $BaTiO_3$, $LiNbO_3$, and $LiTaO_3$

The electronic structure of the ternary d^0 transition-metal oxides is similar in many ways to that of TiO_2. $SrTiO_3$ and $BaTiO_3$ have the ABO_3 perovskite structure, in which the A cation has a valence of 2+ and the B cation is 4+. $LiNbO_3$ and $LiTaO_3$ have the ilmenite structure which is related to corundum, but electronically they are also similar to TiO_2. The electronic structure in the vicinity of E_F is dominated by the Ti and O ions, since the highest filled and lowest empty electronic orbitals on the other cations do not lie close to E_F.

Several single-crystal experimental studies of $SrTiO_3$ and $BaTiO_3$ have been reported.[1] The surface that has received the greatest amount of attention is (100), a model of which is shown in Figure 22. Unlike most

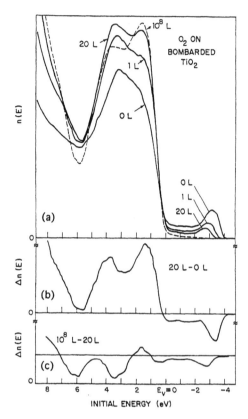

Figure 21
(a) He I UPS spectra for reduced TiO_2 (110) exposed to the amounts of O_2 indicated; (b) UPS difference spectrum for the first adsorbed phase; and (c), difference spectrum for exposures greater than 20 L.

other low-index faces of metal oxides, there are two types of perovskite (100) terminations: one has a BO_2 composition and the other is AO. Both terminations are nonpolar. They are shown in Figure 22, as are two steps between (100) terraces. UPS spectra for stoichiometric $SrTiO_3$ (100) and $BaTiO_3$ (100) are very similar to that shown in Figure 17 for TiO_2, except for somewhat different relative amplitudes of the bonding and nonbonding valence-band components. No intrinsic surface states are found in the bulk bandgap in either UPS or EELS spectra.

The nature of surface defects has been studied more extensively on $SrTiO_3$ than on $BaTiO_3$, although the results are basically the same for both oxides. Bombardment of the surface produces net reduction, with the preferential formation of O-vacancy point defects. Filled bandgap electronic states similar to those for TiO_2 are produced, and the development of those states is similar, although not identical, to that on rutile. The EELS

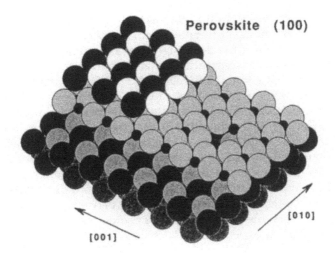

Figure 22
Model of the perovskite ABO_3 (100) surface. Small black circles are B cations, large dark circles are A cations. Two steps to other (100) terraces are shown, as is an O-vacancy defect in the BO_2 plane.

spectra for $SrTiO_3$ (100) upon point defect creation are also similar to those for TiO_2. A loss peak at 1 to 2 eV loss energy also appears upon defect creation, although in some studies its amplitude was smaller than that for TiO_2. No definitive atomic picture of the structure of defective $SrTiO_3$ surfaces has yet been obtained.

$SrTiO_3$ is one of the few oxides on which the electronic structure of steps has been investigated.[95,96] UPS spectra of nominally flat (100) surfaces were compared with those from stepped surfaces cut slightly off of the (100) orientation. The only differences observed between the two surfaces were slight changes in peak widths and intensities, and it was concluded that there were no significant differences in the electronic structure of the two surfaces on the scale that can be resolved in UPS.

Several theoretical calculations have been performed of the electronic structure of $SrTiO_3$ (100).[6,79,97-100] Early LCAO calculations predicted that the stoichiometric surface would possess a high density of intrinsic defect states in the bulk bandgap. When experiments did not find those states, the theory was modified, as for TiO_2, to include coulomb repulsion between d electrons on surface cations and other effects not included in the original calculations. This removed the surface states from the bandgap. DV-Xα cluster calculations of the $SrTiO_3$ (100) surface predicted a slight narrowing of the bandgap at the surface, but no true bandgap surface states.

The changes in electronic structure when a surface O ion is removed from $SrTiO_3$ (100) were considered by using the cluster approach. Occupied states occur in the bulk bandgap, similar to the case for TiO_2.

Population analysis shows that the electrons at the defect site partially populate the 3d orbitals of the adjacent Ti cations, and the electronic charge density is found to extend slightly into vacuum above the surface.

A band structure approach was also applied to both surface O vacancies and Ti adatoms on $SrTiO_3$ (100).[100,101] The results obtained are somewhat different from the cluster results. Removal of a surface O ion did not produce discrete surface states in the bandgap; it only gave a slight narrowing of the bandgap at the surface because of a surface resonance at the bottom of the conduction band. Even when a subsurface O ion was removed and when relaxation of ions adjacent to the defect was included, no bandgap states were produced. The only situation that would produce bandgap surface states was the addition of a Ti adatom to the surface. The discrepancy between these predictions and experimental observations may be another example of the failure of band theory to explain the electronic structure of transition-metal oxides.

3. LiNbO$_3$ and LiTaO$_3$

In these oxides the Li ions are Li^+, and the transition-metal ions are Nb^{5+} and Ta^{5+}, both having a d^0 configuration. Their bulk bandgaps are 3.5 and 4 eV, respectively. Both UPS and EELS measurements have been performed, and both annealed and UHV-fractured or scraped surfaces have been used.[102-105] Figure 23 shows He II UPS spectra for UHV-fractured $LiNbO_3$ ($10\bar{1}2$) surfaces both before and after Ar^+-ion bombardment, and for the ion-bombarded surface after exposure to O_2. [See Section V.C below for a discussion of the ($10\bar{1}2$) surface structure.] Spectra for $LiTaO_3$ ($10\bar{1}2$) surfaces are similar. Stoichiometric surfaces only exhibit emission from the O 2p valence band, with no emission in the bulk bandgap. Upon electron or ion bombardment, a band of occupied defect states appears in the bandgap. XPS spectra of the Nb 3d core level show unequivocally that the defect states are associated with reduction of the Nb cations. Whereas the stoichiometric surface only shows emission from Nb^{5+} ions, increasing particle bombardment first gives emission corresponding to Nb^{4+}, and ultimately also Nb^{3+}, ions. As for other transition-metal oxides, the bandgap states thus presumably correspond to electrons partially occupying Nb 4d (or Ta 5d) orbitals at surface O-vacancy sites. Defect state transitions are also seen in EELS spectra from both $LiNbO_3$ and $LiTaO_3$. Ion bombardment produces loss features between 2.3 and 3.5 eV in both oxides; these presumably correspond to transitions between the occupied bandgap defect states and a low-lying excited state (a normally empty defect level in band terminology). Even UHV-fractured surfaces exhibit those loss peaks, although much smaller in amplitude, indicating that a small density of O-vacancy point defects is present even on fractured surfaces.

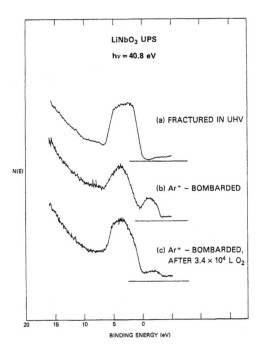

Figure 23
He II UPS spectra for UHV-fractured LiNbO₃ (a) before and (b) after Ar⁺-ion bombardment and (c) after subsequent exposure to O₂. (From Ritz, V.H.; Bermudez, V.M. *Phys. Rev. B.* 1981, 24, 5559. With permission.)

4. V₂O₅

V₂O₅ is the maximal valency oxide of vanadium. It has a layered orthorhombic structure in which the V ions are octahedrally coordinated with O ions, but one of the V–O distances is significantly shorter than the others. Being a layered structure, the only stable face is the basal (001), and single crystals cleave well along that plane. However, V₂O₅ loses oxygen very easily, and even bombardment with the low energy electron beam used in LEED reduces the surface. It has thus been very difficult to study the electronic structure of stoichiometric V₂O₅. However, UPS spectra have been taken of UHV-cleaved surfaces, and Figure 24 shows such spectra for the stoichiometric (001) surface before and after both electron and Ar⁺-ion bombardment.[106] The only emission from the stoichiometric surface is from the O 2p valence band, with no V 3d or bandgap emission present. Upon electron or ion bombardment, however, a distinct peak appears in the bandgap region that corresponds to reduction of the surface and partial population of the V 3d orbitals on surface cations. No EELS measurements have been made of empty-state electronic structure because of damage of the surface by the incident-electron beam.

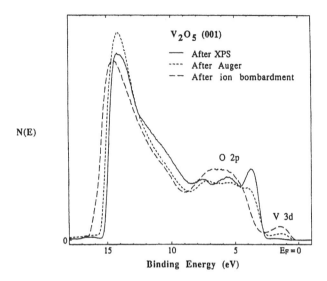

Figure 24
He I UPS spectra of the cleaved V_2O_5 (001) surface (solid curve), after bombardment with 2 keV electrons (short dashed curve), and after subsequent bombardment with 500 eV Ar^+ ions (long dashed curve).

As for all transition-metal oxides, the adsorption properties of V_2O_5 depend on whether or not the surface is reduced. The stoichiometric (001) surface is inert to most molecules, although CO can slightly reduce the surface through interaction with the weakly bound surface O ions to form surface carbonate; on ion-bombarded surfaces, where many of the surface O ions have been removed, CO does not interact. Molecules that do interact with the reduced V_2O_5 (001) surface include O_2 and SO_2, although both interactions are weak.

C. Low-Z d^n Oxides

Oxides of the transition metals at the beginning of the 3d series have a relatively simple electronic structure. The magnetic exchange effects that become dominant in the high-Z magnetic insulators are small here. However, even for such metallic oxides as V_2O_3 and TiO_x, there is a significant localized component to the bonding, so that a band theory approach is not wholly satisfactory. The low-Z d^n oxides are not particularly important as minerals since they are lower oxides of the respective transition metals; in nature, they would oxidize to the maximal valency oxide of that metal atom. However, they are important in understanding the electronic structure of transition-metal oxides in general, so some consideration will be given to them here.

1. Ti_2O_3

Although an interesting material in its own right, the importance of the corundum oxide Ti_2O_3 has been in the light it has shed on the properties of defective rutile surfaces, particularly concerning chemisorption. Ti_2O_3 cleaves well along the $(10\bar{1}2)$ plane, which is illustrated in Figure 25. The corundum structure has one third of its possible octahedral cation sites empty, and the $(10\bar{1}2)$ plane is the one that contains those sites. Therefore, cleaving the crystal along that plane breaks the smallest number of interionic bonds. The cation electron configuration in Ti_2O_3 is Ti^{3+} $3d^1$, although pairing of adjacent Ti cations along the c-axis produces a semiconductor with a 0.1-eV bandgap. Single crystals are available, and there have been several studies of the electronic structure of stoichiometric and of defective Ti_2O_3 $(10\bar{1}2)$.[107-111]

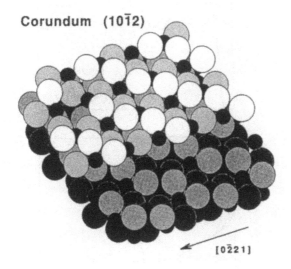

Corundum $(10\bar{1}2)$

$[0\bar{2}21]$

Figure 25
Model of the corundum $(10\bar{1}2)$ surface, including a $[0\bar{2}21]$ step to another $(10\bar{1}2)$ terrace, and an O-vacancy defect.

Figure 26 presents UPS spectra of UHV-cleaved Ti_2O_3 $(10\bar{1}2)$ taken at two different photon energies. The emission between 4 and 12 eV is from the predominantly O 2p valence band, and the emission from 0 to 1.5 eV corresponds to the filled Ti 3d a_1 orbitals. Resonant UPS measurements on Ti_2O_3 show clearly that the 0 to 1.5 eV emission corresponds to Ti 3d orbitals.[112] The dashed curves in Figure 26 are Gaussians used to fit the two spectra in order to determine the changes that occur as a function of photon energy. The increase in the intensity at the lower edge of the valence band with increasing photon energy is due to the Ti 3d hybridization in that region of the valence band and the increase in cross section for photoemission from d orbitals at higher photon energies. These UPS spectra are in basic agreement with the expected bulk electronic structure of Ti_2O_3, and there is no clear evidence for intrinsic surface states on this surface. Resonant photoemission experiments do show an increase in emission between 1.5 and 4 eV at some energies,[113] although bulk calculations indicate that there is an absolute bulk bandgap in that energy region. However, possible complications from different final-state screening effects could give rise to emission there, so no definitive conclusions can be drawn.

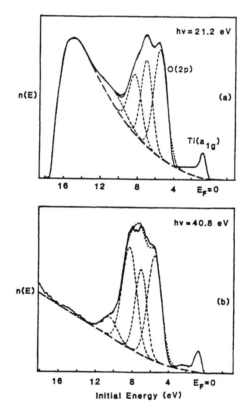

Figure 26

(a) He I and (b) He II UPS spectra for UHV-cleaved Ti_2O_3 ($10\bar{1}2$). Solid lines are the experimental data, long dashed lines are assumed backgrounds, short dashed curves are Gaussians used to fit the O 2p band, and the dotted curves are the resultant fits. (From Kurtz, R.L.; Henrich, V.E. *Phys. Rev. B.* 1982, *26*, 6682. With permission.)

When Ti_2O_3 surfaces are reduced by particle bombardment, the changes in electronic structure are much less dramatic than for TiO_2. Like TiO_2, the structure in the valence band is smeared because of disorder, and the band moves slightly to higher binding energy. But no new emission features appear in the spectrum; the amplitude of the Ti 3d intensity merely increases and broadens slightly.

The striking similarity between stoichiometric Ti_2O_3 and reduced TiO_2 can be seen by comparing Figure 26 with the dashed spectrum in Figure 17. The similarity also extends to the joint density-of-states measured by EELS. The major feature in the EELS spectrum of stoichiometric Ti_2O_3 is a peak at a loss energy of 1.5 to 2 eV.[107] In terms of the band picture, this energy would correspond to a transition from the filled a_1 band to the lowest-lying empty Ti 3d band. However, again the band picture may not adequately represent the excited state of the electronic system, and it has been suggested that this might be better regarded as a collective plasma excitation than as an interband transition.[114] The other features in the loss spectrum of Ti_2O_3 are very similar to those for TiO_2.

Unfortunately, there are no reliable electronic structure calculations of any sort for even bulk Ti_2O_3, let alone its surfaces. This has seriously hampered further attempts to separate any possible surface features from experimental spectra.

The similarity of the electronic structure of stoichiometric Ti_2O_3 with that of reduced TiO_2 has led to several studies of the interaction of molecules with the Ti_2O_3 ($10\bar{1}2$) surface.[1] While water was found to dissociate at O-vacancy defects on TiO_2 surfaces, it was not possible from studies on TiO_2 alone to separate the role played by the electronic vs. the structural components of the defect. The UHV-cleaved Ti_2O_3 ($10\bar{1}2$) surface, however, contains surface Ti^{3+} cations without an appreciable density of structural defects. Therefore, it was possible to separate electronic and geometric components to the water/surface interaction. H_2O was found to adsorb molecularly on the stoichiometric Ti_2O_3 surface, with little interaction with the Ti 3d electrons. This is thus a donor/acceptor interaction, and it shows that Ti 3d electrons *alone* are not sufficient for dissociation of H_2O. Reduced Ti_2O_3 surfaces do dissociate H_2O, as shown in Figure 27. The UPS spectra in Figure 27a show that the interaction with Ti 3d electrons is very weak, which is characteristic of a donor/acceptor or acid/base reaction. That the adsorption is dissociative is apparent from the UPS difference spectra in Figure 27b, which show the appearance of the 3σ and 1π molecular orbitals of adsorbed OH^- for even very small H_2O exposures. Thus, the geometric properties of point defects play a crucial role in the dissociation of water on titania surfaces.

Stoichiometric Ti_2O_3 surfaces were also used to study the interaction of SO_2 with titanias. On TiO_2 the interaction with SO_2 was clearly a redox one via electrons at defect sites; in that respect it is similar to the interaction of O_2 with TiO_2. However, exposure of the UHV-cleaved Ti_2O_3 ($10\bar{1}2$) surface to SO_2 produces the strongest interaction seen for any molecule on any metal oxide. Figure 28 shows UPS spectra for that system. Interaction with the Ti 3d electrons is immediate and intense; by 10 L exposure the near-surface Ti^{3+} ions are almost completely oxidized. Defects play no

H₂O on Sputtered Ti₂O₃ (10$\bar{1}$2)

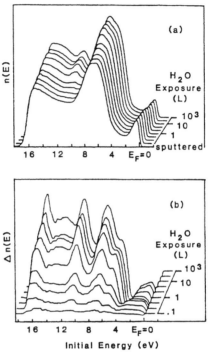

Figure 27

(a) He I UPS spectra for H_2O exposure of defective Ti_2O_3 (10$\bar{1}$2). (b) Difference spectra for the data in (a), aligned to the upper edge of the O 2p valence band. (From Kurtz, R.L.; Henrich, V.E. *Phys. Rev. B.* 1982, *26*, 6682. With permission.)

role at all in this reaction, which continues to the highest exposures used. Thus, the SO_2/Ti_2O_3 interaction is better described as corrosion than chemisorption. SO_2 is a much stronger oxidant for titanias than is O_2.

There is one additional lower oxide of Ti that is important in understanding the electronic structure of the Ti–O system, but little is known about its surface properties. $TiO_{x=1}$ has the rocksalt crystal structure, with all of the Ti cations nominally Ti^{2+}. However, this is not a very stable crystal structure, and there is a wide range of stoichiometry over which the compound can exist. For compositions close to x = 1, the structure distorts to monoclinic, and about 15% of both the cation and anion sites are vacant. This produces a very porous structure that has been shown to readily absorb oxygen, oxidizing the Ti ions to Ti^{4+}.[115] UPS measurements give a filled density-of-states similar to that of Ti_2O_3, although there is a much larger amount of emission at E_F, which indicates that the oxide is metallic.

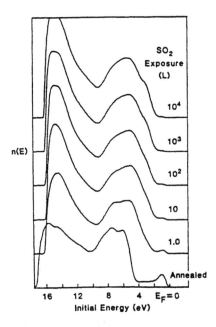

Figure 28
UPS difference spectra for UHV-cleaved Ti$_2$O$_3$ (10$\bar{1}$2) exposed to SO$_2$ up to 10^4 L. (From Smith, K.E.; Henrich, V.E. *Phys. Rev. B.* 1985, 32, 5384. With permission.)

2. Vanadium Oxides

The V–O system contains an even larger number of stable bulk oxide phases than does Ti–O, and several of these have been studied by surface science techniques on well-characterized surfaces.[1] In addition, all of the Magnéli phases that exist in the Ti–O system also exist for V–O. Upon reduction of the maximal valency oxide V$_2$O$_5$, the first major lower oxide structure reached is V$_6$O$_{13}$. It is a mixed-valence oxide with a formal valence of V$^{4.33+}$ (d$^{0.67}$). It is metallic at high temperatures, but undergoes a metal/insulator transition at 150 K.[116] It has a layered structure and cleaves well along its basal (001) plane. UPS measurements of that face exhibit a weak band of emission slightly below E$_F$, as shown in Figure 29. The O 2p valence band is similar to that of V$_2$O$_5$. Angle-resolved UPS measurements show little dispersion of the V 3d orbitals, indicating that those electrons are more localized in V$_6$O$_{13}$ than in the lower oxides VO$_2$ and V$_2$O$_3$. LEED studies have shown that V$_2$O$_5$ (001) will transform to V$_6$O$_{13}$ (001) as the sample is reduced by electron bombardment.

Further reduction of the O/V ratio gives VO$_2$, the rutile structure oxide of vanadium that exhibits the most dramatic metal–insulator transitions in the system. Little is known about its surface electronic structure, although one UPS experiment has been performed on small UHV-fractured single

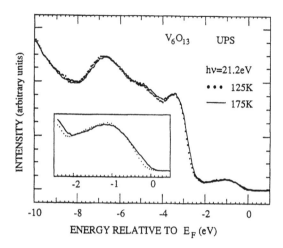

Figure 29
Valence-band He I UPS spectra of UHV-cleaved V_6O_{13} (001) below (solid curve) and above (dotted curve) the metal–insulator transition. The inset shows the region just below E_F in greater detail.

crystals.[117] Those measurements showed a higher density-of-states at E_F than for V_6O_{13} in the metallic phase and a distinct decrease in emission at E_F below the metal–insulator transition at 340 K.

Further bulk reduction gives the corundum oxide V_2O_3. It is a close relative of Ti_2O_3, although each V cation has two 3d electrons. V_2O_3 is also metallic at high temperatures and undergoes a metal–insulator transition at about 150 K. Like Ti_2O_3, it cleaves along the $(10\bar{1}2)$ plane, and UHV-cleaved single-crystal samples have been studied with a variety of surface-sensitive techniques.[1] He I and He II UPS spectra of V_2O_3 $(10\bar{1}2)$ are shown in Figure 30. The O 2p valence band in V_2O_3 is somewhat wider in energy than in the other vanadias, and the V 3d emission is, of course, larger in amplitude because of the greater concentration of V 3d electrons. In an energy-band picture, the V 3d emission comes from partially filled a_1 and e_π bands. Here in particular, however, an energy-band picture may not be appropriate for even the ground-state electronic structure. The presence of metal–insulator transitions in the V–O system indicates that electron correlation effects are large. So, even though the oxides are metallic above certain temperatures, there is a strong localized component to their electronic structure.

Unlike Ti_2O_3, the UPS spectra for UHV-cleaved V_2O_3 $(10\bar{1}2)$ do not agree with the expected bulk density-of-states. Optical measurements indicate that there should be a 1.5-eV gap between the V 3d and O 2p bands, which is much larger than that actually observed. (Based upon the UPS spectra, there may not even be a gap between those bands.) Chemisorption experiments on UHV-cleaved surfaces indicate that the emission

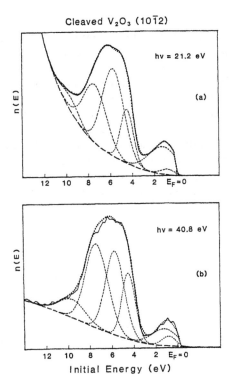

Cleaved V_2O_3 $(10\bar{1}2)$

Figure 30
(a) He I and (b) He II UPS spectra for UHV-cleaved V_2O_3 $(10\bar{1}2)$. Compare with Figure 26. (From Kurtz, R.L.; Henrich, V.E. *Phys. Rev. B.* 1983, *28*, 6699. With permission.)

between the V 3d and O 2p bands is not due to surface defects resulting from the cleaving process. Intrinsic surface states could be present, but, as discussed above in connection with Ti_2O_3, electron correlation effects may well broaden the V 3d emission. There have as yet been no theoretical calculations of the electronic structure of V_2O_3 surfaces.

As for other metal oxides, oxygen vacancies are the dominant type of surface point defect. Changes in the UPS spectra are also similar for V_2O_3 and Ti_2O_3, with an increase in the cation d emission, but no new features appearing in the spectra. They can thus be thought of as an increase in the density of d electrons on cations adjacent to defects.

The most reduced stable oxide of vanadium is $VO_{x\approx1}$. Its properties are similar to those of $TiO_{x\approx1}$ in that there is a wide stoichiometry range, and a large number of both cation and anion sites are vacant in the bulk. While there have been no experimental measurements of the surface properties of VO_x, the electronic structure of an idealized, defect-free VO (100) surface has been calculated.[119,120] Several differences from the bulk electronic structure were predicted, including a narrow intrinsic surface state.

3. Cr_2O_3

In spite of the importance of chromium oxides, little work has been done on well-characterized single-crystal samples. The most stable oxide, Cr_2O_3, which has the corundum structure, has received a small amount of attention recently, and Figure 31 shows an He II photoemission spectrum from the $(10\bar{1}2)$ cleavage face.[121] The emission between 0 and 3 eV is primarily from the Cr 3d electrons, with O 2p emission dominating the spectrum between 3 and 11 eV. The solid curve is a configuration–interaction cluster calculation of the bulk electronic structure, and the UPS spectra, which are taken at a very surface-sensitive photon energy, are consistent with the bulk predictions. Thus, no obvious surface states are present on the stoichiometric surface, although more work will be necessary before the surface electronic structure can be well characterized.

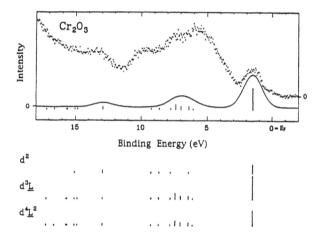

Figure 31
Experimental valence-band UPS spectrum of UHV-cleaved Cr_2O_3 $(10\bar{1}2)$ (dotted curve), compared with the configuration–interaction cluster calculation (solid line). Final states in the CI theory are decomposed into configuration components. (From Li, X.; Henrich, V.E.; Saitoh, T.; Fujimori, A. *Mat. Res. Soc. Symp. Proc.* 1993, *307*, 205. With permission.)

Although the Cr^{3+} cations in Cr_2O_3 have only three d electrons, and it is thus usually classified as a low-Z transition-metal oxide, Cr_2O_3 is an antiferromagnetic insulator like the higher-Z rocksalt oxides discussed below. Analysis of its resonant photoemission properties suggest that it lies at the border between Mott–Hubbard and charge-transfer insulators.[11]

Some photoemission measurements have been performed of the changes in surface properties of Cr_2O_3 upon point defect creation.[122] Unlike all other transition-metal oxides that have been studied, the Cr_2O_3 $(10\bar{1}2)$ surface appears to become *Cr deficient* under Ar^+-ion bombardment. The reason for this is not clear, and more measurements will be required to thoroughly characterize defects on Cr_2O_3 surfaces.

D. High-Z d^n Oxides and Magnetic Insulators

Many of the oxides of the metals in the high-Z half of the 3d-transition-metal series are antiferromagnetic insulators. As discussed in Section II.D above, energy-band approaches appear to be inappropriate for describing even their ground-state electronic structure. At present, the configuration–interaction cluster method seems to give the best description of their properties and excited-state spectra. Since there is still controversy concerning their bulk electronic structure, there have been very few studies of surface electronic structure on well-characterized samples.

One of the major differences between the electronic structure of the low-Z and high-Z transition-metal oxides is that exchange interactions in the magnetic insulators produce large changes in the energy ordering and location of the various 3d orbitals; see Figure 5 above. As the number of 3d electrons on the cations increases, there is stronger overlap in energy of the metal 3d and O 2p orbitals, making it more difficult to separate those components in photoemission spectra. As the covalent ground-state and final-state screening effects become larger, so-called satellite peaks in photoemission spectra become more pronounced. Note in Figure 31, however, that even in Cr_2O_3 satellite emission near 13 eV is significant.

1. MnO

The only oxide of manganese whose surface has been considered is the rocksalt antiferromagnetic insulator MnO.[123-126] Figure 32 shows an He II UPS spectrum of UHV-cleaved MnO (100), along with analogous spectra of CoO and NiO. All of these oxides cleave along the (100) plane, as do other rocksalt compounds. The features labeled π and σ indicate the region of the valence bands where the O 2p contribution is dominant, but the cation 3d wavefunctions extend throughout the band. The sharper features A and B are primarily cation 3d emission that is screened by $O^{2-} \to$ ligand charge transfer.

The UPS spectra for MnO (100) are consistent with the bulk electronic structure, and no intrinsic surface states have been identified. However, the cleaved MnO (100) surface is not thought to be as nearly a truncation of the bulk crystal structure as are most other cleaved oxide surfaces. It does not exhibit good LEED patterns, and it is necessary to polish and anneal the surface in order to obtain reasonable patterns. While the cleaved (100) surfaces of other antiferromagnetic insulators are quite inert toward molecular adsorption, cleaved MnO (100) surfaces are as reactive as are ion-bombarded ones. This suggests that a high density of defects is created in the cleavage process.

There is some uncertainty surrounding the origin of the small peak near 9 eV in the UPS spectrum of MnO (100) in Figure 32. Its location is similar to that of bulk electronic structure features such as that in Cr_2O_3 discussed above. However, no such structure is predicted in configuration–interaction calculations of bulk MnO. Also, its amplitude is found to

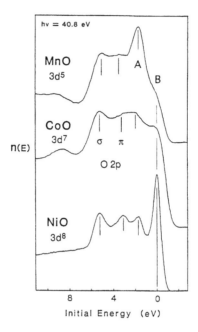

Figure 32
He II UPS spectra of UHV-cleaved (100) surfaces of MnO, CoO, and NiO. The spectra have been aligned at the uppermost emission feature B.

be quite sensitive to surface treatment.[127] Resonant photoemission measurements show that the peak is of Mn 3d character, and it is thought to be associated with Mn vacancies on the surface. But that interpretation is tentative at best.

Ion bombardment of the MnO (100) surface produces slight reduction upon the creation of O-vacancy point defects. However, no new features appear in the UPS spectra, and other changes are consistent with a slight increase in cation 3d charge. The similarity of the spectra for cleaved and reduced surfaces reinforces the interpretation that there is at least disorder, if not reduction, on the cleaved surface. No theoretical calculations have been performed for any type of MnO surface.

2. Oxides of Iron

The oxides of iron are some of the most important transition-metal-oxide minerals. Wustite (Fe_xO), hematite (α-Fe_2O_3), and magnetite (Fe_3O_4) are extremely important components of Earth's crust and play important roles in geochemical reactions. Unlike Mn, iron has only two stable oxidation states: Fe^{2+} and Fe^{3+}. The Fe ions in magnetite exist in both valence states. As with the other high-Z transition-metal oxides, there is still uncertainty as to the nature of the bulk electronic structure of iron oxides.

Most experimental studies have therefore concentrated on bulk properties, with little attention paid to surfaces.

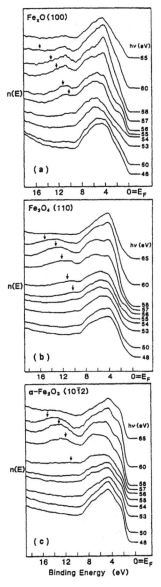

Figure 33

Angle-integrated UPS spectra as a function of photon energy for UHV-cleaved (a) Fe_xO (100), (b) Fe_3O_4 (110), and (c) α-Fe_2O_3 (10$\bar{1}$2). The predicted location of the $M_{23}M_{45}M_{45}$ Auger emission is indicated by the arrows. (From Lad, R.J.; Henrich, V.E. *Phys. Rev. B.* 1989, *39*, 13478. With permission.)

Figure 33 presents angle-integrated UPS spectra of UHV-fractured samples of Fe_xO, α-Fe_2O_3, and Fe_3O_4.[128,129] Fe_xO has the rocksalt crystal structure, and, like TiO_x and VO_x, it has a range of stoichiometry. It is, in fact, very difficult to grow stoichiometric FeO. All of the Fe cations in Fe_xO are Fe^{2+}. α-Fe_2O_3 has the corundum structure, with all of its Fe ions Fe^{3+}. Fe_3O_4 has the inverse spinel structure, in which two thirds of the cation sites are octahedrally coordinated with O ions and the other third are tetrahedrally coordinated. The tetrahedral sites are all occupied by Fe^{3+} ions, while the octahedral sites are equally divided between Fe^{2+} and Fe^{3+} ions. The surfaces used were those that fracture most easily: Fe_xO (100), α-Fe_2O_3 ($10\bar{1}2$), and Fe_3O_4 (110). The valence bands in each of the oxides are about 10 eV wide, with Fe 3d and O 2p orbitals intermixed throughout the band.

One of the important uses of resonant photoemission has been to separate the Fe 3d and O 2p contributions to the electronic structure of iron oxides. For a low photon energy such as 30 eV, the cross section for emission from p orbitals is larger than that for d orbitals, so UPS spectra weight the O 2p emission more heavily. Figure 34a shows such spectra for all three oxides, and the similarities in band shape are apparent. Fairly complete isolation of the Fe 3d orbitals can be accomplished by taking differences between UPS spectra above and below the large Fe 3p → 3d resonance at about 55 eV; this has been done in Figure 34b. Across such a narrow photon energy range, there is no appreciable change in the cross section for O 2p emission, so the difference spectra contain only the Fe density-of-states. Interpretation of the features in the Fe 3d emission requires comparison with configuration–interaction calculations, one of which is shown for Fe_2O_3 in Figure 34b.

Comparing the Fe^{2+} density-of-states for Fe_xO and the Fe^{3+} density in α-Fe_2O_3 in Figure 34b, several differences can be seen. The Fe^{3+} band is slightly narrower, presumably because of the stability of the half-filled 3d shell in the $3d^5$ configuration. When another electron is added to make an Fe^{2+} ion, the band broadens with the appearance of a shoulder at the upper edge of the band. This shoulder is characteristic of the Fe^{2+} configuration. There is also a narrow emission peak that dominates the upper half of the band.

Resonant photoemission has also been used to partially separate the Fe^{2+} and Fe^{3+} contributions to the valence-band density-of-states for Fe_3O_4. It turns out that the resonant profiles (i.e., the intensity of cation 3d emission vs. photon energy) for Fe_xO and α-Fe_2O_3 are slightly different. If it is assumed that the resonant profiles for Fe^{2+} and Fe^{3+} in Fe_3O_4 are the same as those in the reference compounds, then it is possible to find pairs of photon energies across the resonance that emphasize one cation valence state over the other when differences between the spectra are taken. (That assumption has not been proved independently, but, as will be shown below, the results indicate that the separation of valence states is at least partially achieved.) Figure 35 presents difference spectra for

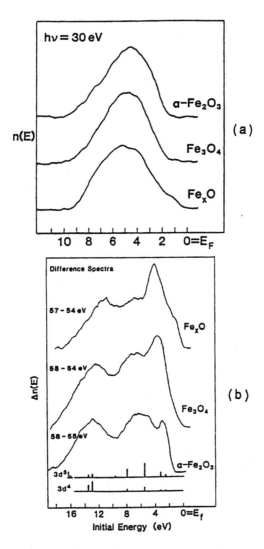

Figure 34
(a), UPS spectra for UHV-cleaved Fe_xO (100), Fe_3O_4 (110), and α-Fe_2O_3 $(10\bar{1}2)$ for $\hbar\omega = 30$ eV, which emphasizes the O 2p emission. (b), UPS difference spectra taken for photon energies just above and just below the Fe 3p $\rightarrow$ 3d resonance, which emphasizes the Fe 3d-derived final states. The vertical lines in (b) represent the relative intensities of the final states calculated in a configuration–interaction approximation. An inelastic background has been removed from each spectrum. (From Lad, R.J.; Henrich, V.E. *J. Vac. Sci. Tech. A.* 1989, *7*, 1893 and *Phys. Rev. B.* 1989, *39*, 13478. With permission.)

Fe_3O_4 taken in three different ways. The top difference spectrum shows the total Fe emission and is the same as the corresponding curve in Figure 34b. The other two differences have been taken between energies that emphasize either the Fe^{2+} or the Fe^{3+} emission. The resultant separation

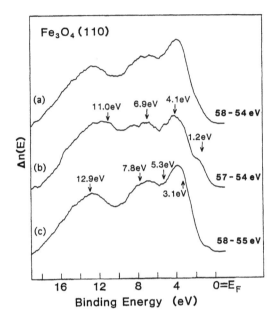

Figure 35

Estimates of the Fe 3d-derived final states in Fe_3O_4 (110), determined by the difference between valence spectra at photon energies of (a) 58 and 54 eV, (b) 57 and 54 eV, and (c) 58 and 55 eV. An inelastic background was removed from each spectrum. The arrows indicate the binding-energy locations corresponding to the Fe^{2+} and Fe^{3+} maxima measured from Fe_xO and α-Fe_2O_3, respectively. (From Lad, R.J.; Henrich, V.E. *J. Vac. Sci. Tech. A.* 1989, 7, 1893. With permission.)

gives densities-of-states that are very similar to those in the monovalent oxides; the arrows show the location of peaks in the spectra for Fe_xO and α-Fe_2O_3. To test the validity of the separation, difference spectra were also taken after the Fe_3O_4 surface had been exposed to O_2; the results are shown in Figure 36. As expected chemically, there is no interaction with the Fe^{3+} ions, since they are already in their highest oxidation state. There is a strong interaction, however, with Fe^{2+} ions; notice the elimination of the shoulder on the upper edge of the band that is characteristic of Fe^{2+}. O_2 therefore interacts via a (presumably dissociative) redox reaction with surface Fe^{2+} ions.

Although the UPS spectra discussed above were taken in a very surface-sensitive region of photoelectron kinetic energy, they have been interpreted primarily in terms of bulk electronic structure. This is because the bulk structure in not yet well enough understood to permit clear identification of any surface states. The oxidation experiment on Fe_3O_4 discussed above clearly shows that the spectra are seeing surface densities-of-states, but those densities are also consistent with the bulk electronic structure. As for many other metal oxides, it is not clear yet just

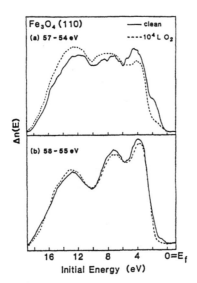

Figure 36
Comparison of the Fe 3d-derived final states measured from the clean Fe_3O_4 (110) surface and following exposure to 10^4 L of O_2: (a) 57–54 eV difference spectrum emphasizing Fe^{2+}-derived states, and (b) 58–55 eV difference spectrum accentuating the Fe^{3+}-derived states. (From Lad, R.J.; Henrich, V.E. *J. Vac. Sci. Tech. A.* 1989, *7*, 1893. With permission.)

how the surface electronic structure differs from that of the bulk. Although the bulk electronic structure of the Fe oxides was one of the earliest cases considered by configuration–interaction calculations, no calculations have yet been performed of the surface electronic properties of any of the iron oxides.

Even if the bulk electronic structure is not known accurately, it is still possible to identify extrinsic defect surface states, and this has been done for α-Fe_2O_3 (0001) and (10$\bar{1}$2) surfaces.[129-132] An example of the changes that occur as an ion-bombarded Fe_2O_3 (0001) surface is annealed to restore the surface stoichiometry is shown in Figure 37. Ion bombardment reduces the surface, and thus some of the normally Fe^{3+} cations become 2+ or lower. That is manifested in the shoulder of emission at the upper edge of the band, which was discussed above in connection with Fe_xO and Fe_3O_4. The amplitude of the shoulder is sensitive to adsorbates, as expected for Fe^{2+} cations. A vestige of it remains even on the most stoichiometric surface that could be achieved.

3. CoO

The only cobalt oxide whose electronic properties have been studied on well-characterized single-crystal samples is CoO.[133-137] It has the rocksalt structure and, like MnO and NiO, is an antiferromagnetic insulator.

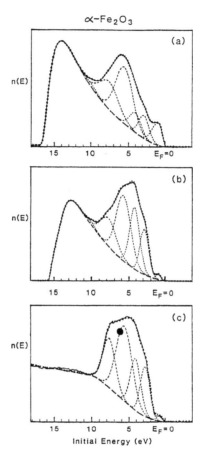

Figure 37

(a) He I UPS spectrum of Ar⁺-ion-bombarded a-Fe₂O₃ (0001); (b) He I and (c) He II spectra for that surface after annealing to 1100 K. (From Kurtz, R.L.; Henrich, V.E. *Phys. Rev. B.* 1987, *36*, 3413. With permission.)

The Co^{2+} ions have a $3d^7$ ground-state configuration. According to band theory, CoO should be metallic because of its partially filled 3d subshell; it is one of the classic examples of the failure of band theory to describe transition-metal oxides that CoO is in fact a very good insulator. An angle-integrated UPS spectrum of UHV-cleaved CoO (100) is shown in Figure 32. Peak A is dominated by emission from Co 3d majority-spin t_{2g} orbitals, while peak B comprises largely Co 3d electrons from the majority-spin e_g and minority-spin t_{2g} orbitals; all of that emission is screened by transfer of electrons from the O^{2-} ligands. However, the entire valence band contains a mixture of both O 2p and Co 3d wavefunctions.

The CoO (100) surface exhibits extremely interesting behavior upon the creation of O-vacancy defects.[133] Ion bombardment produces first a broadening of the valence band because of the partial population of normally empty Co 3d orbitals on cations at defect sites, and then a shoulder appears at the upper edge of the valence band as the defect density is increased; this behavior is similar to that for most other transition-metal oxides. For all other metal oxides, however, surface reduction produces an increase in the electrical conductivity of the surface. While the electrons at defects are largely localized on ions adjacent to the defect, there is still an increased probability of their hopping to neighboring sites, thus increasing conductivity. The slightly reduced CoO (100) surface, on the other hand, has *lower* surface conductivity than does the stoichiometric surface. When electrons are added to the Co^{2+} ions at defects, they first populate the last empty minority-spin t_{2g} orbital, producing the relatively stable $3d^8$ configuration in which three of the four exchange- and ligand-field-split subbands are filled. Thus, electrons at defects on CoO (100) appear to be *more* localized than those on the stoichiometric surface. Increasing the degree of surface reduction by bombarding with higher energy Ar^+ ions does ultimately result in an increase in surface conductivity as lower-valent Co ions or atoms are produced.

A resonant photoemission study has been performed of the changes in interionic bonding that occur at O-vacancy point defect sites on CoO (100) surfaces.[134] This is possible because of the difference in resonant behavior between ionic and charge-transfer states in photoemission. In the configuration–interaction description, the ground-state electronic configuration of CoO can be written as

$$\psi_g = a\,|\,3d^7> + b\,|\,3d^8\underline{L}> \tag{12}$$

where the first term represents the ionic component of the bonding and the second term the covalent part. The photoemission final-state wavefunction is then

$$\psi_f = \alpha\,|\,3d^6> + \beta\,|\,3d^7\underline{L}> \tag{13}$$

The $3d^6$ final-state term can *only* be derived from the $3d^7$ ground-state term; the $3d^7\underline{L}$ term can be derived from either of the two ground-state terms. Theoretical and experimental work on resonant photoemission in the configuration–interaction picture has shown that $3d^{n-1}$ final-state features resonate more strongly than do final states that involve ligand charge transfer to the cations. Thus, that part of the photoemission spectrum which resonates is proportional to the amount of ionic bonding in the ground state. It is found that when the surface is partially reduced, the strength of the resonance in the valence band is reduced, presumably because of the change from ionic to more covalent bonding at surface O-vacancy defects such as the one shown in Figure 6. Exposure of the reduced surface to O_2 partially restores the strength of the valence-band resonance as O_2 dissociatively adsorbs at surface defect sites.[138]

4. NiO

One of the most important of the rocksalt transition-metal oxides is bunsenite, NiO. It was one of the first materials to be described as a Mott insulator, although recent work has shown that it is in reality a charge-transfer insulator. There have been several surface spectroscopic studies of the UHV-cleaved NiO (100) surface,[77,135,139-143] although much of the interpretation of the data has been in terms of bulk electronic structure. An angle-integrated He II UPS spectrum is shown in Figure 32. The Ni^{2+} ions in NiO have a $3d^8$ ground state, which is relatively stable since three of the four subbands are filled. NiO is also antiferromagnetic. The emission peaks A and B are composed largely of the screened $3d^8\underline{L}$ final states. There is no evidence in any of the surface-sensitive spectroscopic results for intrinsic bandgap surface states on NiO (100).

There has, however, been one report of a surface-related feature in the empty density-of-states above the bandgap.[77,140] Figure 38 shows He I and He II UPS spectra (solid curves) for UHV-cleaved NiO (100), along with an electron-induced secondary-electron spectrum (dashed curve). A peak appears in both the He I and secondary-electron spectra that is completely absent in the He II spectrum. It is not, in fact, emission from any filled state, but instead corresponds to a fixed final-state kinetic energy 6.9 eV above the vacuum level of NiO. Such peaks occur when excited electrons lose energy through inelastic collisions and preferentially occupy maxima in the empty density-of-states as they thermalize. The peak in NiO is close to the location of the bulk Ni 4p level.[144] However, the peak is unduly sensitive to surface treatment. There is a strong electron emission angular dependence to the feature, and ion bombardment reduces its intensity much more than that of any of the valence-band UPS features. No more recent studies of this feature have been performed, so its origin is still somewhat uncertain.

There has been one theoretical LCAO calculation of the electronic structure of the NiO (100) surface.[145] It predicts that intrinsic surface states will be present near the band edges, but they would only result in a slight narrowing of the bandgap at the surface that would probably be very difficult to observe experimentally.

The effects of ion bombardment on the electronic structure of NiO (100) surfaces have been studied by using surface spectroscopies. Figure 18 shows He II UPS spectra before and after 500 eV Ar^+-ion bombardment.[77] The equivalent of about 15% of a monolayer of O ions has been removed here, and the persistence of LEED patterns suggests that the O vacancies are primarily in the surface plane. Since the minority-spin t_{2g} orbitals are filled in Ni^{2+}, any additional electrons must begin to populate the minority-spin e_g levels. This gives rise to a distinct shoulder on the upper edge of the valence band, even for small defect densities. This behavior is consistent with that for MnO and CoO. The reduced NiO (100) surface is much more highly conducting than is the stoichiometric surface.

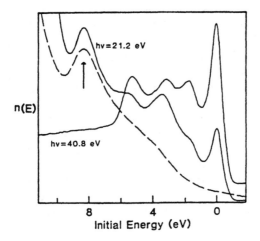

Figure 38
Angle-integrated He I and He II UPS spectra of the valence levels (solid curves) and an electron-beam-induced secondary-electron spectrum (dashed curve; $E_i = 100$ eV) for UHV-cleaved NiO (100). (From McKay, J.M.; Henrich, V.E. *Phys. Rev. Lett.* 1984, *53*, 2343. With permission.)

Although surface O vacancies are formed on NiO by ion bombardment, they are not the equilibrium type of effect. Annealing to high temperatures in UHV and quenching to room temperature has been found to produce Ni vacancies on the surface.[146] However, no surface science studies have been performed of thermally generated surface defects on NiO (100).

The adsorption properties of NiO are similar to those of most other transition-metal oxides in that the stoichiometric surface is inert to most molecules, and chemisorption, whether acid/base or redox, occurs primarily at surface defects. However, CO has been found to react with stoichiometric NiO (100) surfaces in a manner similar to that on V_2O_5 (001). For high CO exposures, a shoulder appears on the upper edge of the valence band that is similar to that in Figure 18 for ion bombardment. In this case it is believed that CO removes surface O ions through the formation of CO_2, and that the O-vacancy defect sites thus formed become sites for the adsorption of additional CO. However, the possibility of surface carbonate formation, which is believed to be the product for CO adsorption on V_2O_5, could occur for NiO also.

5. Copper Oxides

The most important applications of copper oxides at present involve high-temperature superconductors, and most of the surface science work has thus been on those compounds. Since they are of no real geologic interest, we will not consider them here. Only binary Cu oxides will be discussed.

The Cu⁺ cations in the cuprite oxide Cu_2O have a filled-shell $3d^{10}$ configuration similar to that of ZnO. The unit cell is cubic, with O ions at the center and corners of the unit cell and Cu ions occupying four of the eight tetrahedral interstitial sites. This is a somewhat unusual structure in that each Cu ion has only two O nearest neighbors, and each O ion has four tetrahedrally oriented Cu nearest neighbors. UPS, XPS, and LEED experiments have been conducted on both the (100) and (111) planes.[147] The (100) surface is polar and is thus relatively unstable; the (111) surface is nonpolar, consisting of a plane of Cu ions sandwiched between two planes of O ions. He II UPS spectra for the (111)–(1 × 1) surface and the (100)–(3√2 × √2) R 45° surfaces are shown in Figure 39. The Cu 3d orbitals between 1 and 4 eV overlap the O 2p orbitals, whose emission dominates from 5 to 8 eV.[149] The spectra of both surfaces depend upon their history, but the effect is more pronounced for the polar (111).

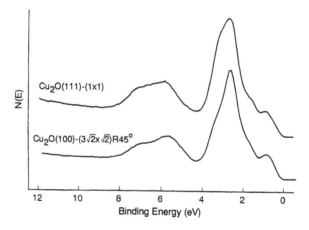

Figure 39
He II UPS spectra for Cu_2O (111) and (100) surfaces reconstructed as indicated. (From Cox, D.F.; Schulz, K.H. Personal communication.)

The monoclinic oxide CuO is truly a transition-metal oxide, since its Cu^{2+} ions have a $3d^9$ configuration. It is an antiferromagnetic semiconductor with a bandgap of 1.4 eV.[149] While most studies have been performed on oxidized copper thin films, one resonant UPS measurement has been made on single crystals.[150] Figure 40 compares the UPS spectra taken on both types of surface. Also shown in Figure 40 is the result of a cluster calculation for $[CuO_4]^{6-}$.[151] The unscreened $3d^8$ final states contribute most to the emission at binding energies above 8 eV, with the screened $3d^9\underline{L}$ emission lying closer to E_F. This makes CuO a charge-transfer insulator like NiO and CoO. While no specific surface states were identified on CuO, the spectra did indicate that the surface lost oxygen with time after cleaving in UHV.

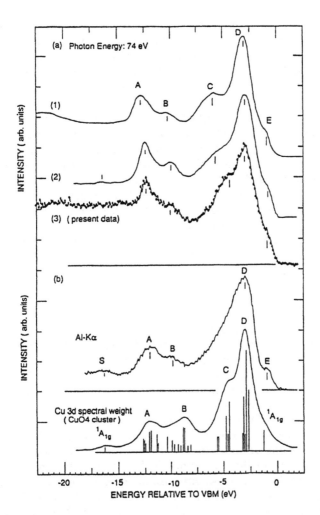

Figure 40
(a) Valence-band UPS spectra taken with $\hbar\omega = 74$ eV for (1) and (2) CuO on oxidized Cu, and (3) UHV-cleaved single-crystal CuO (face unspecified); and (b) $Al_{K\alpha}$ XPS for single-crystal CuO compared with the results of a calculation of the Cu 3d spectral weight for a $[CuO_4]^{6-}$ cluster. (From Shen, Z.X.; List, R.S.; Dessau, D.S.; Parmigiani, F.; Arko, A.J.; Bartlett, R.; Wells, B.O.; Lindau, I.; Spicer, W.E. *Phys. Rev. B.* 1990, *42*, 8081. With permission.)

6. MoO₃ and MoO₂

Molybdenum exhibits at least five stable valence states in different compounds. The maximal valency oxide of Mo is MoO_3, a layered ortho-rhombic structure in which the cations are Mo^{6+}. MoO_3 is an insulator, and its (010) basal plane has been found to be very easily reduced, rather like that of V_2O_5 discussed above. This has created problems in attempts to study its surface properties, since electron or ion beams, or even near-bandgap ultraviolet radiation, readily reduce it to a lower oxide.[152,153] There are a great number of lower molybdenas, since the material forms stable crystallographic shear structures having Magnéli planes, and $Mo_{18}O_{52}$, $Mo_{17}O_{47}$, and the $Mo_nO_{(3n-1)}$ structures Mo_9O_{26}, Mo_8O_{23}, Mo_7O_{20}, Mo_6O_{17}, Mo_5O_{14}, and Mo_4O_{11} are all known to exist.[154]

The stoichiometric MoO_3 (010) surface is quite inert. The valence band in UPS spectra is 7 eV wide, with no emission seen in the bulk bandgap. However, reduced surfaces exhibit a significant amount of bandgap emission; these states are largely of Mo 4d character. Angle-resolved UPS measurements of reduced surfaces showed that the Mo 4d states were almost dispersionless in directions parallel to the surface, but disperse by more than 1 eV in the direction normal to the surface.[153] They are thus unlike most other O-vacancy defect surface states on metal oxides, since they are extended in the direction normal to the surface. The reason for this is thought to be related to the crystallographic shear plane structures that form upon reduction.

7. Other Metal Oxides

The surfaces of three oxides of the 5d-transition-metal series have also been investigated on well-characterized single-crystal samples: ReO_3, WO_3, and Na_xWO_3. All three have basically the ABO_3 perovskite structure, in which the W or Re ions occupy the octahedral B sites. The 12-fold-coordinated A site is vacant in the first two, while the Na^+ ions occupy some of those sites in Na_xWO_3.

Stoichiometric WO_3 is an insulator containing all W^{6+} $5d^0$ cations. The Na atoms donate their 3s electrons to the W 5d orbitals in Na_xWO_3, which results in metallic behavior for x > 0.3. Several UPS measurements have been performed for these compounds.[155-158] Angle-integrated spectra for UHV-cleaved WO_3 (100) and three different compositions of Na_xWO_3 are shown in Figure 41. The dominant change in the spectra upon the addition of Na is the appearance and growth of the W 5d bandgap emission, although there are changes in the shape of the O 2p valence band as well, probably as a result of changes in W–O hybridization. While no evidence has been found for intrinsic surface states on WO_3 (100), there may be such states between the O 2p and W 5d bands in Na_xWO_3. Theoretical calculations of the electronic structure of WO_3 (100) also do not predict the existence of any intrinsic surface states.[159]

As with most other metal oxides, surface O-vacancy defects can be produced by a variety of surface treatments.[160-162] Under electron or ion bombardment, a broad band of defect states appears in the bulk bandgap of WO_3 (100). In Na_xWO_3 there is some question as to whether or not the surface is depleted of Na compared with the bulk.

In ReO_3 the Re^{6+} ions have a $5d^1$ electronic configuration, which makes ReO_3 metallic. The (100) surface has been studied by UPS.[163,164] The valence

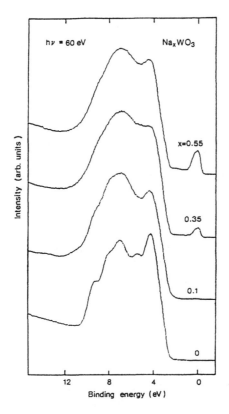

Figure 41
UPS spectra of the valence-band region of Na_xWO_3 (100) as a function of x. (From Hollinger, G.; Petrosa, P.; Doumerc, J.P.; Himpsel, F.J.; Reihl, B. *Phys. Rev. B.* 1985, 32, 1987. With permission.)

band is wider in ReO_3 than in WO_3, and there is no gap between the Re 5d and O 2p bands as there is in Na_xWO_3. No surface states were identified. DV-$X\alpha$ cluster calculations have been performed for both the stoichiometric and reduced ReO_3 (100) surfaces,[6,165] and the calculated density-of-states is in qualitative agreement with experiment.

Surface-sensitive spectra have also been taken on single crystals of RuO_2,[166,167] IrO_2,[167] and ThO_2.[168] RuO_2 surfaces were found to be rather complicated and highly nonstoichiometric. There was some indication that RuO_3 was present in the surface region as well. UPS spectra on UHV-scraped samples exhibit a strong, sharp emission peak about 1 eV below E_F, which is attributed to emission from the Ru 4d orbitals (Ru^{4+} has a $4d^4$ configuration), and a broad band of emission from about 3 to 11 eV that is thought to correspond predominantly to O 2p electrons. Spectra from IrO_2 exhibit a similar sharp peak due to Ir 5d emission, although the O 2p band in IrO_2 is about 12 eV wide. ThO_2 has the same cubic fluorite structure as uraninite. UPS spectra of the nonpolar (111) surface showed that the O 2p valence band had a shape very similar to that in most other metal oxides, and there was no evidence for intrinsic surface states in the bulk bandgap.

VI. SIO$_2$

One of the yawning voids in oxide surface science is the area of silicates. In spite of the tremendous importance of SiO$_2$ and other silicates in geology and geochemistry, there have been virtually no surface science studies of the surface electronic structure of even quartz. Probably the main reason for this is the difficulty of working with highly insulating materials; most of the UPS and many XPS studies that have been reported are on very thin silica films (<50 Å). However, because of the importance of SiO$_2$, we will discuss the general features of its electronic structure, even though there is little to say specifically about the surface.

The Si atoms in SiO$_2$ are tetrahedrally coordinated with O ions, a local bonding environment similar to that in ZnO. Figure 42 shows the results of an energy-band calculation of the bulk electronic structure of α-quartz.[169] The emission between 0 and –3 eV is from the nonbonding region of the valence band, which is primarily of O 2p origin. The emission between –4.5 and –9 eV corresponds to the bonding O 2p orbitals, hybrid-ized with Si 3sp^3 wavefunctions. Note that there is a gap between the bonding and nonbonding O 2p bands;[169,170] this gap, which does not occur for metal oxides, is clearly visible in some photoemission spectra from SiO$_2$ films.[171] The full O 2p bandwidth is about 9 eV in this calculation. The O 2s band lies between –17 and –19 eV. The Si 3s,p conduction band extends upward from 5.8 eV. The calculated bandgap of 5.8 eV is quite a bit smaller than the measured value of 8.9 eV, but that is expected from the local-density approximation used. Since quartz is an insulator, the Fermi level would lie somewhere in the bulk bandgap.

Experimental spectra of the valence band and the O 2s level for thin, amorphous SiO$_2$ films are shown in Figure 43.[172] The solid curve is an He II UPS spectrum, the dot-dashed curve is an Al$_{K\alpha}$ XPS spectrum, and the dashed curve is a soft X-ray emission spectrum. The different energies and techniques weight various regions of the band in different ways, but all results are in general agreement. The full O 2p bandwidth is about 11 eV, slightly wider than the calculation, but the separation into bonding and nonbonding regions is clearly visible. The most surface sensitive of these spectra is the He II UPS, which heavily weights the nonbonding O 2p orbitals. However, no studies have been performed to identify surface states on this or any silica.

Theoretical calculations have also been performed of the bulk elec-tronic structure of other forms of SiO$_2$.[173] The gap between the bonding

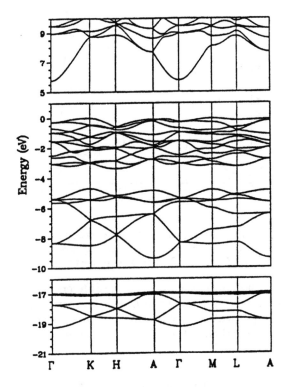

Figure 42
Energy-band structure of α-quartz. (From Binggeli, N.; Troullier, N.; Martins, J.L.; Chelikowski, J.R. *Phys. Rev. B.* 1991, *44*, 4771. With permission.)

and nonbonding O 2p bands is present for β-quartz, β-tridymite, α- and β-crystobalite, keatite, and coesite. For the tetragonal, high-density stishovite phase, however, no gap exists, and the valence-band density-of-states looks more like that of metal oxides.

One group has reported experimental studies of single-crystal α-quartz that addressed the size of the bandgap on the (10$\bar{1}$0) surface.[174,176] Figure 44a shows an EELS spectrum of an SiO$_2$ (10$\bar{1}$0) surface that exhibits (1 × 1) LEED patterns. The bandgap determined from this spectrum is 8.6 eV, only slightly smaller than the bulk value; this value is consistent with combined UPS and IPS measurements taken on thin SiO$_2$ films.[175] Figure 44b presents the EELS spectrum from that surface after reduction by ion bombardment. The two loss features at 5.5 and 7.6 eV are attributed to surface O vacancies. (Recall that in EELS one is measuring only transition energies; these features are not to be confused with emission of electrons from occupied bandgap surface states.)

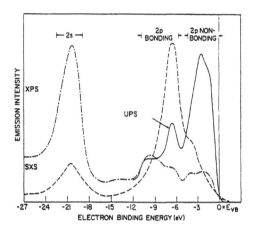

Figure 43
Emission spectra for amorphous SiO_2: He II UPS spectrum (solid curve); $Al_{K\alpha}$ XPS spectrum (dot-dashed curve); and soft X-ray emission spectrum (dashed curve). (From DiStefano, T.H.; Eastman, D.E. *Phys. Rev. B.* 1971, 27, 1560. With permission.)

Beyond this, nothing is known about the differences between the bulk and surface electronic structure of SiO_2. This is a very important area for future research, particularly considering the importance of silicate minerals in both terrestrial and atmospheric geochemistry.

ACKNOWLEDGMENT

The author owes a tremendous debt of thanks to P.A. Cox, who taught him many of the things in this chapter in the course of writing our metal-oxide surface book.

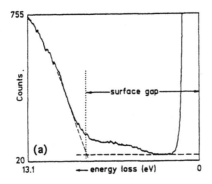

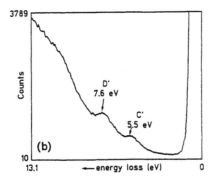

Figure 44
EELS spectra for (a) α-quartz (10$\bar{1}$0)-(1 × 1) surface, and (b) α-quartz after Ar⁺-ion bombardment. (From Bart, F.; Gautier, M.; Durand, J.P.; Henriot, M. *Surf. Sci.* 1992, *274*, 317. With permission.)

REFERENCES

1. Henrich, V.E.; Cox, P.A. *The Surface Science of Metal Oxides.* Cambridge University Press: Cambridge, 1994.
2. Cox, P.A. *The Electronic Structure and Chemistry of Solids.* Oxford University Press: Oxford, 1987.
3. Lee, V.C.; Wong, H.S. *J. Phys. Soc. Jpn.* 1978, *45*, 895.
4. Cox, P.A. *Transition Metal Oxides: An Introduction to Their Electronic Structure and Properties.* Clarendon Press: Oxford, 1992.
5. Satoko, C.; Tsukada, M.; Adachi, H. *J. Phys. Soc. Jpn.* 1978, *45*, 1333.
6. Tsukada, M.; Adachi, H.; Satoko, C. *Prog. Surf. Sci.* 1983, *14*, 113.
7. Goodenough, J.B. *Prog. Solid State Chem.* 1972, *5*, 145.
8. Terakura, K.; Oguchi, T.; Williams, A.R.; Kübler, J. *Phys. Rev. B.* 1984, *30*, 4734.
9. Fujimori, A.; Minami, F. *Phys. Rev. B.* 1984, *30*, 957.

10. Sawatzky, G.A. *Core Level Spectroscopy in Condensed Systems.* J. Kanamori; A. Kotani; Eds. Springer-Verlag Series in Solid State Sciences, Vol. 81. Springer-Verlag: Berlin, 1988.
11. Zaanen, J.; Sawatzky, G.A.; Allen, J.W. *Phys. Rev. Lett.* 1985, *55*, 418.
12. Woodruff, D.P.; Delchar, T.A. *Modern Techniques of Surface Science.* Cambridge University Press: Cambridge, 1986.
13. Zangwill, A. *Physics at Surfaces.* Cambridge University Press: Cambridge, 1988.
14. Ibach, H.; Ed. *Electron Spectroscopy for Surface Analysis.* Springer-Verlag: Berlin, 1977.
15. Kunz, C. *Photoemission in Solids*; Topics in Applied Physics, Vol. 27; L. Ley; M. Cardona; Eds. Springer-Verlag: Berlin, 1979; pp. 299–348.
16. Davis, L.C. *J. Appl. Phys.* 1986, *59*, R25.
17. Henrich, V.E. *Surf. Sci.* 1976, *57*, 385.
18. Henrich, V.E.; Dresselhaus, G.; Zeiger, H.J. *Phys. Rev. B.* 1980, *22*, 4764.
19. Henrich, V.E.; Dresselhaus, G.; Zeiger, H.J. *Phys. Rev. Lett.* 1976, *36*, 158.
20. Underhill, P.R.; Gallon, T.E. *Solid State Commun.* 1982, *43*, 9.
21. Cox, P.A.; Williams, A.A. *Surf. Sci.* 1986, *175*, L782.
22. Garrone, E.; Zecchina, A.; Stone, F.S. *Philos. Mag. B.* 1980, *42*, 683.
23. He, J.W.; Møller, P.J. *Phys. Stat. Sol. B.* 1986, *133*, 687.
24. Lee, Y.C.; Tong, P.; Montano, P.A. *Surf. Sci.* 1987, *181*, 559.
25. Onishi, H.; Egawa, C.; Aruga, T.; Iwasawa, Y. *Surf. Sci.* 1987, *191*, 479.
26. Tjeng, L.H.; Vos, A.R.; Sawatzky, G.A. *Surf. Sci.* 1990, *235*, 269.
27. Stoneham, A.M.; Sangster, M.J.L. *Philos. Mag. B.* 1981, *43*, 609; Stoneham, A.M.; Sangster, M.J.L.; Tasker, P.W. *Phil. Mag. B.* 1981, *44*, 603.
28. Satoko, C.; Tsukada, M.; Adachi, H. *J. Phys. Soc. Jpn.* 1978, *45*, 1333.
29. Russo, S.; Noguera, C. *Surf. Sci.* 1992, *262*, 245.
30. LaFemina, J.P.; Duke, C.B. *J. Vac. Sci. Tech. A.* 1991, *9*, 1847.
31. Shluger, A.L.; Grimes, R.W.; Catlow, C.R.A.; Itoh, N. *J. Phys. Condens. Matter.* 1991, *3*, 8027.
32. Tasker, P.W.; Duffy, D.M. *Surf. Sci.* 1984, *137*, 91.
33. Henrich, V.E.; Kurtz, R.L. *J. Vac. Sci. Tech.* 1981, *18*, 416.
34. He, J.W.; Møller, P.J. *Chem. Phys. Lett.* 1986, *129*, 13.
35. Protheroe, A.R.; Steinbrunn, A.; Gallon, T.E. *J. Phys. C: Solid State Phys.* 1982, *15*, 4951.
36. Protheroe, A.R.; Steinbrunn, A.; Gallon, T.E. *Surf. Sci.* 1983, *126*, 534.
37. Ohuchi, F.S.; Kohyama, M. *J. Am. Ceram. Soc.* 1991, *74*, 1163.
38. Altman, E.I.; Gorte, R.J. *Surf. Sci.* 1989, *216*, 386.
39. Gautier, M.; Duraud, J.P.; Pham Van, L.; Guittet, M.J. *Surf. Sci.* 1991, *250*, 71.
40. Gignac, W.J.; Williams, R.S.; Kowalczyk, S.P. *Phys. Rev. B.* 1985, *32*, 1237.
41. Pisani, C.; Causà, M.; Dovesi, R.; Roetti, C. *Prog. Surf. Sci.* 1987, *25*, 119.
42. Causà, M.; Dovesi, R.; Pisani, C.; Roetti, C. *Surf. Sci.* 1989, *215*, 259.
43. Ciraci, S.; Batra, I.P. *Phys. Rev. B.* 1983, *28*, 982.
44. Gay, R.R.; Nodine, M.H.; Henrich, V.E.; Zeiger, H.J.; Solomon, E.I. *J. Am. Chem. Soc.* 1980, *102*, 6752.
45. Margoninski, Y.; Eger, D. *Phys. Lett.* 1976, *59A*, 305.
46. Margoninski, Y.; Eger, D. *J. Electron Spectrosc.* 1978, *13*, 337.
47. Onsgaard, J.; Barlow, S.M.; Gallon, T.E. *J. Phys. C: Solid State Phys.* 1979, *12*, 925.
48. Ebina, A.; Takahashi, T. *Surf. Sci.* 1978, *74*, 667.
49. Froitzheim, H.; Ibach, H. *Z. Phys.* 1974, *269*, 17.
50. Dorn, R.; Lüth, H.; Büchel, M. *Phys. Rev. B.* 1977, *16*, 4675.
51. Margoninski, Y. *Surf. Sci.* 1980, *94*, L167.
52. Lee, D.H.; Joannopoulos, J.D. *Phys. Rev. B.* 1981, *24*, 6899.
53. Ivanov, I.; Pollmann, J. *Phys. Rev. B.* 1981, *24*, 7275.
54. Tsukada, M.; Miyazaki, E.; Adachi, H. *J. Phys. Soc. Jpn.* 1981, *50*, 3032.
55. Sekine, R.; Adachi, H.; Morimoto, T. *Surf. Sci.* 1989, *208*, 177.
56. Duke, C.B. *The Chemical Physics of Solid Surfaces and Heterogeneous Catalysis.* D.A. King; D.P. Woodruff; Eds. Elsevier: Amsterdam, 1988; pp. 69–118.

57. Watson, R.E.; Perlman, M.L.; Davenport, J.W. *Surf. Sci.* 1982, *115*, 117.
58. Göpel, W.; Bauer, R.S.; Hansson, G. *Surf. Sci.* 1980, *99*, 138.
59. Göpel, W.; Brillson, L.J.; Brucker, C.F. *J. Vac. Sci. Tech.* 1980, *17*, 894.
60. Göpel, W.; Lampe, U. *Phys. Rev. B.* 1980, *32*, 6447.
61. Rubloff, G.W.; Lüth, H.; Grobman, W.D. *Chem. Phys. Lett.* 1976, *39*, 493.
62. Sherwood, P.M.A. *Phys. Rev. B.* 1990, *41*, 10151.
63. deFrésart, E.; Darville, J.; Gilles, J.M. *Appl. Surf. Sci.* 1982, *11/12*, 637.
64. Erickson, J.W.; Semancik, S. *Surf. Sci.* 1987, *187*, L658.
65. Cox, D.F.; Fryberger, T.B.; Semancik, S. *Surf. Sci.* 1989, *224*, 121.
66. Cavicchi, R.; Tarlov, M.; Semancik, S. *J. Vac. Sci. Tech. A.* 1990, *8*, 2347.
67. Egdell, R.G.; Eriksen, S.; Flavell, W.R. *Surf. Sci.* 1987, *192*, 265.
68. Semancik, S.; Fryberger, T.B. *Sensors Actuators B.* 1990, *1*, 97.
69. Themlin, J.M.; Sporken, R.; Darville, J.; Caudano, R.; Gilles, J.M.; Johnson, R.L. *Phys. Rev. B.* 1990, *42*, 11914.
70. Egdell, R.G.; Eriksen, S.; Flavell, W.R. *Solid State Commun.* 1986, *60*, 835.
71. Munnix, S.; Schmeits, M. *Phys. Rev. B.* 1983, *27*, 7624.
72. Munnix, S.; Schmeits, M. *Solid State Commun.* 1982, *43*, 867.
73. Munnix, S.; Schmeits, M. *Phys. Rev. B.* 1986, *33*, 4136.
74. Munnix, S.; Schmeits, M. *Surf. Sci.* 1983, *126*, 20.
75. Cox, D.F.; Fryberger, T.B.; Semancik, S. *Phys. Rev. B.* 1988, *38*, 2072.
76. Sadeghi, H.R.; Henrich, V.E. *J. Catal.* 1988, *109*, 1.
77. McKay, J.M.; Henrich, V.E. *Phys. Rev. B.* 1985, *32*, 6764.
78. Munnix, S.; Schmeits, M. *Phys. Rev. B.* 1984, *30*, 2202.
79. Morin, F.J.; Wolfram, T. *Phys. Rev. Lett.* 1973, *30*, 1214.
80. Wolfram, T.; Ellialtioglu, S. *Appl. Phys.* 1977, *13*, 21.
81. Kasowski, R.V.; Tait, R.H. *Phys. Rev. B.* 1979, *20*, 5168.
82. Tsukada, M.; Satoko, C.; Adachi, H. *J. Phys. Soc. Jpn.* 1978, *44*, 1043.
83. Tsukada, M.; Satoko, C.; Adachi, H. *J. Phys. Soc. Jpn.* 1979, *47*, 1610.
84. Munnix, S.; Schmeits, M. *Phys. Rev. B.* 1983, *28*, 7342.
85. Zhang, Z.; Jeng, S.P.; Henrich, V.E. *Phys. Rev. B.* 1991, *43*, 12004.
86. See, A.K.; Bartynski, R.A. *J. Vac. Sci. Tech. A.* 1992, *10*, 2591.
87. Cox, P.A.; Egdell, R.G.; Flavell, W.R.; Kemp, J.P.; Potter, F.H.; Rastomjee, C.S. *J. Electron Spectrosc.* 1990, *54/55*, 1173.
88. Wang, C.R.; Xu, Y.S. *Surf. Sci.* 1989, *219*, L537.
89. Kobayashi, H.; Yamaguchi, M. *Surf. Sci.* 1989, *214*, 466.
90. Munnix, S.; Schmeits, M. *J. Vac. Sci. Tech. A.* 1987, *5*, 910.
91. Munnix, S.; Schmeits, M. *Phys. Rev. B.* 1985, *31*, 3369.
92. Rohrer, G.S.; Henrich, V.E.; Bonnell, D.A. *Science.* 1990, *250*, 1239.
93. Rohrer, G.S.; Henrich, V.E.; Bonnell, D.A. *Mater. Res. Soc. Symp. Proc.* 1991, *209*, 611.
94. Rohrer, G.S.; Bonnell, D.A. *Surf. Sci.* 1991, *247*, L195.
95. Brookes, N.B.; Law, D.S.L.; Padmore, T.S.; Warburton, D.R.; Thornton, G. *Solid State Commun.* 1986, *57*, 473.
96. Raiker, G.N.; Muryn, C.A.; Hardman, P.J.; Wincott, P.L.; Thornton, G.; Bullett, D.W.; Dale, P.A.D.M.A. *J. Phys. Condens. Matter.* 1991, *3*, S357.
97. Wolfram, T.; Kraut, E.A.; Morin, F.J. *Phys. Rev. B.* 1973, *7*, 1677.
98. Wolfram, T.; Ellialtioglu, S. *Theory of Chemisorption*; J.R. Smith; Ed. Springer-Verlag: Berlin, 1980; pp. 149–181.
99. Tsukada, M.; Satoko, C.; Adachi, H. *J. Phys. Soc. Jpn.* 1980, *48*, 200.
100. Toussaint, G.; Selme, M.O.; Pecheur, P. *Phys. Rev. B.* 1987, *36*, 6135.
101. Selme, M.O.; Toussaint, G.; Pecheur, P. *Non-Stoichiometric Compounds: Surfaces, Grain Boundaries and Structural Defects*; J. Nowotny; W. Weppner; Eds. Kluwer: Dordrecht, 1989; pp. 173–186.
102. Courths, R. *Ferroelectrics.* 1980, *26*, 749.
103. Ritz, V.H.; Bermudez, V.M. *Phys. Rev. B.* 1981, *24*, 5559.
104. Cháb, V.; Kubátová, J. *Appl. Phys. A.* 1986, *39*, 67.

105. Kasper, L.; Hüfner, S. *Phys. Lett.* 1981, *81A*, 165.
106. Zhang, Z.; Henrich, V.E. *Surf. Sci.* 1994, *321*, 133.
107. Kurtz, R.L.; Henrich, V.E. *Phys. Rev. B.* 1982, *25*, 3563.
108. Smith, K.E.; Henrich, V.E. *Phys. Rev. B.* 1985, *32*, 5384.
109. Kurtz, R.L.; Henrich, V.E. *Phys. Rev. B.* 1982, *26*, 6682.
110. Smith, K.E.; Henrich, V.E. *Phys. Rev. B.* 1988, *38*, 5965.
111. Smith, K.E.; Henrich, V.E. *Phys. Rev. B.* 1988, *38*, 9571.
112. Smith, K.E.; Henrich, V.E. *Solid State Commun.* 1988, *68*, 29.
113. McKay, J.M.; Mohamed, M.H.; Henrich, V.E. *Phys. Rev. B.* 1987, *35*, 4304.
114. Bianconi, A.; Stizza, S.; Bernardini, R. *Phys. Rev. B.* 1981, *24*, 4406.
115. Henrich, V.E.; Zeiger, H.J.; Reed, T.B. *Phys. Rev. B.* 1978, *17*, 4121.
116. Kawashima, K.; Ueda, Y.; Kosuge, K.; Kachi, S. *J. Crystal Growth.* 1974, *26*, 321.
117. Bermudez, V.M.; Williams, R.T.; Long, J.P.; Reed, R.K.; Klein, P.H. *Phys. Rev. B.* 1992, *45*, 9266.
118. Kurtz, R.L.; Henrich, V.E. *Phys. Rev. B.* 1983, *28*, 6699.
119. Laks, B.; Gonçalves da Silva, C.E.T. *Surf. Sci.* 1978, *71*, 563.
120. Laks, B.; Gonçalves da Silva, C.E.T. *J. Phys. C: Solid State Phys.* 1977, *10*, L99.
121. Li, X.; Henrich, V.E.; Saitoh, T.; Fujimori, A. *Mat. Res. Soc. Symp. Proc.* 1993, *307*, 205.
122. Li, X. Ph.D. thesis, Yale University: New Haven, CT, 1994 (unpublished).
123. Lad, R.J.; Henrich, V.E. *Phys. Rev. B.* 1988, *38*, 10860.
124. Lad, R.J.; Henrich, V.E. *J. Vac. Sci. Tech. A.* 1988, *6*, 781.
125. Fujimori, A.; Kimizuka, N.; Akahane, T.; Chiba, T.; Kimura, S.; Minami, F.; Siratori, K.; Taniguchi, M.; Ogawa, S.; Suga, S. *Phys. Rev. B.* 1990, *42*, 7580.
126. Hermsmeier, B.; Osterwalder, J.; Friedman, D.J.; Sinkovic, B.; Tran, T.; Fadley, C.S. *Phys. Rev. B.* 1990, *42*, 11895.
127. Jeng, S.P.; Lad, R.J.; Henrich, V.E. *Phys. Rev. B.* 1991, *43*, 11971.
128. Lad, R.J.; Henrich, V.E. *J. Vac. Sci. Tech. A.* 1989, *7*, 1893.
129. Lad, R.J.; Henrich, V.E. *Phys. Rev. B.* 1989, *39*, 13478.
130. Kurtz, R.L.; Henrich, V.E. *Phys. Rev. B.* 1987, *36*, 3413.
131. Lad, R.J.; Henrich, V.E. *Surf. Sci.* 1988, *193*, 81.
132. Hendewerk, M.; Salmeron, M.; Somorjai, G. *Surf. Sci.* 1986, *172*, 544.
133. Mackay, J.L.; Henrich, V.E. *Phys. Rev. B.* 1989, *39*, 6156.
134. Jeng, S.P.; Henrich, V.E. *Solid State Commun.* 1990, *75*, 1013.
135. Shen, Z.X.; Shih, C.K.; Jepsen, O.; Spicer, W.E.; Lindau, I.; Allen, J.W. *Phys. Rev. Lett.* 1990, *64*, 2442.
136. Brookes, N.B.; Law, D.S.L.; Warburton, D.R.; Wincott, P.L.; Thornton, G. *J. Phys. Condens. Matter.* 1989, *1*, 4267.
137. Shen, Z.X.; Allen, J.W.; Lingberg, P.A.P.; Dessau, D.S.; Wells, B.O.; Borg, A.; Ellis, W.; Kang, J.S.; Oh, S.J.; Lindau, I.; Spicer, W.E. *Phys. Rev. B.* 1990, *42*, 1817.
138. Jeng, S.P.; Zhang, Z.; Henrich, V.E. *Phys. Rev. B.* 1991, *44*, 3266.
139. Cox, P.A.; Williams, A.A. *Surf. Sci.* 1985, *152/153*, 791.
140. McKay, J.M.; Henrich, V.E. *Phys. Rev. Lett.* 1984, *53*, 2343.
141. Sakisaka, Y.; Akimoto, K.; Nishijima, M.; Ouchi, M. *Solid State Commun.* 1977, *24*, 105.
142. Oh, S.J.; Allen, J.W.; Lindau, I.; Mikkelsen, J.C., Jr. *Phys. Rev. B.* 1982, *26*, 4845.
143. Thuler, M.R.; Benbow, R.L.; Hurych, Z. *Phys. Rev. B.* 1983, *27*, 2082.
144. Sawatzky, G.A.; Allen, J.W. *Phys. Rev. Lett.* 1984, *53*, 2339.
145. Lee, V.C.; Wong, H.S. *J. Phys. Soc. Jpn.* 1981, *50*, 2351.
146. Roberts, M.W.; Smart, R.St.C. *J. Chem. Soc. Faraday Trans. 1.* 1984, *80*, 2957.
147. Schulz, K.H.; Cox, D.F. *Phys. Rev. B.* 1991, *43*, 1610.
148. Cox, D.F.; Schulz, K.H. (personal communication).
149. Ghijsen, J.; Tjeng, L.H.; van Elp, J.; Eskes, H.; Westerink, J.; Sawatzky, G.A.; Czyzyk, N.T. *Phys. Rev. B.* 1988, *38*, 11322.
150. Shen, Z.X.; List, R.S.; Dessau, D.S.; Parmigiani, F.; Arko, A.J.; Bartlett, R.; Wells, B.O.; Lindau, I.; Spicer, W.E. *Phys. Rev. B.* 1990, *42*, 8081.
151. Eskes, H.; Tjeng, L.H.; Sawatzky, G.A. *Phys. Rev. B.* 1990, *41*, 288.

152. Firment, L.E.; Ferretti, A. *Surf. Sci.* 1983, *129*, 155.
153. Firment, L.E.; Ferretti, A.; Cohen, M.R.; Merrill, R.P. *Langmuir.* 1985, *1*, 166.
154. Vincent, H.; Marezio, M. *Low-Dimensional Electronic Properties of Molybdenum Bronzes and Oxides.* C. Schlenker; Ed. Kluwer: Dordrecht, 1989; p. 49.
155. Egdell, R.G.; Innes, H.; Hill, M.D. *Surf. Sci.* 1985, *149*, 33.
156. Höchst, H.; Bringans, R.D.; Shanks, H.R. *Phys. Rev. B.* 1982, *26*, 1702.
157. Benbow, R.L.; Hurych, Z. *Phys. Rev. B.* 1978, *17*, 4527.
158. Hollinger, G.; Petrosa, P.; Doumerc, J.P.; Himpsel, F.J.; Reihl, B. *Phys. Rev. B.* 1985, *32*, 1987.
159. Bullett, D.W. *J. Phys. C: Solid State Phys.* 1983, *16*, 2197.
160. Fleisch, T.H.; Zajac, G.W.; Schreiner, J.O.; Mains, G.J. *Appl. Surf. Sci.* 1986, *26*, 488.
161. Bringans, R.D.; Höchst, H.; Shanks, H.R. *Phys. Rev. B.* 1981, *24*, 3481.
162. Langell, M.A.; Bernasek, S.L. *Phys. Rev. B.* 1981, *23*, 1584.
163. Fujimori, A.; Minami, F.; Akahane, T.; Tsuda, N. *J. Phys. Soc. Jpn.* 1980, *49*, 1820.
164. Hollinger, G.; Himpsel, F.J.; Martensson, N.; Reihl, B.; Doumerc, J.P.; Akahane, T. *Phys. Rev. B.* 1983, *27*, 6370.
165. Tsukada, M.; Tsuda, N.; Minami, F. *J. Phys. Soc. Jpn.* 1980, *49*, 1115.
166. Atanasoska, Lj.; O'Grady, W.E.; Atanasoski, R.T.; Pollak, F.H. *Surf. Sci.* 1988, *202*, 142.
167. Daniels, R.R.; Margaritondo, G.; Georg, C.A.; Lévy, F. *Phys. Rev. B.* 1984, *29*, 1813.
168. Ellis, W.P.; Boring, A.M.; Allen, J.W.; Cox, L.E.; Cowan, R.D.; Pate, B.B.; Arko, A.J.; Lindau, I. *Solid State Commun.* 1989, *72*, 725.
169. Binggeli, N.; Troullier, N.; Martins, J.L.; Chelikowski, J.R. *Phys. Rev. B.* 1991, *44*, 4771.
170. Dovesi, R.; Pisani, C.; Roetti, C.; Silvi, B. *J. Chem. Phys.* 1987, *86*, 6967.
171. Fischer, B.; Pollak, R.A.; DiStefano, T.H.; Grobman, W.D. *Phys. Rev. B.* 1977, *15*, 3193.
172. DiStefano, T.H.; Eastman, D.E. *Phys. Rev. Lett.* 1971, *27*, 1560.
173. Ching, W.Y. *Structure and Bonding in Noncrystalline Solids.* G.E. Walrafen; A.G. Revesz; Eds. Plenum Press: New York, 1986; pp. 77–99.
174. Bart, F.; Gautier, M.; Durand, J.P.; Henriot, M. *Surf. Sci.* 1992, *274*, 317.
175. Azizan, M.; Baptist, R.; Brenac, A.; Chauvet, G.; Nguyen Tan, T.A. *J. Phys. (Paris).* 1987, *48*, 81.
176. Bart, F.; Gautier, M.; Jollet, F.; Durand, J.P. *Surf. Sci.* 1994, *306*, 342.

Chapter **3**

APPLICATION OF X-RAY ABSORPTION SPECTROSCOPY FOR SURFACE COMPLEXATION MODELING OF METAL ION SORPTION

Kim F. Hayes and Lynn E. Katz

CONTENTS

0-8493-8351-X/96/$0.00+$.50
© 1996 by CRC Press, Inc.

I. INTRODUCTION

The reaction of metal ions with mineral surfaces in natural environments, including soils, sediments, and aquifers, typically causes a significant fraction of the ions to sorb to the solids, often lowering the metal ion aqueous phase concentration well below the solubility limits of the solid phases that may form. A number of processes can occur during sorption, including absorption by diffusion into the solid matrix or adsorption at the surface/water interface by surface complexation, surface polymerization, and surface precipitation. The type of sorption process which occurs is highly dependent on the quantity and type of mineral phases present. Solution conditions, such as pH, ionic strength, metal ion concentration, and the presence and concentration of other sorbing species, also play a major role in determining the extent and type of operative sorption process.

Over the past two decades, surface complexation models (SCMs) have emerged as one of the most promising methods to model metal ion sorption because of their ability to account for the pH, ionic strength, and surface-coverage dependence of sorption. These models are based on the premise that reactions between surface functional groups and sorbing species are analogous to aqueous solution complexation reactions between metal ions and complexing ligands. SCM reactions can be written to distinguish between inner- and outer-sphere surface complexes, surface polymerization, and surface precipitation. Surface coverage, ionic strength, and pH dependence of metal ion sorption can also be rationally understood and modeled in terms of SCM reactions. In spite of the attractiveness of the SCM approach, it is not without certain drawbacks. For example, before appropriate surface reactions, stoichiometry, and equilibrium constants can be selected, the controlling reaction processes and surface species must be known or assumed. Fortunately, recent advances in identification of sorbed species by X-ray absorption spectroscopy (XAS) have made it possible to overcome this limitation for many systems of interest.

This chapter reviews the type of spectroscopic information that is needed to use SCMs and recent spectroscopic studies which have provided this information. Because the most extensive testing of the SCM approach has come from studies of divalent cation sorption to metal oxide, hydroxide, and oxyhydroxide sorbents, this chapter mainly covers that literature. By way of introduction, a brief review of divalent metal cation sorption and the dependence of sorption processes on solution conditions is given. Empirical modeling approaches which require conditional constants, such as the linear partitioning coefficient, K_d, are discussed, in view of their current widespread use in contaminant transport models, to point out their limitations compared with SCMs. Background information on the theoretical framework of the SCM approach is also provided to introduce the number and type of parameters which must be evaluated for applying SCMs. As an example, the $Co(II)/\alpha-Al_2O_3$ system has been selected to illustrate how both spectroscopic and equilibrium sorption data are used to implement various versions of the SCM approach over a wide range of conditions in pH, ionic strength, and surface coverage. Finally, a comparison among three SCMs, the nonelectrostatic model (NEM), the diffuse layer model (DLM), and the triple layer model (TLM) is given illustrating the relative capabilities of the three SCMs for predicting sorption over wide ranges of solution conditions.

II. SORPTION BEHAVIOR OF DIVALENT CATIONS

A large body of literature exists which summarizes the dependence of metal ion sorption on the properties of mineral oxides, metal ions, and solution conditions (see, e.g., James and Parks;[1] Hayes;[2] Dzombak and Morel;[3,4] Schindler and Stumm;[5] Davis and Kent;[6] Stumm;[7] McBride[8]). A brief overview of divalent metal ion sorption behavior is given here to establish the range and types of surface reactions that SCMs must be capable of representing.

In general, sorption of divalent metal cations on metal oxides, hydroxides, and oxyhydroxides (hereafter collectively referred to as metal oxides) increases with pH (Figure 1). This results from either inner- or outer-sphere surface complexation formation of divalent metal ions (Me^{2+}) with surface hydroxyls ($\equiv$SOH) as depicted in Figure 2 and described in the following reactions:

$$\equiv SOH + Me^{2+} = \equiv SOMe^+ + H^+ \qquad \text{(inner sphere)} \qquad (1)$$

$$\equiv SOH + Me^{2+} = \equiv SO^--Me^{2+} + H^+ \quad \text{(outer sphere)} \qquad (2)$$

As these reactions illustrate, divalent metal ions are in direct competition with protons (or other cationic species), and, hence, the extent of sorption is highly dependent on solution pH. Because surface hydroxyls tend to dissociate sorbed protons as the solution pH increases, metal ion sorption is favored at higher pH values. While the above reactions account for the observed dependence of metal ion sorption as a function of the pH, the relative pH range and extent of divalent cation sorption depend on which metal oxide and particular divalent cation is present in the system. For example, Cr(III) sorbs at a lower pH than Co(II) which sorbs at a lower pH than Sr(II) on corundum (Figure 1A). Comparing two different oxides, it is found that Co(II) sorbs at a lower pH on corundum than quartz (Figure 1B). Metal cations which sorb at lower pH on a given metal oxide have a greater affinity for the surface. Alternatively, surfaces which sorb metal cations at lower pH have a greater affinity for a given sorbing cation. In general, the relative affinity of a series of divalent cations for a given metal oxide or a series of metal cations for a given metal oxide, such as those shown in Figure 1, can be related to the acid–base properties of the metal oxide surface hydroxyls and sorbing metal cations. By considering the origin of the relative acid–base properties of metal oxides surface hydroxyls and divalent cations, it is possible to rationalize the sorption trends observed.

A. Relative Affinity of Surface Hydroxyls for Divalent Cations

When metal oxides are exposed to water, surface hydroxyls are formed. Spectroscopic studies have shown that a number of different

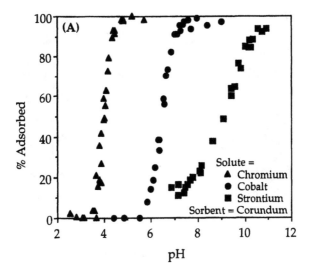

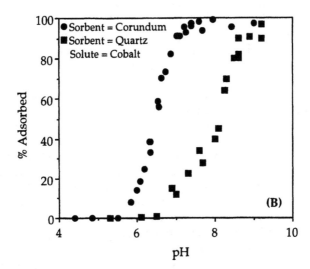

Figure 1
General dependence of metal ion sorption on pH for oxide minerals: (A) effect of solute properties; (B) effect of sorbent properties.

types of surface hydroxyls can be present on a particular surface. These groups can be distinguished from one another by the number of lattice metal atoms coordinated to a given surface hydroxyl.[5,9] From the standpoint of reactivity with divalent metal cations, two general types of surface hydroxyls can be distinguished: (1) terminal surface hydroxyls which are coordinated to only one metal atom from the crystal lattice ($\equiv$S–OH) and

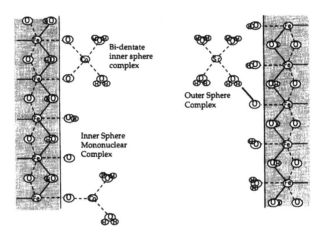

Figure 2
Inner- and outer-sphere surface complexation of divalent metal ions.

(2) bridging surface hydroxyls which are typically coordinated by either two or three metal atoms from the lattice (($\equiv$S)$_2$–OH or ($\equiv$S)$_3$–OH) (Figure 3).[8] Bridging hydroxyls, which tend to act like Bronsted acids and disso-ciate protons in water, are thought to have little or no tendency to form complexes with sorbing metal ions. Terminal hydroxyls, on the other hand (with multiple lone pairs of electrons on the hydroxyl oxygen), can have significant Lewis base properties (i.e., they can share their lone pairs of electrons) and may form strong complexes with sorbing metal ions, espe-cially upon the loss of the hydroxyl proton.

The relative Lewis base properties of terminal surface hydroxyls for a series of metal oxides can be interpreted in terms of the valence coor-dination number ratio (VCNR), the ratio of the valence charge of a lattice atom to the number of oxygen atoms coordinated by the lattice metal.[8] Terminal surface hydroxyls with lower values of the VCNR are expected to have greater Lewis base properties than those with higher values. For example, in the case of the terminal oxygens of quartz SiO$_2$, Si has a valence of IV and is coordinated to four oxygens giving a VCNR of $^4/_4$ = +1. This implies that an Si surface hydroxyl experiences an effective charge of +1. In the case of a terminal oxygen in an aluminum oxide like corun-dum (α-Al$_2$O$_3$), the Al has a valence of III and is coordinated to six oxygens. Therefore, a charge of 3+ is shared equally by six oxygens, resulting in a VCNR of $^3/_6$ = +0.5. In this case, an Al surface hydroxyl experiences an effective charge of +0.5. Because the attraction of terminal oxygen electrons by Si is greater than Al in the respective oxides, the Si terminal hydroxyl is less likely to donate its electrons to form an inner-sphere complex. As a result, terminal Al hydroxyls tend to be stronger Lewis base sites than Si hydroxyls and to form stronger complexes with metal ions. This is thought to be the reason that metal cations like Co(II)

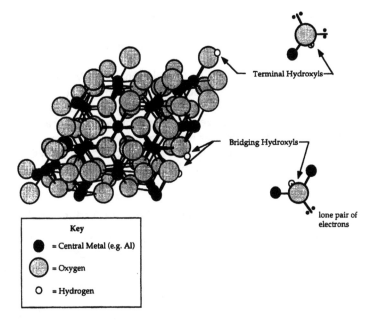

Figure 3
Representation of terminal and bridging surface hydroxyl groups on an oxide mineral.

TABLE 1.

pH$_{PZC}$ and Valence-Coordination Number Ratio
for Various Metal Oxides

Group	VCNR[a]	pH$_{PZC}$[b]	Metal oxide
≡Si–OH	+4/4 = 1.00	2.9	SiO$_2$
≡Ti–OH	+4/6 = 0.67	5.8	TiO$_2$
≡Fe–OH	+3/6 = 0.50	8.5	α-Fe$_2$O$_3$
≡Al–OH	+3/6 = 0.50	9.1	α-Al$_2$O$_3$

[a] After McBride.[8]
[b] After Davis and Kent.[6] pH$_{PZC}$ values given for comparative purposes only. Actual values can vary depending on preparation and methods used.

have a higher affinity for corundum than quartz and sorb at lower pH even though comparison of the pH$_{PZC}$ (Table 1) indicates that the surface of corundum is more positively charged at any particular pH (Figure 1).

In contrast, Bronsted acidity, the ability to give up a proton, is correlated with higher rather than lower values of VCNR. Higher values of VCNR indicate a greater attraction of the surface oxygen electrons by the lattice metals resulting in a weaker attraction of the surface oxygens for the proton. As a result, metal oxides like quartz with a relatively high value of the VCNR tend to have surface hydroxyls which dissociate protons more easily compared with those with lower VCNR, like corundum. Analogously, metal ions would tend to be more easily dissociated and be less strongly complexed by surface hydroxyls on quartz compared with corundum.

While generalizations based on correlations with VCNR discussed above are simplifications of the reactivity of the surface, they provide a basis for understanding the relative affinity of a series of metal oxides for a given divalent metal cation. In general, metal oxides which have a lower value of the VCNR have a greater tendency to form stronger complexes compared with oxides having higher values. Values of the VCNR for various metal oxides are given in Table 1 for comparison.

B. Relative Affinity of Divalent Metal Cations for Surface Hydroxyls

Although the acid–base properties of surface hydroxyls determine the relative affinity of various oxide minerals for divalent cations, the properties of divalent metal cations themselves determine the relative affinity they have for a particular metal oxide. In terms of sorption characteristics, two classes of divalent metal ions can be distinguished: (1) those that sorb relatively strongly and form inner-sphere complexes and (2) those that sorb relatively weakly and form outer-sphere complexes.[10]

Strongly sorbing metal ions are found in the middle of the periodic table and include the divalent transition (e.g., Zn, Cu, Ni, Co, Fe, and Mn) and other heavy metal cations (e.g., Cd and Pb). For these divalent cations, sorption at low metal ion surface coverage (below 5%) has been found to occur through inner-sphere coordination of the metal ions with surface hydroxyls (see spectroscopic section below). Several different correlations have been put forth to explain the relative ability of these cations to form inner-sphere complexes with terminal surface hydroxyls. The well-known Irving–Williams series gives the relative ability of transition-metal ions to form coordinative complexes with organic Lewis bases.[11] Based on that scaling, the following selectivity order with Cu at the apex is expected:

$$Zn^{2+} < Cu^{2+} > Ni^{2+} > Co^{2+} > Fe^{2+} > Mn^{2+}$$

Alternatively, the ability to form inner-sphere complexes with surface hydroxyls has been found to be related to the ability of the aqueous species to form hydroxide complexes. The analogy of surface complexation formation with hydroxide complex formation can be seen by comparing the following two reactions:

$$Me^{2+} + H_2O = Me(OH)^+ + H^+ \quad \text{(hydroxide complex formation)} \quad \textbf{(3)}$$

$$Me^{2+} + \equiv SOH = \equiv SOMe^+ + H^+ \quad \text{(inner-sphere complex formation)} \quad \textbf{(4)}$$

TABLE 2.

pK Values of Metal Hydroxide Formation

Divalent cation	pK Value
Cu^{2+}	7.7
Pb^{2+}	7.7
Zn^{2+}	9.0
Co^{2+}	9.7
Ni^{2+}	9.9
Cd^{2+}	10.1

Data taken from Reference 12.

Assuming a correlation between surface complexation and hydroxide complexation exists, the following surface affinity order would be expected, based on the first hydrolysis constants given in Table 2:

$$Cu^{2+} \approx Pb^{2+} > Zn^{2+} > Co^{2+} > Ni^{2+} > Cd^{2+}$$

Qualitative agreement between the Irving–Williams series and that based on the value of the first hydrolysis constant has been observed for inner-sphere coordination of surface hydroxyls by divalent transition and other heavy metal ions. For example, in selectivity studies conducted on α-FeOOH and $Al(OH)_3$, the order $Cu > Pb > Zn > Ni > Co > Cd$ has been observed.[13,14] This is qualitatively consistent with the predictions based on the hydrolysis constants. It is also consistent with the Irving–Williams series for divalent transition metal cations with $Zn^{2+} < Cu^{2+} > Ni^{2+} > Co^{2+}$.

The second class of divalent metal ions, those which form relatively weak outer-sphere complexes, include the divalent alkaline earth ions (e.g., Be, Mg, Ca, Sr, and Ba). For these cations, the relative affinity to form complexes with surface hydroxyls can be correlated with the ionic radii. This is because the ionic radius gives the relative tendency of these cations to form electrostatic bonds, the most important aspect of outer-sphere sorption. Ions of the same valence but with smaller ionic radii have a greater charge density and so can more effectively form an ion–pair complex. Based on the ionic radii (shown below in parentheses in nanometers), the following selectivity trends are expected for the alkaline earth ions, assuming they form purely electrostatic, ion–pair complexes:[15]

$$Mg^{2+}(0.072) > Ca^{2+}(0.100) > Sr^{2+}(0.118) > Ba^{2+}(0.135)$$

The above sequence has been observed in studies of the sorption of alkaline earth adsorption on γ-Al_2O_3 where the sorption of $Mg^{2+} > Ca^{2+} > Ba^{2+}$.[16] The trend is also consistent with the expectation based on the expected preference of harder Lewis acids for hard Lewis bases like

TABLE 3.

pK Values of Metal Hydroxide Formation

Alkaline earth ions	pK Value
Mg^{2+}	11.4
Ca^{2+}	12.9
Sr^{2+}	>12.9[a]
Ba^{2+}	>12.9[a]

[a] Values not reported, but expected to be greater than for Ca.

Data taken from Reference 12.

surface hydroxyls.* In terms of acid hardness, $Mg^{2+} > Ca^{2+} > Sr^{2+}$, and so sorption would be expected to follow that order.[9] When soft Lewis base sites are present, however, the order reported above can be reversed. This has been found to be the case for sorption of alkaline earth metal ions by smectite clays, where the well-known Hoffmeister series is observed:

$$Mg^{2+} < Ca^{2+} < Sr^{2+} < Ba^{2+}$$

This order is consistent with sorption at interlayer siloxane ditrigonal cavities in smectites that have soft Lewis acid properties.[9] As seen in the Table 3, the relative order for the formation of outer-sphere complexes, like with the inner-sphere complexes, is also related to the ability to form a metal hydroxide complex. Comparison of Tables 2 and 3 indicates that the first hydrolysis constants for alkaline earth metal ions are considerably lower than the class 1–type metal ions, consistent with the expectation that lower values of the first hydrolysis constants correlate with lower affinity. Regardless of the relative ordering of the alkaline earth metal ions, as a group, they all have lower affinity for surface hydroxyls compared with any of the transition-metal and other heavy metal divalent cations.

C. Metal Ion Sorption Isotherms

Having defined the surface and metal ion properties which determine the relative affinity and types of surface complexes that may form, the impact of changing solution conditions on the extent of metal ion sorption can be more easily understood. The extent of sorption, which refers to the amount of a given ion that partitions between an aqueous and a solid phase, can be represented in terms of a relative surface excess quantity given by Γ, where:

$$\Gamma_i = \frac{n_i}{A} \equiv \left[mol / m^2 \right] \qquad (5)$$

---

* The terms *hard* and *soft* are frequently used to classify metal ions based on their preferences for particular ligands. Soft Lewis acids and bases are generally larger and more polarizable than are hard Lewis acids and bases.

and n_i is the number of moles of species i sorbed per unit mass of material and A is the specific surface area. Sorption behavior for a particular solute/sorbent system over a range of conditions is often represented using an isotherm (constant temperature plot) which relates the sorbed-phase concentration of solute to the equilibrium aqueous-phase concentration. Several types of isotherm shapes have been observed for sorption of divalent metal ions onto metal oxides as a function of pH and metal ion concentration (Figure 4). In all cases, the extent of sorption increases with increasing solute concentration. The exact behavior, however, depends on the pH, range of solution conditions monitored, and the extent of surface coverage. At low concentrations, low coverage, constant pH, and narrow concentration ranges, the extent of sorption on an isotherm plot may exhibit a linear dependence with metal ion concentration (inset of Figure 4A). In such cases, it may be possible to model sorption using a single-parameter isotherm model that is linear with respect to [Me]:

$$\Gamma_{Me} = K_d[Me] \tag{6}$$

where K_d represents the distribution coefficient for the following overall reaction between sorbed and aqueous phase metal ion:

$$Me_{(aqueous)} = Me_{(sorbed)} \tag{7}$$

Because of the dependence of metal ion sorption on solution conditions, a single value for K_d is expected to be applicable only over a very limited range of solution conditions. When a wide range of metal ion concentrations is monitored at constant pH, the sorption isotherm is often highly nonlinear, especially at relatively high metal ion concentrations, where it may level off (Figure 4A). For this type of nonlinear isotherm shape, a two-parameter (S_T, K_L) Langmuir-type model can sometimes be used to model the data:

$$\Gamma_{Me} = \frac{S_T K_L[Me]}{1 + K_L[Me]} \tag{8}$$

where [Me] is the molar concentration of a metal cation in solution, S_T is the total moles of sorption sites per unit surface area, and K_L is the Langmuir constant for the following overall sorption reaction at constant pH:

$$\equiv S + Me = \equiv SMe \tag{9}$$

Because of the strong pH dependence of divalent cation sorption (Figure 4A), however, the Langmuir isotherm parameter K_L, like K_d, will also have only a limited range of applicability.

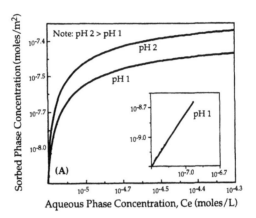

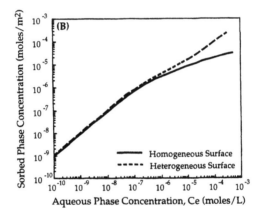

Figure 4

(A) General isotherm behavior for divalent cations at fixed pH and concentrations below the solubility limit (data for two different pH values). (B) Isotherm behavior for sorption on heterogeneous surfaces at fixed pH. (C) Effect of exceeding bulk solution solubility of a metal hydroxide on a sorption isotherm at fixed pH. (D) Effect of increasing solute concentration on a sorption isotherm at fixed pH.

It is often found that nonlinear sorption is even too complicated for a two-parameter Langmuir model to predict the dependence of sorption on $[Me^{2+}]$ over the entire range. An example is the case where two types of surface sites exist that differ in concentration and reactivity (Figure 4B). Under these circumstances, additional parameters may need to be added to the Langmuir-type sorption equation to account for the possibility of multiple site types or heterogeneities in the reactivity of a given site type. The following generalized isotherm illustrates how various additional parameters can be added to the Langmuir isotherm model to represent increasing complexity in isotherm shape:

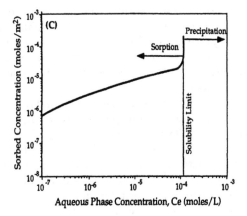

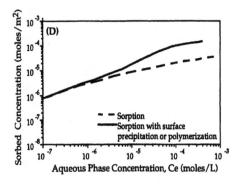

Figure 4 (continued)

$$\Gamma_{Me} = \sum_n \frac{(S_T)_n K_n^{\beta_n} [Me]^{\beta_n}}{\left(1 + K_n [Me]^{\delta_n}\right)^{\gamma_n}} \qquad (10)$$

In this equation, the summation is over n site types and the factors β, δ, and γ are added to account for the possibility of heterogeneity in the reactivity of each site type. Even these more complicated isotherm models, however, do not specifically account for the strong pH dependency of divalent metal cation sorption and would, in general, be limited to constant pH conditions unless specifically optimized over a range of pH values. Furthermore, the parameters generated provide little understanding of the sorption process and make it difficult to extrapolate to other systems or conditions outside the range of calibration.

Additional features may arise in the shape of the isotherm that are not specifically taken into account in the isotherm models. For example, at high solute concentrations, the aqueous phase metal ion concentration may approach and even eventually exceed the solubility limit of oxide, hydroxide, or oxyhydroxide solid phases that it may form. Furthermore, the solubility limits of disordered phases usually are not known accurately, and the possibility exists that homogenous precipitation may result even when not anticipated.

Because the extent of sorption is often calculated from a mass balance approach in which only the aqueous phase concentration has been determined analytically, precipitation is often indistinguishable from sorption. When precipitation occurs, from the standpoint of sorption analyses based on concentrations measured only in the aqueous phase, the sorbed and precipitated phases would be plotted as sorbed amounts in the isotherm plot. Hence, the occurrence of precipitation can complicate the interpretation of sorption isotherms, particularly as the metal ion concentration approaches solubility limits. In most cases, homogenous precipitation of a metal oxide phase would be obvious, appearing as an abrupt increase in the amount "sorbed" at the aqueous-phase concentrations exceeding the solubility. In many cases, however, the increase in sorption relative to the general isotherm is not as abrupt as depicted in Figure 4C. Rather, the relative increase in sorption capacity is more gradual, occurring over an order of magnitude in surface concentration (Figure 4D). Although homogeneous precipitation of disordered phases may be one cause, alternative explanations for this behavior include the formation of surface polymers and surface precipitates. The forms of the isotherm models discussed above have no theoretical framework and are ill-equipped to account for change in isotherm shape which results from such behavior.

A discussion of the various types of isotherms that might be applied to metal ion sorption depending on the range of coverage, can be found in the review by Kinniburgh.[17] Examples of metal ion sorption on metal oxides at constant pH, where the range of surface coverage data is wide enough to observe the overall trends discussed above, include work by Honeyman,[18] Morel and co-workers,[4,19,20] Charlet and Manceau,[21] and Katz and Hayes.[22,23]

D. Metal Ion Sorption Edges

As discussed above, metal ion sorption is strongly pH dependent. As a result, it is common to collect data under conditions where the metal ion and solid concentrations are kept constant and the pH is varied. When these data are plotted as percent of the total metal sorbed vs. pH, sorption typically increases over a narrow pH range from less than 10% of the total to more than 90% (Figure 1). The characteristic sigmoidal curve that results is generally referred to as the *sorption edge*. Often, it is the ability to predict the changes in the shape and location of the sorption edge that is desired in modeling metal ion sorption rather than the shape of the constant pH isotherm discussed above.

For a given metal cation, the location of the sorption edge is dependent on the total metal ion and total surface hydroxyl concentration in the system. The concentration of surface hydroxyl sites can be increased either by utilizing a solid material with higher surface area or by increasing the amount of solid material in the system. For a given type of metal oxide, increasing the concentration of surface hydroxyls by adding more solid generally causes the location of the sorption edge to shift to a lower pH (Figure 5A). On the other hand, increasing the solution concentration of total metal ion while holding the total number of surface sites constant either has no effect on the position of the edge at very low coverage or causes it to shift to the right and become less steep with increasing coverage (Figure 5B). This behavior can be rationalized to a first approximation by assuming that the acid–base properties of metal oxide surface hydroxyls and metal ion sorption can be represented by the equilibrium and mass balance expressions shown in Table 4. Based on the mass balance expression shown in Table 4, an expression can be established among pH, equilibrium constants, the total metal concentration, Me_T, and the total site concentration, S_T. Starting with the definition of the fractional sorption,

$$\frac{[\equiv SOMe^+]}{Me_T} = \frac{[\equiv SOMe^+]}{[Me^{2+}] + [\equiv SOMe^+]} = \left(\frac{[H^+]}{K_{Me}[\equiv SOH]} + 1\right)^{-1} \tag{11}$$

and substituting

$$[\equiv SOH] = \alpha_1 S_T \tag{12}$$

where α_1 is given by:

$$\alpha_1 = \left(\frac{[H^+]}{K_{a1}} + 1 + \frac{K_{a2}}{[H^+]} + \frac{K_{Me}(Me_T - [\equiv SOMe^+])}{[H^+]}\right)^{-1} \tag{13}$$

leads to an expression for fractional sorption given by:

$$\frac{[\equiv SOMe^+]}{Me_T} = \left(\frac{[H^+]}{K_{Me}\alpha_1 S_T} + 1\right)^{-1} \tag{14}$$

For the case of 50% sorption where fractional sorption equals 0.5, $Me_T - [\equiv SOMe^+] = 0.5$,

$$\alpha_1 = \left(\frac{[H^+]}{K_{a1}} + 1 + \frac{K_{a2}}{[H^+]} + \frac{0.5K_{Me}}{[H^+]}\right)^{-1} \tag{15}$$

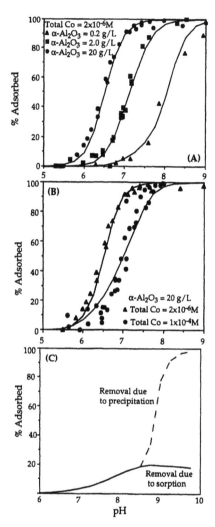

Figure 5
(A) Effect of increasing the concentration of reactive sites on the location of pH adsorption edges. (B) Effect of increasing the concentration of total metal ion on the shape of pH adsorption edges. (C) Effect of exceeding the aqueous phase solubility of a metal ion on the shape of pH adsorption edge.

and the overall expression for fractional sorption reduces to:

$$\frac{\left[\equiv SOMe^{+}\right]}{Me_{T}} = 0.5 = \left(\frac{\left[H^{+}\right]}{K_{Me}\alpha_{1}S_{T}} + 1\right)^{-1} \qquad (16)$$

TABLE 4.

Mass Law and Mass Balance Equations for
Surface Complexation Models

Mass Law Equations for Proton and Metal Ion Sorption

$\equiv SOHO_2^+ = \equiv SOH + H^+$	K_{a1}
$\equiv SOH = \equiv SO^- + H^+$	K_{a2}
$\equiv SOH + Me^{2+} = \equiv SOMe^+ + H^+$	K_{Me}

Mass Balance Equations for Total Metal and Surface Sites

$$S_T = [\equiv SOH_2^+] + [\equiv SOH] + [\equiv SO^-] + [\equiv SOMe^+]$$
$$Me_T = [\equiv SOMe^+] + [Me^{2+}]$$

Using Equation 16, the effects of changing Me_T concentrations on α_1 can be used to assess the direction of the shift of the fractional sorption edge. For example, at very low Me_T to S_T ratios changing Me_T at constant S_T has no effect on the position of the pH edge. This can be rationalized in terms of the effect of sorption on the distribution of uncomplexed surface species. Because $[\equiv SOMe^+]$ is so small under these conditions, α_1 is invariant for changes in Me_T and the pH edge at 0.5 fractional sorption would not shift according to Equation 16.

If Me_T is comparable to S_T, an increase in Me_T (for constant S_T) requires a concomitant increase in pH for the fractional sorption to remain at 0.5 (Figure 5B). Under these conditions, the surface distribution of uncomplexed sites is affected, α_1 decreases, and the pH of 0.5 fractional sorption must increase for the term $((K_{Me}\alpha_1 S_T)^{-1}[H^+])$ to remain constant and equal to the value of 1 according to Equation 16. Likewise when S_T is increased at constant Me_T, the pH must decrease for the term $((K_{Me}\alpha_1 S_T)^{-1}[H^+])$ to remain constant and equal to the value of 1. Hence, the sorption edge must shift to lower pH to maintain the fractional sorption at 0.5 (Figure 5A).

The concentration ratio of Me_T to S_T under which the above trends occur and the degree of the shifts in the position and shape of the edge depend on the relative values of the equilibrium constants, K_{a1}, K_{a2}, and K_{Me} (Table 4). Furthermore, when the relative concentration of metal ion in solution becomes high, other processes which contribute to metal ion removal from solution may affect the slope of the sorption edge. For example, if the bulk solubility of the metal ion is exceeded, the "apparent" sorption edge determined by measuring the concentration of metal ion in the aqueous phase following solid–liquid separation appears to increase dramatically (Figure 5C). Alternatively, as surface coverage becomes quite high, the sorption process may change from simple sorption of monomeric complexes to the formation of surface polymers or precipitates. This, too, can result in significantly more metal ion sorbed over a narrow pH range leading to a steeper slope of the sorption edge, as suggested by the change in stoichiometry shown in the following polymer or surface precipitation reactions:[23]

Surface Polymer Reactions

$$\equiv SOH + 2Me^{2+} + 2H_2O = \equiv SOMe_2(OH)_2^+ + 3H^+ \qquad (17)$$

$$\equiv SOH + 4Me^{2+} + 5H_2O = \equiv SOMe_4(OH)_5^{2+} + 6H^+ \qquad (18)$$

Surface Precipitate Reaction

$$\equiv SOH + Me^{2+} + 2H_2O = \equiv SO[Me(OH)_2\,(S)] + 2H^+ \qquad (19)$$

Each of the above reactions has a reaction stoichiometry of greater than one proton released per metal ion sorbed. Hence, any one of these reactions compared with the monomer reaction with one proton released per metal ion sorbed would cause an increase in the slope of the sorption edge.

Although not specifically mentioned above, another complexity in pH sorption edge behavior arises at very low surface coverages (less than 0.1%). For example, at very low surface coverage, changes in Me_T for fixed S_T can cause the sorption edge to shift to higher pH, even though surface concentrations are well below the saturation limit (see, e.g., Honeyman[18]). While at higher coverage such shifts with increasing Me_T make sense based on saturation of the number of surface hydroxyl sites according to the discussion above, at low coverage this is hard to reconcile if it is assumed that only one type of reactive surface hydroxyl site exists. In fact, shifts in the sorption edge at very low coverage are usually attributed to the presence of a small number of very reactive surface hydroxyl sites.[18] Assuming the presence of such sites, the shift at very low coverage can then also be rationalized by the saturation, in this case, of the small number of reactive sites with increasing Me_T. As a result, at least two types of surface hydroxyl sites are often needed to account for sorption of divalent cations from very low to moderate surface coverage, one site type at a relatively low concentration (1% of the total) which has a relatively greater affinity for divalent cations and another site type with lower affinity and in greater concentration (99%) which controls sorption in the low to moderate coverage range (0.1 to 10%).

In addition to the effects of Me_T or S_T on metal ion sorption, the presence of competing ions can also influence the extent of sorption, depending on the type of surface complex formed. For example, for weakly sorbing metal cations like the alkaline earth cations discussed above, changes in the background electrolyte concentration can significantly reduce the amount of sorption (Figure 6A). On the other hand, for cations with a strong affinity for surface hydroxyl sites, like the transition-metal and heavy metal cations, changing the background electrolyte concentrations has little or no

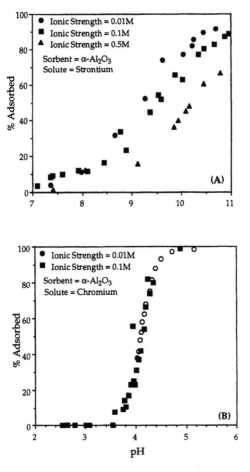

Figure 6
Effect of increasing the concentration of background electrolyte on the location of the pH adsorption edge for (A) a weakly sorbing cation (Sr on α-Al$_2$O$_3$) and (B) for a strongly sorbing cation (Cr on α-Al$_2$O$_3$).

effect on sorption (Figure 6B). Reactions describing these differences in affinity have been formulated as follows:

Inner Sphere

$$\equiv\text{SOH} + \text{Me}^{2+} = \equiv\text{SOMe}^+ + \text{H}^+ \qquad K_{IS} \qquad (20)$$

Outer Sphere

$$\equiv\text{SOH} + \text{Me}^{2+} = \equiv\text{SO}^-\text{-Me}^{2+} + \text{H}^+ \qquad K_{OS} \qquad (21)$$

$$\equiv SOH + Na^+ = \equiv SO^--Na^+ + H^+ \qquad K_{Na} \qquad (22)$$

In general, $K_{Me}^{IS} \gg K_{Me}^{OS} \gg K_{Na}$. As a result, when a univalent electrolyte cation like Na is present at concentrations which are several orders of magnitude in excess of the total metal ion concentration, it may be able to displace a weakly sorbed, outer-sphere cation but never a strongly sorbed cation.

It is clear based on the above discussion that trends in divalent cation sorption are highly dependent on the solution conditions, particularly pH, and that as surface coverage increases, multiple reaction processes may occur simultaneously. Selection of the types of surface reaction and complexes present under various reaction conditions, however, requires independent knowledge of their existence. While it is possible to speculate about the existence of the types of reactions just presented, it is not possible to justify one reaction or another or a combination of reactions purely by fitting sorption data. As any modeling exercise will show, as the number of reactions and parameters with the appropriate conditional form are increased, the fits to data will improve. In fact, the types of reactions presented above have been confirmed by recent spectroscopic studies to determine the structure and composition of sorbed complexes as a function of solution conditions. The next section provides an overview of these studies.

III. STRUCTURE OF SURFACE COMPLEXES

A. Types of Surface Complexes

A variety of spectroscopic techniques are available which may be applied *in situ* to study the structure of metal ion surface complexes at the mineral–water interface. The techniques include Raman and Fourier transform infrared spectroscopy (FTIR); magnetic resonance spectroscopies including electron spin or paramagnetic resonance (ESR or EPR), electron-nuclear double resonance (ENDOR), electron spin-echo envelope modulation (ESEEM), and nuclear magnetic resonance (NMR); Mossbauer spectroscopy (MOSS); and XAS. A recent review that details the advantages and limitations of each of these methods can be found in Brown.[24] Based on the *in situ* spectroscopic studies of ion sorption at mineral–water interfaces that have been conducted to date, a clearer picture of the types of reaction processes and surface complexes that may result during sorption is emerging. The most important results from the studies with respect to SCMs are summarized in this section. Because the most definitive *in situ* determination of surface structure and composition of sorbed species has come from the results of XAS, emphasis on these studies is given here (Table 5).

TABLE 5.

XAS Studies on Oxide Surfaces

System	Coverages	Type of complex	Ref.
Se(IV) and Se(VI) on goethite	High	Inner and outer sphere, respectively	25
Pb(II) on γ-alumina	High	Inner sphere, monodentate, multinuclear (clusters)	26
Co(II) on γ-alumina and rutile	Moderate to high	Inner sphere, multinuclear (clusters)	27, 28
Pb(II) on goethite	High	Inner sphere, multinuclear	29
Cr(III) on goethite and HFO	High	Inner sphere, bidentate, multinuclear (clusters) and solid solutions	21
Ur(VI) on silica and montmorillonite	High	Inner sphere, bidentate, mono- and multinuclear	30
Co(II) on calcite		Solid solution	31
Np(V) on goethite	High	Inner sphere, mononuclear	32
Cd(II) on alumina	High	Inner sphere, precipitation	33
Se(IV) on alumina	High	Inner sphere, mononuclear	33
As(VI) on ferrihydrite, goethite, lepidocrocite, and akaganeite	High	Inner sphere, bidentate (monodentate also on ferrihydrite)	34
Cr(III) on silica	High	Inner sphere, monodentate, multinuclear (clusters)	35
Sr(II) on goethite, silica, alumina, montmorillonite, hectorite, and kaolinite	High	Outer sphere, mononuclear	36
Co(II) on kaolinite	Moderate to high	Inner sphere, bidentate mono- and multinuclear (clusters)	37, 38

To date, XAS has been used to determine the identity, location, and coordination environment for a number of metal ions sorbed to various oxide minerals (Table 5). Because XAS data analysis allows the coordination number and bonding distances for different coordination shells surrounding a sorbing metal ion out to about 6 Å to be determined, it is possible to distinguish among inner-sphere or outer-sphere surface complexes, surface polymers and precipitates, and solid solution formation (Figures 2 and 7). For example, for outer-sphere coordination, the sorbing cation remains completely hydrated, no surface metal atoms enter into the near coordination environment, and the XAS spectrum looks identical to the aqueous-phase metal ion spectrum. When inner-sphere complexes form, however, water is displaced from the inner coordination shell of the sorbing cation allowing surface metal atoms from an oxide sorbent to reside within about 3 or 4 Å of the sorbing cation. The presence of these surface metal ions leads to the appearance of spectral features in the XAS spectrum known as "beat patterns." Likewise, it is possible to distinguish mononuclear surface species from multinuclear species based on the presence or absence of additional metal atoms in the near coordination environment surrounding the sorbing cations. When surface polymers or precipitates form, bridging hydroxyls connect adjacent sorbing metal cations reducing the distance between them. The appearance of spectral beat patterns, indicative of additional metal atoms within 6 Å of a sorbing metal ion, distinguishes multinuclear species from mononuclear ones. Hence, it is possible to distinguish among the variety of near coordination environments (Figures 2 and 7) for metal ions sorbed at mineral–water interfaces by XAS.

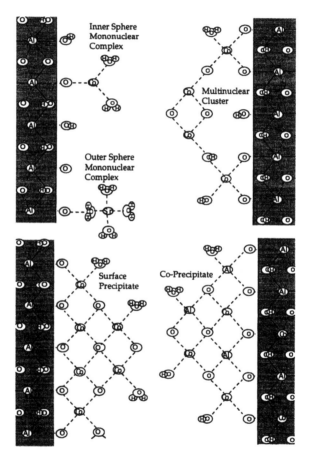

Figure 7
Types of surface complexes observed using XAS.

B. Applications of X-Ray Absorption Spectroscopy

1. *Evidence for Inner- and Outer-Sphere Monomer Complexes*

The first successful application of XAS to study surface complex formation at the mineral–water interface was conducted by Hayes et al.[25]

They showed that XAS could be used to distinguish between inner- and outer-sphere surface complexes based on the differences between the XAS of sorbed selenate (SeO_4^{2-}) compared with sorbed selenite (SeO_3^{2-}) on goethite. The appearance of two surface Fe atoms in the near coordination environment of sorbed selenite at 3.38 Å and the absence of Fe in the XAS of sorbed selenate clearly demonstrated that selenite was bonded directly to the surface as an inner-sphere complex, whereas selenate was separated from the surface by at least one layer of water, indicative of outer-sphere surface complex formation.

Waychunas et al.[34] have also recently employed XAS to study anion sorption on iron oxides. They studied the sorption of the strongly sorbing anion, arsenate (AsO_4^{2-}), by three FeOOH crystalline polymorphs, goethite, akaganeite, and lepidocrocite, and a poorly crystalline ferrihydrite. Consistent with the selenite studies, the XAS data analysis supported the formation of inner-sphere, bidentate surface complexes of arsenate on all the iron oxide minerals. Monodentate species, however, were observed in the study of arsenate sorption to ferrihydrite, particularly in samples in which the As/Fe ratios were low. While both of the above studies were conducted with divalent anions instead of cations, they have clearly demonstrated the potential of the XAS for distinguishing among sorbed complexes *in situ*.

Recent XAS studies with metal cations have indicated the usefulness of this technique for these sorbates as well. For example, XAS studies of cation sorption show that transition-metal and other heavy metal cations sorb as inner-sphere complexes on metal oxide sorbents at low surface coverage. Chisholm-Brause et al.[27] have shown that Co(II) forms inner-sphere, bidentate surface complexes with TiO_2 and inner-sphere multidentate complexes with γ-Al_2O_3. XAS studies with Pb^{2+} have led to similar conclusions, indicating that at low coverage Pb^{2+} sorbs as an inner-sphere monodentate complex with γ-Al_2O_3 and as an inner-sphere complex with goethite.[26,29] In addition, based on XAS results, inner-sphere, mononuclear bidentate surface complexes have been proposed for UO_2^{2+} on hydrous iron oxide.[30] Results from studies of strongly sorbing cations having other than divalent charge have also confirmed that inner-sphere complexes may form. In a recent study of neptunium(V), permissive evidence has been reported for the inner-sphere coordination of the species $NpO_2(H_2O)_5^+$ species with goethite.[32] In other recent studies, Cr^{3+} was found to form unidentate, inner-sphere surface complexes with the terminal hydroxyls on SiO_2,[35] and multidentate complexes on Fe oxide minerals.[21] XAS studies by Dent et al.[30] have suggested that monomeric or multinuclear species of U(VI) may form on montmorillonite and silica depending on the pH and metal ion concentration. Thus, all data to date support the hypothesis that transition-metal and other heavy metal cations form inner-sphere surface complexes on metal oxides at low coverages.

No XAS data have yet been published which provide unequivocal support for the hypothesis that weakly sorbing alkaline earth metal cations form only outer-sphere surface complexes with metal oxides. The main data supporting this hypothesis come from ionic-strength dependence studies which show that univalent cations like Na^+ can cause desorption of weakly sorbed alkaline earth cations from metal oxides, but not strongly sorbed transition metal cations.[10] Hayes and co-workers[36] have, however, collected XAS data for Sr^{2+} on a variety of metal oxide and clay minerals (α-FeOOH, SiO_2, γ-Al_2O_3, montmorillonite, hectorite, and kaolinite) and have seen no difference in the XAS of the cation sorption samples compared with the aqueous solution samples, suggesting that the Sr^{2+} remains hydrated upon sorption. One problem with XAS studies of Sr^{2+} is that the primary and secondary hydration shells around the Sr cation are apparently not well ordered, causing the XAS data to be severely damped beyond the second coordination shell of about 3 to 4 Å, even for aqueous samples. This has made the basis for the absence of near coordination features compared with the aqueous phase spectrum speculative at this point. Nonetheless, the data do not refute the expectation that Sr^{2+} remains hydrated and forms outer-sphere complexes upon sorption. More work in this area is needed.

2. Evidence for Multinuclear Surface Complexes

In the case of cations which sorb strongly to metal oxide surfaces, when surface coverage exceeds about 10%, XAS data support the contention that multinuclear species begin to form. For example, Chisholm-Brause et al.[26] have shown that multinuclear Pb^{2+} surface complexes form at surface coverage in excess of 10% during sorption by γ-Al_2O_3 and that the number or size of multinuclear species increases with increasing surface concentration. Similar findings have been observed for sorption of Co(II) by γ-Al_2O_3 and TiO_2.[27] The underlying surface structure has also been found to influence the polymerization process. Chisholm-Brause[28] has found that the size of the clusters formed on TiO_2 were smaller than those formed on γ-Al_2O_3 for equivalent surface coverages, concluding that the size of multinuclear species that form is dependent on the structure and juxtaposition of surface hydroxyls. XAS studies of Co(II) sorption on clay surfaces are also consistent with these findings.[37,38] For example, O'Day et al.[37] compared the sorption of Co(II) on kaolinite and α-SiO_2 and found that the size of cobalt surface clusters was larger on kaolinite compared with quartz. Recently, Papelis and Hayes[39] have found that multinuclear clusters of Co(II) also form on the surface hydroxyl edge sites of montmorillonite at high surface coverage. They were able to distinguish the multinuclear species that formed on the surface hydroxyl edge sites from outer-sphere complexes that formed with interlayer siloxane ditrigonal sites by selectively desorbing Co(II) from the interlayer sites with Na^+ ion concentrations above 0.1 M.

Evidence for multinuclear species formation has also been reported for Cr(III) sorption at high surface coverages on SiO_2, goethite, and hydrous ferric oxide.[21,35] In the case of Cr(III) sorption on silica, the growth of γ-CrOOH-type structure in small patches on the surface was indicated. On goethite, however, the better template match of surface hydroxyls with α-CrOOH structure resulted in the epitaxial growth of an α-CrOOH coating, followed by a phase transition to γ-CrOOH-type structure as the coating expanded outward from the surface. These studies point to the importance of the juxtaposition of reactive surface hydroxyls to the manner in which surface multinuclear species form and evolve. In general, all of the higher coverage studies of cation sorption illustrate that, as apparent surface coverages exceed about 10%, surface polymers and precipitates begin to form on many different types of mineral oxides. This common critical coverage value is probably due, in part, to the reactive site density of about 2 to 3 sites/nm^2 on average for many metal oxides.[6]

Surface precipitation or polymerization of alkaline earth metal cations is not generally expected because of the higher solubility of metal oxides and hydroxides of these metals and their weaker sorption. Because of the weaker sorption, surface coverages usually will not exceed 10% at reasonable pH values in the absence of bulk phase precipitation. In view of the weak sorption, no XAS studies have been attempted that would address the issue of the effects of surface coverage on alkaline earth sorption.

As all of the above studies indicate, XAS can be used to identify the type of complex formed at surfaces and the location of the complex with respect to the surface. Indeed, the results suggest that the type of complex formed is dependent on the sorbing ion, the mineral surface characteristics, and the surface coverage of the mineral. In the next section, the way in which this information can be used in surface complexation modeling will be demonstrated after a brief review of SCMs.

IV. SURFACE COMPLEXATION MODELING OVERVIEW

A. Constraint Equations and Model Parameters

The results summarized in the previous section illustrate that XAS can be used to select among different reaction processes which occur during sorption of metal cations to mineral surfaces. Surface complexation models (SCMs) require accurate description of the reaction phenomena described in the previous sections and have been used in conjunction with proposed surface reactions to elucidate the effects that solution conditions have on sorption. SCMs assume that the surface is composed of functional groups which act as ligands that can form ion–pair and coordination complexes with aqueous-phase solutes. The models vary in the way surface reactions are formulated and the depiction of the interfacial region.

For each surface species postulated, a chemical equilibrium expression is developed in terms of the components, and a minimum set of reactants from which the species can form. Each of the equilibrium constants for the species formation reaction incorporates some type of activity correction. The particular form of the activity correction differs among the various SCMs and depends on the assumptions regarding the structure of the interfacial region, the type of surface complexes allowed, and the resulting relationship between charge and potential.[2,6,40]

All of the SCMs can be formulated mathematically to yield a set of simultaneous equations which are readily solved using any number of numerical models. Computer programs are available which, at a minimum, combine surface complexation with solution speciation modeling (e.g., MINEQL,[41] HYDRAQL,[42] MINEQL+[43]). In general, SCMs include a set of equations describing the thermodynamic, stoichiometric, and electrostatic constraints on the surface, interfacial region, and bulk solution, along with a set of equations expressing the relationship between surface charge and potential imposed by the particular model. For metal oxide systems, these equations and constraints include

1. Mass law expressions for protonation/deprotonation reactions of each type of surface hydroxyl group (usually for a given metal oxide just one hydroxyl site type with amphoteric properties is assumed);

2. Specification of the total number of surface hydroxyl groups of each type;

3. A mass law expression for each complexation reaction between a surface hydroxyl group and sorbing ion, incorporating activity corrections due to electrostatic interactions;

4. A mass law expression for each aqueous-phase complexation, precipitation, and redox reaction, incorporating activity corrections due to electrostatic interactions;

5. Charge–balance equations for each sorption plane in the particular model based on the number of ionized species sorbed at that location;

6. Expressions relating the interfacial charge to the potential for each sorption plane in a given model; and

7. Electroneutrality condition for the bulk solution and surface.

B. Description of Several Surface Complexation Models

A variety of SCMs have been used for predicting sorption of metal ions and other species on metal oxides (Table 6 and Figure 8) including the non-electrostatic model (NEM) (Figure 8a), the diffuse layer model (DLM) (Figure 8b), the constant capacitance model (CCM) (Figure 8c), and the triple layer model (TLM) (Figure 8d). The primary differences among these models relate to the description of the interlayer region, specifically the location of sorbed species with respect to the surface and

TABLE 6.
Surface Complexation Reactions and Model Parameters

	Mass Law Constant	NEM Location of Complex	CCM Location of Complex	DLM Location of Complex	TLM Location of Complex
Protolysis Reactions					
$SOH + H^+ = SOH_2^+$	K^+	o-plane	o-plane	o-plane	o-plane
$SOH = SO^- + H^+$	K^-	o-plane	o-plane	o-plane	o-plane
Surface Complexation Reactions					
$\equiv SOH + Me^{2+} = \equiv SOMe^+ + H^+$	K_{Me}	o-plane	o-plane	o-plane	o- or β-plane
$\equiv SOH + L^- = \equiv SL + OH^-$	K_L	o-plane	o-plane	o-plane	o- or β-plane
$\equiv SOH + Cat^+ = \equiv SO^- - Cat^+ + H^+$	K_{Cat}	—	—	—	β-plane
$\equiv SOH + An^- + H^+ = \equiv SOH_2^+ - An^-$	K_{An}	—	—	—	β-plane
Charge/Potential Relationships					
$\sigma_{o\text{-plane}}$		N/A	$C_1\psi_o$	—	$(\psi_o - \psi_\beta)/C_1$
$\sigma_{\beta\text{-plane}}$		N/A	—	—	$-(\psi_\beta - \psi_d)/C_2$
$\sigma_{d\text{-plane}}$		—	—	$0.1174\sqrt{I}$ $\sinh(zF\psi_d/2RT)$	$0.1174\sqrt{I}$ $\sinh(zF\psi_d/2RT)$
Model Parameters		K^+, K^-, N_s	K^+, K^-, N_s, C_1	K^+, K^-, N_s	$K^+, K^-, K_{Cat}, K_{An},$ N_s, C_1, C_2

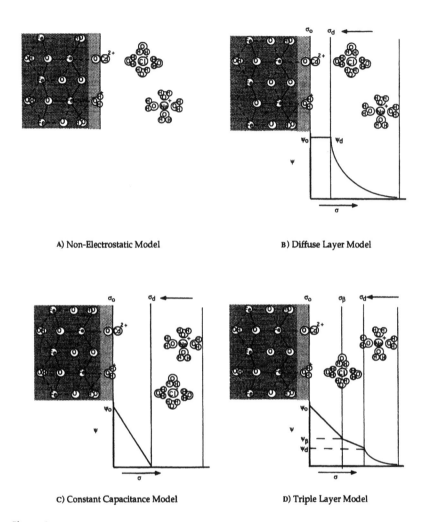

Figure 8
(A) Schematic of the nonelectrostatic version of the surface complexation model. (B) Schematic of the diffuse layer version of the surface complexation model. (C) Schematic of the constant capacitance version of the surface complexation model. (D) Schematic of the triple-layer version of the surface complexation model.

the description of charge–potential relationships across the interfacial region. The simplest SCMs are the NEM and DLM in which only one sorption plane is allowed. In the case of the NEM, the electrostatic effects associated with the buildup of interfacial charge and potential due to sorption are ignored, while in the DLM, the Gouy–Chapman diffuse layer model is used to describe the relationship between surface charge and potential in the interfacial region. For both the DLM and NEM, three parameters are needed to characterize the acid–base properties of the surface hydroxyl groups: the site density, N_S, and the surface protolysis constants, K^+ and K^-. Another commonly used SCM which only has one sorption plane is the CCM. This model differs, however, from the NEM and DLM in that it requires one additional parameter, the inner-layer capacitance, C_1, and is restricted to relatively high and constant ionic strength conditions.[40] Because the NEM, DLM, and CCM have only one sorption plane, they cannot distinguish among the relative affinity of sorbed species except by varying the value of the sorption constant. Although this has not limited the ability of these models to predict the behavior of sorbing cations under constant ionic strength conditions, it remains to be demonstrated whether or not these simplistic models of the interfacial region can account for the competitive effects of background electrolytes on weakly sorbing vs. strongly sorbing cations (Figure 6).[4,44] Another limitation of these three SCMs is their inability to provide accurate descriptions of the changes in the interfacial charge and potential due to ion sorption, either because they have no electrostatic prediction capabilities (NEM), must be used only for a constant ionic strength (CCM), or are only applicable to low ionic strength conditions (DLM). Hence, even though they may provide reasonable descriptions of cation sorption under a limited set of conditions, one-plane SCMs will not, in general, provide adequate enough predictions of surface charge and potential to assess their impact on colloidal properties like coagulation which depend on these properties.

In view of the above limitations, more complex sorption models incorporating the full Gouy–Chapman–Stern–Grahame representation of the interfacial region, in which inner- and outer-planes of sorption are allowed (Figure 8d), were developed. The first application of the Gouy–Chapman–Stern–Grahame model to metal oxide surfaces took the form of the TLM.[45] The TLM differs from the one-plane models by having two sorption planes for the formation of surface complexes. In the original interpretation of this model, only H^+ and OH^- were selected to sorb at the inner-surface plane while all other sorbing ions were restricted to the outer sorption plane.[45] More recently, Hayes and co-workers[10,46] illustrated that by placing strongly sorbing cations at the inner-surface plane and weakly sorbing cations at the outer-layer plane (Figure 8d), the TLM could be used to predict the effects of changes in ionic strength on metal ion sorption. This has led to the current interpretation of the TLM that placement of sorbing ions at the inner plane best represents inner-sphere complexes while outer-plane placement corresponds best to outer-sphere complexes. The additional complexity and more realistic depiction of the interfacial region of the TLM, however, comes with a cost. Four additional SCM parameters are required for the TLM compared with the NEM and DLM, an inner- and outer-layer capacitance, C_1 and C_2, and two reaction constants to specifically account for the effects of background electrolytes on surface protolysis, K_{Cat} and K_{An} (Table 6).

SCMs are also distinguished by the method employed to account for activity corrections for surface species caused by non-ideal conditions when the surface becomes charged. In most SCMs, the Davies equation is used for calculating activity coefficients for aqueous-phase species using infinite dilution as the reference state condition. This equation has the form:

$$\ln \gamma_i = -A z_i^2 \left(\frac{I^{1/2}}{1 + I^{1/2}} - bI \right) \tag{23}$$

where γ_i is the activity coefficient of species i, z_i is the charge on species i, A is $1.82 \times 10^{-6}(\varepsilon T)^{-3/2}$, and b is equal to $0.3 z_i^2 A$. This model is valid for ionic strengths up to approximately 0.5 M.[12] At concentrations above this level, it is necessary to incorporate interactions among neighboring ions into models describing activity corrections using such methods as the Brønsted–Guggenheim specific ion interaction model or the Pitzer equations.[47,48]

The form of the activity correction for surface species is directly related to the molecular hypotheses regarding the structure of the interface. Almost all of the SCMs employ a zero charge reference state in which the activity coefficients for surface species comprise a chemical and an electrostatic component:

$$\gamma_i = \gamma_i^0 \exp(zF\Psi/RT) \tag{24}$$

where γ_i^0 is the chemical portion and the exponential term with the model-dependent surface potentials and $\exp(zF\Psi/RT)$ is the electrostatic portion. Typically, γ_i^0 values for surface species are assumed to be equal, and, hence, they cancel out. For example, the mass law expression for proton sorption using any of the SCMs is:[49]

$$K = \frac{[SOH][H^+]\gamma_{SOH}\gamma_{H^+}}{[SOH_2^+]\gamma_{SOH^{2+}}} \tag{25}$$

and since γ_{SOH}^0 and $\gamma_{SOH_2^+}^0$ are assumed equal, the resulting expression for the mass law is:

$$K = \frac{[SOH][H^+]\gamma_{H^+}}{[SOH_2^+]} \exp(zF\Psi/RT) \tag{26}$$

A slightly different form of the activity corrections for the TLM has been implemented by Hayes and Leckie.[10] In this case, the hypothetical infinite dilution reference state, zero surface charge, and no ionic interactions for both surface and solution species were selected; the standard state for both solution and surface species was defined as 1 mol/L and no ionic interaction. With this version, the mass law expression for the proton sorption reaction reduces to:

$$K = \frac{[SOH][H^+]}{[SOH_2^+]} \exp(zF\Psi / RT) \qquad (27)$$

As can be seen by comparing the two expressions above, the major difference is that the activity corrections to solution species are incorporated directly in the exponential potential term in the version of Hayes and Leckie.[10] Although these mathematical forms appear quite similar, they will lead to different modeling results, particularly when variable ionic strength conditions are being modeled. Therefore, users of the TLM must be sure to use the surface parameters that are consistent with the standard and reference states assumed. In addition, because each of the SCMs highlighted in this section (NEM, DLM, CCM, and TLM) represent and calculate the model surface potentials that appear in the exponential activity correction differently (Table 6), a separate and self-consistent set of surface acidity constants and model parameters must be determined for each version of the model. Model parameters for different SCM versions are not interchangeable. Comprehensive reviews of the manner in which SCM parameters are estimated for metal oxides can be found in the literature.[4,40] The recent study by Katz and Hayes[22] gives the most up-to-date method for obtaining TLM constants from titration and metal ion sorption data. Two theoretical methods have also recently emerged which have the potential for allowing values for surface acidity constants to be estimated independently from titration data.[50-52]

C. Spectroscopy and the Selection of SCM Reactions

Given the various SCM versions, guidelines are needed to select from among the different possible surface reactions and SCMs. Fortunately, independent confirmation of the structure of sorbed surface complexes by XAS is now helping to establish a logical set of selection rules for surface species. For example, at relatively low coverage (below 10%), one of the major issues in selecting an SCM and an appropriate SCM reaction is whether or not a given metal ion forms an inner- or outer-sphere surface complex with surface hydroxyls. In the SCMs, such as the CCM[53,54] and DLM[55,56], only inner-sphere surface complexation is allowed, inasmuch as only one sorption plane at the surface exists. In contrast, TLM analogues of either inner- or outer-sphere surface complexation reactions can be selected, in view of the option of allowing sorption to take place at either the surface plane or a second plane farther out from the surface.[10,57] All models, however, have been successful to some degree in predicting sorption behavior. For example, the CCM has been widely used to model the sorption of relatively weakly sorbing alkaline earth and more strongly sorbing transition-metal divalent cations on a variety of metal oxide surfaces.[44] Likewise, the DLM has been used successfully to model sorption of alkaline earth and transition-metal cations at low coverages to metal oxides.[4,56] Similarly, the TLM has found widespread application in modeling the sorption of divalent cations.[10,22,23,45,57,58]

One might ask, if all models work equally well in predicting divalent cation sorption using one or two surface complexation reactions, why do we need independent spectroscopic confirmation of the types of complexes? The answer lies in the fact that, when modeling studies are conducted over wide ranges in surface coverage, ionic strength, or in more complex multiple sorbate systems, only limited modeling success has been achieved. When sorption studies cover a wider range of conditions or have been conducted with multiple and competitively sorbing solutes, the differences between weakly sorbing and strongly sorbing cations become more evident and the choice of appropriate reactions more restrictive. For example, Hayes[2] using the TLM found that it was not possible to model the ionic strength dependence of the sorption of weakly sorbing cation Ba^{2+} unless an outer-sphere surface reaction was selected. Even the TLM, however, overpredicted the effects of the background electrolyte unless the surface constants for the electrolyte were reduced from the values obtained independently from surface titrations. More recent studies have found better prediction of competition between background electrolyte ions and weakly sorbing alkaline earth cations when both inner- and outer-sphere surface complexes were allowed.[59] This suggests that the difficulty that Hayes and Leckie[10] had may have been due to the assumption that only outer-sphere complexes were present. In fact, it may be that mixed populations of inner- and outer-sphere surface complexes reside on metal oxide surfaces depending on solution pH and ionic strength conditions for this class of cations.

Spectroscopic evidence that more weakly sorbing divalent ions may sorb as both inner- and outer-sphere complexes has recently come to light from two apparently conflicting XAS studies of selenate sorption. Manceau and Charlet[60] have recently completed a study which indicates that selenate forms inner-sphere complexes with goethite, in contrast to the original XAS finding of Hayes et al.[25] who found only outer-sphere coordination of selenate on goethite. The apparently contradictory results, however, may be attributable to the different experimental conditions of the two studies. In the study by Manceau and Charlet, the sorption experiments were conducted at 0.1 M $NaNO_3$, while Hayes et al. conducted their study at 0.01 M $NaNO_3$.[59] Hence, it may be that the relative population of inner- and outer-sphere surface complexes were reversed when the ionic strength went from 0.01 to 0.1 M with the higher ionic strength favoring the formation of inner-sphere surface complexation. Although these hypotheses need to be confirmed, these results reaffirm the need for *in situ* spectroscopy studies to provide guidance for selecting appropriate surface complexation reactions.

The above discussion suggests that the more complex representation of the interfacial region provided by the TLM may be required to distinguish weakly and strongly bonded metal ions with separate sorption planes. Because of the restriction in the NEM, DLM, and CCM that all sorbates must reside at the surface plane, these models are less flexible in accounting for the differences in the affinity of inner- and outer-sphere complexes and the relative effects of background electrolytes. As such they are likely to be less successful in modeling the competition among surface species when both types of surface complexes are present.

In addition to distinguishing between inner- and outer-sphere surface complexes, proper selection of the type of surface species is often necessary to predict the sorption of strongly sorbing metal cations over wide ranges in coverage. For example, Farley et al.[19] have modified the DLM by allowing for the formation of a solid solution between the sorbing metal and surface at high coverage. Using this modified version of the DLM, Cd^{2+} sorption and Cr^{3+} sorption on ferrihydrate have been successfully modeled from low to high coverage.[19-21] In spite of the inconsistency of the modified DLM with XAS data (the DLM model assumes solid solution formation instead of the formation of surface polymers and precipitates indicated by XAS), the mathematical form of the modification has worked well for predicting sorption at high coverage. Unlike the solid solution DLM, Katz and Hayes[23] have recently modified the TLM to include surface polymers and surface precipitate reactions, rather than solid solution formation, and have been successful in accounting for sorption of Co(II) on γ-Al_2O_3 over a wide range of pH, surface coverage, and solid concentration. In this study, XAS spectroscopic data were used to guide the selection of the multinuclear reactions that were used for modeling of the sorption data. Although Charlet and Manceau[21] conducted an XAS study to guide the selection of SCM parameters for the onset of solid solution at high coverage, the study by Katz and Hayes[23] was the first to use an SCM that had surface polymer and precipitation reactions which were consistent with XAS data.

All of the above studies with the DLM and TLM indicate that SCMs can be modified with the guidance of XAS data to account for sorption of strongly sorbing cations over wide ranges in surface coverage. The following section compares the ability of three SCMs, the NEM, DLM, and TLM, to fit the data described in Katz and Hayes[22,23] and to demonstrate how XAS data is used to guide the selection of SCM reactions.

V. ILLUSTRATION OF SURFACE COMPLEXATION MODELING

A. Description of the Protocol Used to Model Divalent Metal Ion Sorption

The major elements of surface complexation modeling involve selection of an SCM, e.g., the NEM, DLM, CCM, or TLM described above, and the determination of the necessary metal oxide surface parameters for the selected SCM. For application of any of the SCM versions, estimation of the sorbent surface area, surface hydroxyl site density, and values for the two surface protolysis, K^+ and K^-, are required. In the case of the NEM and DLM, this is the only information that is needed for characterizing the acid–base surface properties. When the CCM is selected, an additional interfacial capacitance parameter, C_1, is required. When the TLM is selected, four additional parameters are required, two interfacial capacitances, C_1 and C_2, and two electrolyte surface protolysis reaction constants, K_{An} and K_{Cat}. Table 6 above summarizes the SCM surface parameters required to the use the various SCMs.

After determining the necessary surface parameters for modeling proton sorption, SCMs can be used to model sorption behavior of other types of ions. This involves selecting the minimum number of SCM reactions needed to adequately model the sorption data over a wide range of conditions. For modeling divalent metal ion sorption, the proton stoichiometry (the number of protons released per metal ion sorbed), type of surface complex (inner- or outer-sphere), and formation constants for each reaction selected are required. Because a variety of combinations of different sorption reactions and constants may fit various aspects of the sorption data equally well (see, e.g., Westall and Hohl;[61] Hayes et al.;[40] and Katz and Hayes[22]), protocols are needed to insure the best choice of reactions and a more universally accepted set of guidelines to allow reproducibility from one laboratory to another.

Various procedures are available for obtaining the SCM surface parameters, and a number of reviews are available which discuss the techniques for obtaining these parameters for proton and divalent metal cation sorption.[4,22,40,62,63] The approach in this study utilizes acid–base titration data, experimental metal ion sorption data, XAS data, and objective curve-fitting techniques after the protocol established by Katz and Hayes.[22] Specifically, the strategy requires

1. Estimation of the SCM metal oxide surface parameters using surface titration data and the objective curve-fitting routine FITEQL.[64] (A range of valid parameter sets is generated assuming various values of site density, ΔpK_a, and C_1, for TLM, to be later optimized with metal cation sorption data.)

2. Evaluation of ionic strength dependence and XAS studies to guide the selection of divalent cation sorption reactions at low to moderate

coverage. (Past XAS studies suggest that mononuclear species predominate from low to moderate coverage data for strongly sorbing species, 0.1 to 10%,[28] and at all coverages for weakly sorbing species,* Ionic strength dependence studies have been used in the past to distinguish inner- and outer-sphere complex formation.[25,40])

3. Estimation of surface complexation constants for the selected mononuclear surface reactions using one set of moderate-coverage sorption data (0.1 to 10%) in conjunction with objective curve-fitting routine FITEQL[64] (In general, below about 10% surface coverage, divalent cations have not been found to form multinuclear surface species as discussed in the spectroscopic review section above.)

4. Analysis of the predictive capabilities of the ionic strength dependence (If the ionic strength dependence is not well predicted, go back to step 3 and use a different set of valid surface parameters and repeat step 4. Refinement of the range of optimal SCM parameters should be based on fitting of data. Note, however, that each particular model will have some limitations with respect to predicting data over a range of ionic strengths.[22])

5. Analysis of the predictive capabilities of the monomer surface complexation constants over a range of moderate and high surface coverage data (0.1 to 100%). (Further refinement of the optimal SCM parameter range for modeling should be based on the ability to fit a wide range of data. However, modeling with monomeric reactions only is expected to result in an underprediction of higher surface coverage data if multinuclear species are forming.)

6. Evaluation of XAS data to guide the selection of multinuclear SCM reactions and the reaction constants to insure that SCMs predict the formation of these species when they are found to be present at higher surface coverage.

7. Incorporation of multinuclear species formation and surface precipitation reactions into SCMs to accurately account for their formation when such species are present at higher coverages. (A continuum TLM model has been developed which allows surface polymer and precipitate reactions to be specified in accordance with XAS data[22]).

8. Estimation of prediction capabilities of the SCM continuum model over a wide range of surface coverages and conditions (optimization of the continuum model parameters over the whole range of surface coverage using the valid surface parameters from step 5).

B. Results and Discussion

1. Determination of Surface Parameters

The implementation and results of applying the above approach to Co(II) sorption on α-Al$_2$O$_3$ for the NEM, DLM, and TLM are given in this

---

* *Note:* For the purposes of calculating surface coverage, a coverage classification scheme assuming a hydrated radius (H.R.) of Co(II) of 4 Å was utilized in this work.

section. Details for the specific procedures used to collect the data presented below are provided in the Appendix. As discussed above, the first step in SCM involves the determination of valid sets of surface parameters for each SCM for a given solid (step 1 above). For this study, titration data for α-Al_2O_3 (Figure 9), FITEQL,[64] and the procedures described previously by Hayes et al.[40] were used to determine the surface ionization constants. FITEQL was used to obtain the valid sets for log K^+ and log K^- for specific values of N_s ranging from 1 to 10 sites/nm^2 (Tables 7 and 8). For the TLM, a slightly more complex procedure for arriving at a valid set of surface characterization parameters was required. In this case, FITEQL was used to obtain values of log K_{An} and log K_{Cat} for various combinations of assumed values for N_s, ΔpK_a, and C_1.[*22,40] Valid sets of TLM parameters for α-Al_2O_3 for N_S ranging from 1 to 10 sites/nm^2 are summarized in Table 9.

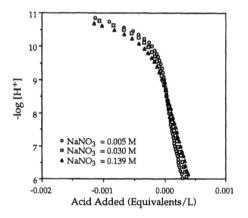

Figure 9
Titration data for α-Al_2O_3. (From Hayes, K.F.; Redden, G.; Ela, W.; Leckie, J.O. *J. Coll. Interf. Sci.* 1991, *142*, 448. With permission.)

TABLE 7.

NEM Fits of Titration Data for α-Al_2O_3

N_s	K^+	K^-	F
1	7.1	−10.5	183.9
2	6.6	−11.0	391.2
3	6.4	−11.2	401
10	5.9	−11.7	316.5

* *Note:* $\Delta pK_\alpha = -(\log K^- + \log K^+)$

TABLE 8.

DLM Fits of Titration Data
for α-Al_2O_3

N_s	K^+	K^-	F
0.75	10.4	-8.1	133.9
0.8	9.6	-8.9	129.6
1	9.0	-9.6	133.7
2	8.3	-10.4	175.7
3	8.1	-10.6	183.0
10	7.5	-11.1	188.9

TABLE 9.

TLM Fits of Titration Data for α-Al_2O_3

C_1	C_2	ΔpK_a	N_s	K_{An}	K_{Cat}	F
0.6	0.2	4	3	10.4	-7.9	123.5
0.8	0.2	2	2	10.1	-7.8	33.1
0.8	0.2	2	3	9.9	-8.0	28.3
0.8	0.2	2	9	9.4	-8.5	23.1
0.8	0.2	4	2	10.1	-7.9	33.2
0.8	0.2	4	3	9.8	-8.1	28.3
0.8	0.2	4	9	9.3	-8.6	23.1
1.1	0.2	0	2	10.0	-7.6	12.9
1.1	0.2	0	3	9.8	-7.8	15.2
1.1	0.2	2	1	10.1	-7.6	14.7
1.1	0.2	2	2	9.6	-8.0	12.9
1.1	0.2	2	9	8.9	-8.7	19.2
1.1	0.2	4	1	10.1	-7.7	15.2
1.1	0.2	4	2	9.6	-8.1	13.0
1.1	0.2	4	3	9.3	-8.3	15.4
1.1	0.2	4	5	9.1	-8.5	17.6
1.1	0.2	4	9	8.8	-8.8	19.3
2.0	0.2	0	3	9.1	-8.5	137.2
2.0	0.2	2	1	9.5	-8.1	50.0
2.0	0.2	2	2	8.9	-8.7	118.8
2.0	0.2	4	2	8.9	-8.7	123.2
2.0	0.2	4	3	8.6	-9.0	142.5

Based on the goodness-of-fit value, F, and visual inspection of the fitting results, it can be seen that the DLM generally provides a better fit to the titration data than the NEM; however, neither gives especially excellent fits to the data (Tables 7 and 8 and Figure 10). Furthermore, for both the DLM and NEM, the value of F generally decreases with decreasing site density, which would appear to indicate that lower N_s values yield better fits to the titration data than higher values. For the DLM, this trend continues until the value of N_s reaches 0.75 sites/nm^2, at which point the F-statistic increases to 133.9. When the value of N_s is reduced below 0.75 sites/nm^2, convergence was not achieved. Based on these results, surface

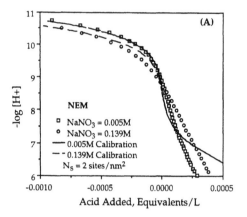

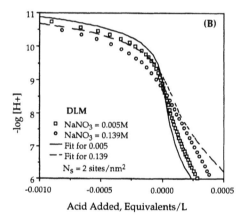

Figure 10
(A) NEM and (B) DLM fits to titration data for α-Al$_2$O$_3$.

constants generated for N$_s$ values between 2 and 10 sites/nm^2 were considered to be valid sets for the DLM and NEM (Tables 7 and 8).

The estimation of valid sets of TLM surface parameters for α-Al$_2$O$_3$ (Table 9 and Figure 11) have been discussed in detail previously and indicate that intermediate values of C$_1$ between 0.8 and 2.0 F/m^2 provide the best fits to the titration data.[22,40] Trends in N$_s$ and ΔpK$_a$ are not as evident for the TLM. The increase in the number of adjustable parameters in the TLM leads to a more complex relationship between N$_s$ and F which depends on the value of C$_1$. In spite of the lack of trends, valid sets of surface parameters for the TLM are found for ΔpK$_a$ between 0 and 6, C$_1$ between 0.8 and 2.0 F/m^2, and N$_s$ between 1 and 10 sites/nm^2 (Table 9).[22,40]

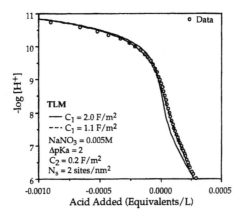

Figure 11
TLM fits to titration data for $\alpha\text{-Al}_2O_3$.

2. Selection of Mononuclear Sorption Reactions for Cobalt(II)

Following surface characterization, the next step in modeling divalent cation sorption is the selection of the appropriate mononuclear species reaction and reaction constant at moderate coverage (step 2 above). In general, inner-sphere reactions are believed to predominate for strongly sorbing metal ions for which changes in ionic strength have little or no effect on the location of the adsorption edge.[10] As shown in Figure 12, the adsorption edges for cobalt $\alpha\text{-Al}_2O_3$ show very little effect of ionic strength, supporting the appropriateness of an inner-sphere reaction. Recent XAS studies also support the selection of an inner-sphere reaction, indicating the presence of Al in the second coordination shell of cobalt sorbed onto alumina at moderate coverage.[28] Based on these results, the following inner-sphere monomer reaction was selected for calibration of the moderate coverage data:

$$\equiv\!SOH + Co^{2+} = \equiv\!SOCo^+ + H^+ \qquad \qquad (28)$$

A stoichiometry of one proton released per metal ion sorbed has been found to be appropriate based on past experimental and modeling studies for strongly sorbing divalent metal ions.[4,10,54,56] Various choices for the proton release should be tested in this step, however, to confirm the selected stoichiometry. In this study, only a one-proton release stoichiometry was found to predict the slope of the pH sorption edges well for the three SCMs tested.

Although we have selected a monodentate inner-sphere surface complex involving just one surface hydroxyl, most XAS studies suggest that bidentate surface complexes form with two surface hydroxyl groups (see

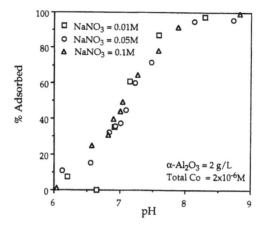

Figure 12
Effect of background electrolyte on cobalt sorption on α-Al$_2$O$_3$.

XAS review section above). Our initial SCM analysis showed that choosing a bidentate complex always led to an underprediction of sorption at higher moderate coverages. Mathematically, this can be traced to the fact that the bidentate surface complex representation results in an equilibrium expression for the surface reaction that is proportional to the square of the concentration of uncomplexed surface hydroxyls, compared with a linear relation for the unidentate complex. By selecting bidentate surface complexes, the surface hydroxyl sites become saturated at much lower surface concentrations, resulting in significant underprediction of sorption as surface coverage increases.

It may be reasoned that the SCM monodentate representation is actually more realistic from a modeling perspective, especially if the formation of a bidentate complex involves coordination of a metal ion to different types of surface hydroxyl sites. For example, based on XAS analysis of sorbed samples and the polyhedral approach for selecting feasible surface sites, it has been proposed that at low coverage, a bidentate complex of Co may be formed on kaolinite by sharing a terminal Al–OH and a nearby bridging Al–OH–Si.[38] Because proton release from bridging hydroxyls is unlikely, only one proton exchangeable terminal hydroxyl site is involved.[50,51] Since titration data from which SCM parameters are determined only characterize the total number of proton exchangeable sites (e.g., terminal surface hydroxyls), a monodentate surface reaction may be a mathematically more accurate representation of the exchange process. In this sense, from a modeling perspective the monodentate surface reaction appears to be more appropriate.

Other XAS studies have also indicated that multidentate complexes may form between terminal hydroxyls in juxtaposition to other non-proton-exchanging bridging surface hydroxyls. For example, Chisholm-Brause[28] originally proposed that Co(II) may form a tridentate complex formation on γ-Al_2O_3 with Co bonding to one terminal hydroxyl at the corner of an AlO_4 tetrahedron and two adjacent bridging hydroxyls exposed on one of the two nonequivalent 111 planes that may be exposed. More recently, however, Brown et al. have instead proposed bidentate surface complex formation of Co(II) with two bridging hydroxyls at the edge of AlO_6 octahedra on the other 111 plane.[65] To date, no compelling thermodynamic arguments support the proposed bidentate bonding of metal ions exclusively to bridging hydroxyls which result in one proton released per metal ion sorbed. Clearly, more work is needed to evaluate the most thermodynamically favorable sites in addition to those that are structurally consistent with crystal structure, proton reaction stoichiometry, and XAS analysis. At present, the best SCM results for strongly sorbing divalent cations have been obtained by postulating monodentate inner-sphere surface reactions which include only one proton exchangeable site in the mass law equations.

3. Monomer Inner-Sphere Surface Reaction Calibration

Divalent metal ion sorption modeling begins by selecting one set of moderate coverage data and fitting this data using the most probable surface complexation reaction based on information from step 2 and FITE-QL (calibration step 3 above). In the case of cobalt sorption on α-Al_2O_3, the most appropriate reaction has been found to be an inner-sphere monomer monodentate reaction with one proton released per metal ion sorbed.* We refer to this procedure as a calibration step because the value of the sorption mass law constant is determined from one set of conditions and then the SCM is subsequently used to predict data at other conditions with this value. While other investigators have used multiple data sets to optimize the sorption mass law constants (e.g., Dzombak and Morel[4]), the single-set calibration approach has the advantage that calibration data can be selected for particular ranges of surface coverage and solid concentrations when only one sorption process is occurring.

Two criteria were imposed for selecting the data set to be used in the calibration step. The principal criterion was that the data should fall in the moderate coverage range where XAS data indicate only monomeric inner-sphere surface species. The second criterion was that the data be collected at the same solid concentration as that used for the surface

---

* *Note:* During the calibration step all reasonable choices for the surface reaction should be tried including monodentate and bidentate complexes, one- or two- proton release reactions or, if appropriate as with the TLM, inner- or outer-sphere surface reactions. Based on our extensive modeling experience with cobalt sorption on corundum, we have found that Equation 28 is the only one which fits the slope of the pH edge sorption data well; outer-sphere, bidentate, or reactions with two protons released per metal ion sorb do not provide reliable fits to the calibration data. In addition, we have assumed that it is best to start by using the minimum number of reaction types, for example inner- or outer-sphere, and only increasing the number based on poor quality fits or independent confirmation that more than one reaction type is likely.

ionization parameters evaluated in the previous section as a precautionary step to avoid any effects that particle concentration may have on sorption equilibria. Based on these criteria, 20 g/L α-Al_2O_3 and 2×10^{-6} M total cobalt data were used for the calibration. The data from this sorption pH edge represent approximately 0.12 to 1.2% surface coverage. Based on previous XAS studies, this range of surface coverages was expected to encompass predominantly mononuclear, inner-sphere surface complexes.[28]

Each of the SCMs was calibrated by finding the best value for the reaction constant, K_{Me}, represented by Equation 28 above using the 20 g/L α-Al_2O_3, 2×10^{-6} M total cobalt data and FITEQL. In the case of the NEM and DLM, the impact of variation of N_s on the calibration value and goodness-of-fit was evaluated (Tables 10 and 11). As expected, higher values of N_s required lower values of the surface complexation reaction to fit the data. For both the NEM and the DLM, reasonable fits to the data were found for all values of N_s between 1 and 10 sites/nm² (Figures 13 and 14). The small values and differences in F indicate very little difference in the ability of the different parameter sets to fit the data.

TABLE 10.

NEM Metal Ion Reaction Constants
for Cobalt Sorption

N_s	K_{Me}	F	K^+	K^-
10	−3.9	1.5	5.9	−11.7
3	−3.2	0.6	6.4	−11.2
2	−2.92	0.3	6.6	−11.0
1	−2.3	0.2	7.1	−10.5

TABLE 11.

DLM Metal Ion Reaction
Constants for Cobalt Sorption

N_s	K_{Me}	F	K^+	K^-
10	−1.7	0.3	7.5	−11.1
3	−1.1	0.3	8.1	−10.6
2	−0.9	0.3	8.3	−10.4
1	−0.8	0.3	9.0	−9.6

A similar calibration procedure for the TLM was implemented for various values of N_s, ΔpK_a and C_1. The results are grouped to provide a convenient means of comparing the impact of N_s, ΔpK_a, and C_1 on the goodness-of-fit (Table 12). In general, for a given value of N_s and C_1, increasing the value of ΔpK_a resulted in a larger value of K_{Me}. Visual inspection of the model fits (Figure 15) and a comparison of the F-statistic for these constants indicate that all fits were similar and acceptable. A comparison of the effects of C_1 for constant N_s and ΔpK_a show that C_1 has

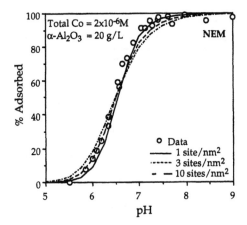

Figure 13
NEM calibration of moderate coverage data for cobalt sorption on α-Al$_2$O$_3$.

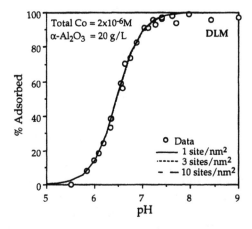

Figure 14
DLM calibration of moderate coverage data for cobalt sorption on α-Al$_2$O$_3$.

only a small effect on the best-fit value of K_{Me}. All combinations, however, give relatively low values of F and again indicate good fits to the data. Likewise, a comparison of different N_s values and two different values of C_1 and ΔpK_a illustrates that all parameter sets represent the data reasonably well (Table 12 and Figure 16). As with the NEM and DLM, lower values of N_s result in higher values of the K_{Me}.

The results of the calibration step above indicate that the monomeric inner-sphere surface reaction and a range of SCM surface parameters can be used to describe a single pH sorption edge. This might lead some to conclude that many sets of SCM parameters will work equally well for modeling divalent cation sorption. It is important to recognize, however,

TABLE 12.

TLM Fits of Moderate Coverage Sorption Data

C_1	ΔpK_a	N_s	$\log K_{Me}$	$\sigma_{\log KMe}$	F
0.8	0	3	−0.6	0.1	0.5
0.8	2	3	−1.0	0.1	0.5
0.8	4	3	−1.1	0.1	0.6
0.8	4	3	−1.1	0.1	0.6
1.1	4	3	−1.3	0.1	0.6
2.0	4	3	−1.4	0.1	0.5
0.8	4	2	−0.9	0.1	0.6
0.8	4	3	−1.1	0.1	0.6
0.8	4	7	−1.5	0.1	0.6
0.8	4	9	−1.6	0.1	0.5
1.1	2	2	−0.8	0.1	0.6
1.1	2	3	−1.0	0.1	0.6
1.1	2	5	−1.2	0.1	0.5
1.1	2	7	−1.4	0.1	0.5
1.1	2	9	−1.5	0.1	0.5
1.1	4	2	−1.1	0.1	0.6
1.1	4	3	−1.3	0.1	0.6
1.1	4	5	−1.4	0.1	0.6
1.1	4	9	−1.7	0.1	0.5

that the evaluation of a set of SCM parameters should not be based on its ability to predict a narrow range, but rather on how well it predicts data over the entire range of conditions for which the model is applicable. Although the calibration step provides guidance on selecting the appropriate surface complexation reaction including reaction stoichiometry, it is the following steps in modeling, the ability to predict a range of sorption data for different ionic strengths and cobalt surface coverages on α-Al_2O_3 that substantially narrow the choice of surface parameters for a given surface reaction and lead to the most effective set of model parameters.

4. SCM Predictions of Variable Ionic Strength

Following the calibration step, each set of valid SCM parameters should be checked to determine how well they predict the effects of ionic strength on sorption (step 4 above). For metal ions which have a high affinity for metal oxide surfaces, sorption is characterized by little or no dependency on ionic strength as discussed above. Hence, each model must be evaluated for its ability to predict the nondependence of sorption with changes in the background electrolyte concentration.

Both the NEM and DLM similarly predict the lack of ionic strength effects on metal ion sorption for the complete range of site density values

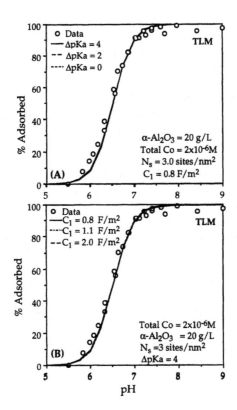

Figure 15
TLM calibration of moderate coverage data for cobalt sorption on α-Al$_2$O$_3$: (A) Effect of ΔpK_a; (B) effect of C_1.

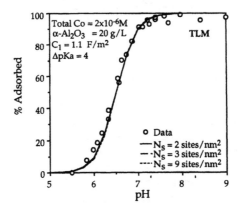

Figure 16
TLM calibration of moderate coverage data for cobalt sorption on α-Al$_2$O$_3$: effect of N_s.

tested. Model predictions of 2 g/L α-Al₂O₃, 2 × 10⁻⁶ M total cobalt data collected at three ionic strengths indicate that there is no significant difference in the location of the pH edges for the different ionic strengths at low site density (Figures 17 and 18). This result is not surprising, however, since neither the NEM nor the DLM incorporates electrolyte surface complexation reactions with surface hydroxyl sites. Hence, no competition between the background electrolyte ions and the cobalt ions for surface sites is expected. In the NEM and DLM, varying ionic strength by changing the background electrolyte concentration only affects model predictions through its impact on the diffuse layer charge in the DLM or aqueous-phase activity corrections in the NEM and DLM. These effects are apparently slight, as indicated by the lack of effect of ionic strength on the predictions of the position of the pH sorption edge. Although not shown, parameter sets with N_s values between 1 and 10 sites/nm² gave equally good predictions of the lack of ionic strength effect on metal ion sorption. Hence, all parameter sets that worked well for the calibration step also worked equally well for predicting the ionic strength dependence of cobalt sorption on α-Al₂O₃ for the NEM and DLM.

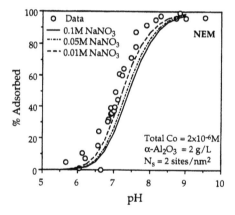

Figure 17
NEM predictions of ionic strength effects for cobalt sorption on α-Al₂O₃.

The TLM was also able to predict the lack of ionic strength effects reasonably well (Figure 19); however, slightly larger shifts in the predictions for different ionic strengths were noticeable, depending on the choice of SCM surface parameters (e.g., see Katz and Hayes[22] for more details). In general, higher N_s values yielded marginally better predictions of the ionic strength effects for site densities above 1.0 site/nm². Predictions of the pH edge based on lower site densities were shifted to higher pH values compared with the data regardless of the value of C_1. C_1 values of 0.8 F/m² or greater were needed to adequately represent the lack of effect of ionic strength on cobalt sorption. The effect of ΔpK_a on the prediction of

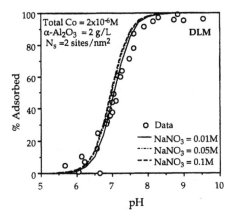

Figure 18
DLM predictions of ionic strength effects for cobalt sorption on α-Al$_2$O$_3$.

the ionic strength effects showed that lower values of ΔpK_a predicted less shift in the pH edge with ionic strength changes. In fact, ΔpK_a values greater than 4 predicted shifts in the pH edge with ionic strength that were larger than the experimental results. In general, higher values of N_s and capacitance yielded slightly better predictions with respect to the data collected at different ionic strengths. Based on an extensive analysis of the prediction of the ionic strength effects, Katz and Hayes[22] concluded that TLM parameter sets with N_s values between 1.0 and 10 sites/nm^2, C_1 values greater than 0.8 up to 2.0 F/m^2, and ΔpK_a values less than 4 were needed to predict the data collected at different ionic strengths reasonably well.

5. SCM Predictions of Variable Surface Coverage

The NEM, DLM, and TLM parameters derived from the 20-g/L data and consistent with ionic strength data were subsequently used to predict sorption for a variety of data which varied over a range of surface coverage (step 5). For this purpose, the following three data sets were selected for which the surface coverage ranges from 0.1 to over 100%: 2 g/L α-Al$_2$O$_3$, 2×10^{-6} M total cobalt; 20 g/L α-Al$_2$O$_3$, 1×10^{-4} M total cobalt; and 2 g/L α-Al$_2$O$_3$, 1×10^{-4} M total cobalt (Table 13).

In the case of the NEM, only parameter sets with the lower values for N_s were able to predict the data collected at surface coverages of less than 2.4% (Figure 20A). For N_s greater than 3 sites/nm^2, the NEM underpredicted the amount of sorption through most of the pH range for the 2 g/L α-Al$_2$O$_3$, 2×10^{-6} M total cobalt sorption edge. As surface coverage increased in the 20 g/L α-Al$_2$O$_3$, 1×10^{-4} M total cobalt sorption edge, the model significantly overpredicted the data (Figure 20B). Surface coverages in this system range from 1.2 to 12% across the sorption edge. Finally, as

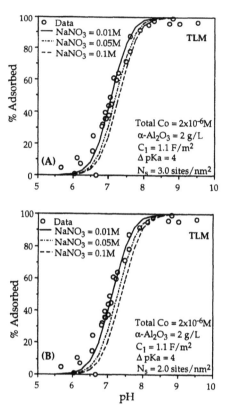

Figure 19

TLM predictions of ionic strength effects for cobalt sorption on α-Al$_2$O$_3$: (A) N$_s$ = 3.0 sites/nm^2. (B) N$_s$ = 2.0 sites/nm^2.

the surface coverage was further increased from 12 to 120% surface coverage in the 2 g/L α-Al$_2$O$_3$, 1 × 10^{-4} M total cobalt data, the model significantly underpredicted the data except for high values of N$_s$ (Figure 20C). The plateau reached at approximately 40% adsorbed when N$_s$ was equal to 1 site/nm^2 was predominantly due to the lower total number of surface hydroxyls at this low value.

Limitations of the NEM were also apparent in the predictions of constant pH isotherm data (Figure 21). Clearly, the model was not capable of describing sorption over the entire isotherm range for data collected at pH 6.9. In fact, the NEM overpredicted isotherm data for surface coverages greater than approximately 1% for all of the site densities. At pH 7.6, the NEM affords reasonable predictions of the data for all but the lowest site density. However, data collected in this range of surface coverage are as high as 80%. It would seem unlikely that monomeric sorption reactions would predominate at these coverages. These results indicate that the NEM is qualitatively capable of predicting some of the trends in the

TABLE 13.

Surface Coverages for Cobalt Sorption pH Edges at 50% Total Co Sorbed

Solid Conc. (g/L)	Total Cobalt Conc. (mol/L)	Coverage (μmol/m²)	Coverage for H.R. = 4 Å (%)	Coverage for N_s = 3 sites/nm² (%)
0.2	2×10^{-6}	0.4	12.11	8.03
2	2×10^{-6}	0.04	1.21	0.8
2	1×10^{-4}	2	60.55	40.15
20	2×10^{-6}	0.004	0.12	0.08
20	1×10^{-4}	0.2	6.05	4.02
20	1×10^{-3}	2	60.55	40.15
100	1×10^{-4}	0.04	1.21	0.80
100	1×10^{-3}	0.4	12.11	8.03

sorption behavior of cobalt to α-Al$_2$O$_3$, e.g., predicting the proper direction of the pH shift in sorption with coverage that is consistent with the data. The NEM, however, has limited quantitative prediction capabilities, since no single site density value can be selected which can accurately predict the position of the pH edge or isotherm data over the range of conditions examined.

Compared with the NEM, the DLM offered improvements in predicting the proper location of the sorption edge over the entire data range (Figure 22). Unlike the NEM, the DLM gave excellent predictions of the 2 g/L α-Al$_2$O$_3$, 2 $\times$ 10^{-6} M total cobalt data for all of the site densities examined. Similar to the NEM, the DLM overpredicted sorption for the 20 g/L α-Al$_2$O$_3$, 1 $\times$ 10^{-4} M total cobalt sorption edge for all site densities. At the even higher surface coverage, the results were similar to the NEM results, with significant underprediction as well. Constant pH isotherm data were also inadequately predicted by the DLM (Figure 23). Although the DLM predicted data both at lower and higher surface coverages of the isotherm, the model significantly overpredicted data for surface coverages from approximately 2 to 10%. Thus, while the DLM provided reasonable predictions for one pH edge and also the ionic strength effects, like the NEM, the model did not predict data collected over a range of solution concentrations at constant pH or pH edges collected over a wide range of surface coverages with one set of parameters.

Because neither the NEM nor DLM consistently underpredicts or overpredicts with respect to surface coverage modeling, improved predictive capabilities across the entire range of coverage with additional multinuclear reactions are unlikely. For example, for these two models, intermediate coverage data were consistently overpredicted while high coverage data were often underpredicted. As a result, even if additional information about the formation of multinuclear surface speciation were independently available from spectroscopic studies, these two models would not be capable of accurately predicting the whole range of coverages. At best, incorporation of multinuclear species in the NEM and DLM would improve the prediction of highest coverage systems, where surface

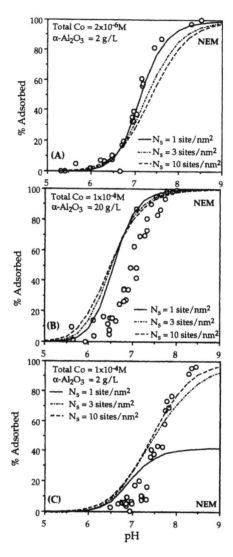

Figure 20
NEM predictions of cobalt sorption edges on α-Al$_2$O$_3$: (A) total Co = 2 × 10^{-6} M, α-Al$_2$O$_3$ = 2 g/l; (B) total Co = 1 × 10^{-4} M, α-Al$_2$O$_3$ = 20 g/L; (C) total Co = 1 × 10^{-4} M, α-Al$_2$O$_3$ = 2 g/L.

precipitation is suspected, while perhaps making the intermediate coverage predictions only slightly worse. Although not attempted here, improvement of the NEM or DLM may be possible by utilizing a two-site model to describe sorption in the low and moderate coverage range and then to incorporate multinuclear sorption reactions to account for sorption at higher coverages. Two-site models are generally needed when coverages

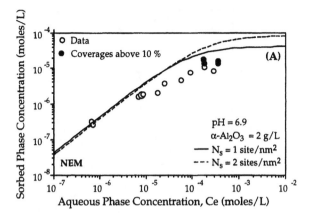

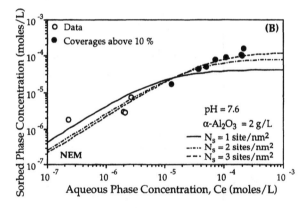

Figure 21
NEM predictions of cobalt sorption isotherms on α-Al_2O_3: (A) pH = 6.9; (B) pH = 7.6.

become extremely low (below 0.1%) to account for the small number of high energy sites that exist on most metal oxides.[18,66]

The results for the TLM, however, are more encouraging than those presented for the DLM and NEM. Katz and Hayes[22] previously conducted an extensive analysis of the TLM for predicting the range of sorption data discussed above for the NEM and DLM. Based on that analysis, it was found that most of the surface parameter sets that worked well for the calibration and ionic strength dependence step did not adequately predict the data over moderate ranges in surface coverage and pH. Two TLM parameter sets, however, did predict sorption data well up to about 12% coverage. These two sets were C_1 = 1.1 F/m^2, N_2 = 3 sites/nm^2, ΔpK_a = 4, and log K_{Me} = –1.2; and C_1 = 1.1 F/m^2, N_2 = 2 sites/nm^2, ΔpK_a = 4, and

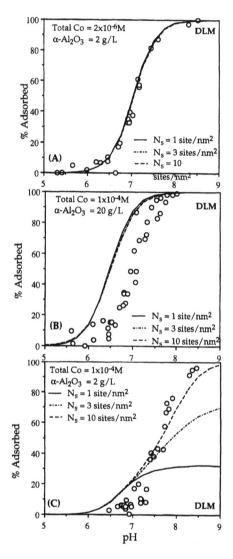

Figure 22
DLM predictions of cobalt sorption edges on α-Al$_2$O$_3$: (A) total Co = 2 × 10^{-6} M, α-Al$_2$O$_3$ = 2 g/L; (B) total Co = 1 × 10^{-4} M, α-Al$_2$O$_3$ = 20 g/L; (C) total Co = 1 × 10^{-4} M, α-Al$_2$O$_3$ = 2 g/L.

log K_{Me} = −1.1. As illustrated, using these parameters, the TLM gives excellent predictions of the 2 g/L α-Al$_2$O$_3$, 2 × 10^{-6} M data, reasonable predictions of the 20 g/L α-Al$_2$O$_3$, 1 × 10^{-4} M data, and underpredicts the 2 g/L α-Al$_2$O$_3$, 1 × 10^{-4} M total cobalt data (Figure 24). These results suggest that the model is capable of predicting sorption data for surface coverages less than approximately 12%. This observation is also evident in the isotherm predictions of data at constant pH (Figure 25).

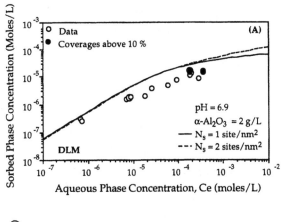

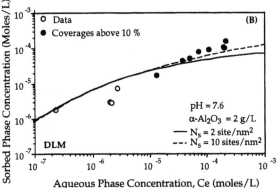

Figure 23
DLM predictions of cobalt sorption isotherms on α-Al$_2$O$_3$: (A) pH = 6.9; (B) pH = 7.6.

Perhaps significantly, unlike the DLM and NEM, the TLM does not overpredict data in the intermediate range between 2 and 10% surface coverage. This suggests that TLM predictions may be improved by incorporating additional surface complexation reactions which would lead to enhanced sorption with increasing coverage in this range. Multinuclear and surface precipitation formation reactions are two such surface complexation reactions which increase the predicted sorption by allowing multiple metal ions to sorb for every surface reaction site occupied. Based on the above analysis for the TLM, the onset of multinuclear reactions should occur at about 12% coverage, if the underprediction at this coverage is an indication of the failure to represent the proper reaction stoichiometry with increasing coverage.

Additional information, however, is needed to verify that the predominant reaction process is changing from monomeric to multinuclear species formation as a function of coverage. XAS provides this type of information.

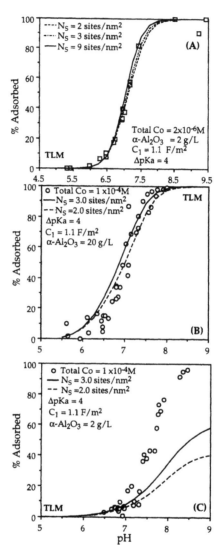

Figure 24
TLM predictions of cobalt sorption edges on α-Al$_2$O$_3$: (A) total Co = 2×10^{-6} M, α-Al$_2$O$_3$ = 2 g/L; (B) Total Co = 1×10^{-4} M, α-Al$_2$O$_3$ = 20 g/L; (C) Total Co = 1×10^{-4} M, α-Al$_2$O$_3$ = 2 g/L.

For example, at lower coverage XAS indicates that inner-sphere monomeric surface complexes form, while at higher coverage multinuclear species formation occurs. For the next step in the TLM modeling effort of cobalt on α-Al$_2$O$_3$, XAS data were consulted to determine the coverage at which multinuclear species begin to form and to identify the species involved for incorporation into SCM reactions.

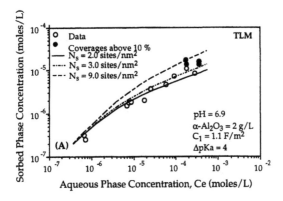

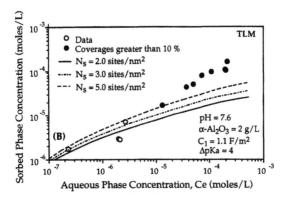

Figure 25
TLM predictions of cobalt sorption isotherms on α-Al$_2$O$_3$: (A) pH = 6.9; (B) pH = 7.6.

6. Extended X-Ray Absorption Fine Structure Studies

In order to test the hypothesis that multinuclear species form at surface coverages in excess of 10%, XAS data were collected. These data were then used to guide the selection of multinuclear SCM reactions to improve the modeling predictions at higher coverages (step 6 above).

XAS spectra can be divided into two analysis regions based on the energy of the absorption edge-jump resulting from ejection of a particular core electron. These regions are the X-ray absorption near edge structure (XANES) and the extended X-ray absorption fine structure (EXAFS). XANES data range from energies about 10 to 20 eV below the edge-jump to 30 to 60 eV above the absorption maximum and can provide information relative to the oxidation state and local geometry of the absorber atom. EXAFS analysis is conducted from 50 to 100 eV above the edge to

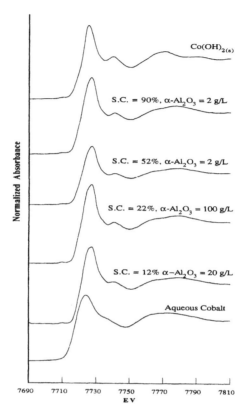

Figure 26
XAS spectra for sorption of varying surface coverages (S.C.) of cobalt on α-Al₂O₃.

500 to 1000 eV beyond the edge and can be used to determine the number and bond distances of atoms in the near coordination environment of the central metal atom.[67]

A comparison of the similarities between XANES spectra of model and unknown samples provides qualitative information about the similarities of the coordination environment of the various compounds. For example, a comparison of the spectra for sorption samples and model compounds indicates that most of the sorption samples have a significantly different coordination environment from either a freshly precipitated Co(OH)₂(s) or a completely hexahydrated Co(II) ion in aqueous solution (Figure 26). With increasing coverage, however, the sorption samples and the precipitated Co(OH)₂(s) appear to become more similar. This provides the first indication that at very high coverage, surface precipitation may be occurring.

Normalized EXAFS spectra are presented in Figure 27. Comparison of the EXAFS data for the sorption samples and model compounds indicates a beat pattern in the sorption samples due to backscattering from

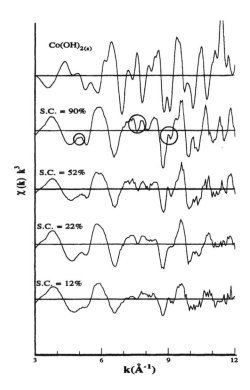

Figure 27
EXAFS for sorption of varying surface coverages (S.C.) of cobalt on α-Al_2O_3.

second-nearest neighbors. The beat pattern indicated by the low amplitude shoulders located at k of approximately 5.0, 7.2, and 9 $Å^{-1}$ are circled for the sample collected at 90% surface coverage (Figure 27). These beat patterns are indicative of backscattering from second-shell neighbor atoms within several angstroms of the cobalt metal ion. It also appears that the magnitude of the beat pattern is dependent on the amount of surface coverage. This indicates that there is an increase in the number of second-nearest neighbor backscattering metal atoms with increasing surface coverage.

The background-subtracted EXAFS data were Fourier transformed (Figure 28) and then back transformed to yield filtered EXAFS spectra. EXAFS analysis indicates that the peak located at 2.08 Å in all of the radial structure functions (RSFs) is due to the presence of oxygen in the ligating shell around cobalt (Table 14). Sorption samples and the hydroxide precipitate contain an additional peak at approximately 3 Å, and the higher sorption samples also contain a peak near 6.3 Å. Analysis of the EXAFS spectra indicates that the peaks located near 3 and 6.3 Å are due to neighboring cobalt atoms in the second and fourth shell surrounding the central metal atom (Table 14). The coordination number associated with

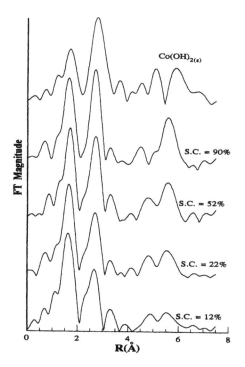

Figure 28

Radial structure functions for sorption of varying surface coverages (S.C.) of cobalt on α-Al$_2$O$_3$.

the Co–Co peak identified at 3.07 Å increased from 0 to 2.49 with increasing coverage. Similarly, the coordination number of the Co–Co peak identified at 6.32 Å increased from 0 to 2.78 with increasing coverage. The presence of second- and fourth-shell neighbor Co–Co interactions indicates the presence of multinuclear sorption complexes. The increase in the coordination number with coverage is consistent with an increase in the size of the multinuclear complexes as the surface coverage increases. Previous EXAFS studies of Co(II) sorption complexes on γ-Al$_2$O$_3$, kaolinite, and quartz surfaces have also shown increasing coordination number of the second- and fourth-shell Co–Co interactions with increasing surface coverage.[27,28,37,38]

Comparing the RSFs for the sorption samples with that of the solid Co(OH)$_{2(s)}$ supports the hypothesis that the cobalt on the surface at high coverage is not present as a well-ordered precipitate. The Co–Co distances obtained were 0.08 and 0.03 Å shorter than found in Co(OH)$_{2(s)}$, perhaps caused by a mismatch with the template formed by the surface hydroxyls and the requirements of an ordered precipitate.

The above analysis was obtained by considering the possible presence of Al in the second shell. The results suggest that between one and two

TABLE 14.

EXAFS Parameters for Sorption Samples and Model Compounds

Coverage (%)	Co-O R Å	N	$\Delta\sigma^2$	ΔE (eV)	Co-Co R (Å)	N	$\Delta\sigma^2$	Co-Al R (Å)	N	$\Delta\sigma^2$	Co-Co R (Å)	N	$\Delta\sigma^2$
12	2.07	5.7	0.0044	-2.044	3.08	1.1	0.0011	3.52	0.54	0.0025	—	—	0.0032
22	2.08	6.1	0.0020	-1.28	3.07	1.24	0.0032	3.48	1.22	0.0025	6.33	1.08	0.0032
52	2.07	6.9	0.0024	0.145	3.07	1.92	0.0028	3.48	1.01	0.0017	6.32	2.08	0.0032
90	2.06	7.5	0.0029	1.026	3.07	2.49	0.0032	3.45	1.98	0.0025	6.32	2.78	0.0032
$Co(OH)_2(s)$[a]	2.097	6.0			3.173	6	0.0032				6.35	6.0	
$Co(NO_3)_2(aq)$[b]	2.108	6.0											

[a] Values from Reference 68.
[b] Values from Reference 69.

Al atoms are present in the second shell of sorbing cobalt (Table 14). The presence of Al in the second shell is direct evidence that cobalt is sorbing as an inner-sphere complex. In addition, the low number of Al atoms indicates that neither a solid solution due to cobalt diffusing into the solid nor a cobalt–aluminum–hydroxide coprecipitate is forming.[27,38] As the surface coverage increases from 12 to 90%, the average coordination number of cobalt atoms in a second shell at 3.07 Å and a fourth shell at 6.32 Å changes from approximately one to nearly three, indicating that an increasing size and/or number of multinuclear clusters form as a function of surface coverage. From a modeling perspective, the changing coordination number is consistent with a transition from mononuclear to binuclear to tetranuclear and, finally, to a surface precipitate. In solution, multinuclear species containing two and four cobalt atoms have been proposed.[70] Therefore, these two species and a surface precipitation reaction were selected for subsequent surface complexation modeling to enhance the prediction capability of the TLM for coverages above 12%.

7. Modeling High Coverage Data Using the Continuum Model

The optimized TLM parameters derived for moderate coverage range were used as the bases for modifying the TLM model to incorporate multinuclear species formation (step 7 above). Katz and Hayes[22] recently developed a continuum version of the TLM which allows the simultaneous consideration of mononuclear, polynuclear, and precipitation surface reactions. Based on the spectroscopic information presented above, the following two inner-sphere surface polymer reactions were employed to attempt to model the highest surface coverage data:

$$\equiv\!SOH + 2Co^{2+} + 2H_2O = \equiv\!SO(Co)_2(OH)_2{}^+ + 3H^+ \qquad (29)$$

and

$$\equiv\!SOH + 4Co^{2+} + 5H_2O = \equiv\!SO(Co)_4(OH)_5{}^{2+} + 6H^+ \qquad (30)$$

In addition, the continuum model allows for the formation of surface precipitates by modifying the activity of a solid phase based on surface coverage due to mononuclear and multinuclear surface complex formation. The activity of the surface precipitating solid phase is allowed to increase from 0 to 1 based on the extent of surface coverage according to the following:

$$\left\{ Me(OH)_{2(s)} \right\} = \frac{\left[Me(OH)_2(s) \right]}{T_s} \qquad (31)$$

in which the parameter T_s, is equal to:

$$T_s = [\equiv SOMe^+] + [\equiv SOMe_2(OH)_2^+] + [\equiv SOMe_4(OH)_5^{2+}]$$
$$+ [Me(OH)_2(s)] \tag{32}$$

According to this definition, the formation of surface complexes makes it thermodynamically feasible for the surface precipitate to form at lower aqueous metal ion concentrations than the homogeneously precipitated solid. As the relative amount of surface precipitate increases, the activity of the metal hydroxide solid phase approaches 1. Under this conceptualization, the polymer species and surface precipitate begin to form a surface coating as surface coverage increases. Hence, this representation is consistent with the spectroscopic evidence presented earlier.

Implementation of the continuum model requires two modifications to the TLM strategy. These modifications include defining T_s in the model and incorporating T_s into the mass law equation for the metal hydroxide surface precipitate (Katz and Hayes[23]). The only additional fitting parameters required for the model are the two surface complexation reaction constants for the two surface polymer reactions given above. All other constants and parameters remain unchanged from the best values for lower surface coverages. A complete listing of the reactions and fitting parameters needed for the continuum model are given in Table 15. Calibration of the polymer model reactions was performed using the constant pH isotherm data collected at pH 7.6. As shown in Figure 29, the model is capable of describing sorption behavior at both moderate and high coverage.

8. Predictions with the TLM Continuum Model

In order to more adequately evaluate the predictive capabilities of the continuum model, the parameters determined from the calibration step were used to predict a range of moderate and high coverage data (step 8). Comparison of the continuum TLM predictions to the data indicates that the model is capable of describing sorption over the whole range of surface coverages examined (Figure 30). Furthermore, as shown in Figure 31, at surface coverages between 60 and 100%, the data predict a distribution of both dimer and tetrameric species. These results are also consistent with the spectroscopic evidence presented previously in which the coordination number of cobalt in the second shell increases with coverage.

Although the continuum model could readily be applied to the NEM or DLM, in this example, such modification is not justified based on the fact that both of these models already overpredicted sorption at moderate coverage. In order for the continuum model to improve predictive capabilities of a model containing only monomer reactions, two criteria are required. First, the model must be capable of describing data throughout

TABLE 15.

Reactions Required for Implementation of the Continuum Model[a]

Reaction	Parameters[b]	log K Values for Co/α-Al$_2$O$_3$ system[c]
Surface Adsorption Reactions[d]		
$\equiv$SOH + Me^{2+} = SOMe$^+$ + H$^+$	K_{Me}	-1.1
$\equiv$SOH + 2Me^{2+} + 2H$_2$O = $\equiv$SOMe$_2$(OH)$_2^+$ + 3H$^+$	K_{P1}	-11.8
$\equiv$SOH + 4Me^{2+} + 5H$_2$O = $\equiv$SOMe$_4$(OH)$_5^{2+}$ + 6H$^+$	K_{P2}	-22.5
Precipitation Reactions		
Me^{2+} + 2H$_2$O = Me(OH)$_2$(s) + 2H$^+$	K_{Sol}	-13.1
Mass Law Expressions for Precipitation		
{Me)OH)$_2$(s)] [H$^+$]^2/T$_s$ [Me^{2+}] = K_{Sol}		
Solution Complexation Reactions[e]		
Me^{2+} + H$_2$O = Me(OH)$^+$ + H$^+$		
Me^{2+} + 2H$_2$O = Me(OH)$_2^0$ + 2H$^+$		
Me^{2+} + 3H$_2$O = Me(OH)$_3^{-1}$ + 3H$^+$		

[a] These parameters are in addition to the parameters required for modeling surface ionization (i.e., surface protolysis, background electrolyte surface complexation, site density, and capacitance values) presented in Table 2.

[b] K_{Me} is the mass law surface complexation constant. K_{Sol} is the mass law constant for formation of the solid hydroxide of the sorbing metal ion.

[c] The surface polymer values reported were calibrated using surface acidity and metal ion sorption constants from Tables 9 and 12 corresponding to N_s = 2 sites/nm^2, ΔpK_a = 4, C_1 = 1.1 F/m^2, C_2 = 0.2 F/m^2.

[d] $\equiv$S(OH) refers to a surface hydroxyl group on the surface of an $\equiv$S(OH)$_3$ solid phase, $\equiv$SOMe$^+$ refers to an inner-sphere mononuclear metal ion surface complex and SOMe$_n$(OH)$_m$ are inner-sphere surface polymers.

[e] See Table A-2 for cobalt hydrolysis constants.

the moderate coverage range and, second, the model must consistently underpredict data at high coverage. Because these criteria were not met by either the NEM or DLM, implementation of a version of the continuum model was not justified. As indicated previously, had these models been modified to include additional surface sites or perhaps additional monomer reactions, it may have been possible to meet these two criteria. However, such modification would have increased the number of fitting

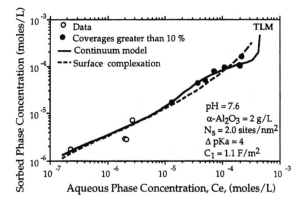

Figure 29
Continuum model calibration of pH 7.6 isotherm data for cobalt sorption on α-Al$_2$O$_3$.

parameters required and, therefore, was not pursued for the purpose of this illustrative example of surface complexation modeling.

VI. SUMMARY AND CONCLUSIONS

This chapter has provided an overview of the type of information that is needed and a protocol for modeling divalent cation sorption at metal oxide–water interfaces using SCMs. An example of the state of the art for incorporating XAS information in respect to SCM of the strongly sorbing divalent cations for metal oxide surfaces was presented. Three SCMs were evaluated for their potential to simulate and predict sorption data over a wide range of solution conditions and surface coverages, the NEM, DLM, and TLM. XAS data were consistent with the formation of inner-sphere surface complexes for sorption of cobalt(II) to α-Al$_2$O$_3$. A single mononuclear inner-sphere reaction, however, was unable to account for sorption except over a limited range of conditions for all of the models. The TLM was found to be the only model suitable for modification based on independent XAS information that indicated the formation of multinuclear reactions at high surface coverage. Because surface polymer and surface precipitation reactions can only lead to increased prediction of sorption relative to the monomer reaction, they would only make overpredictions worse at intermediate coverages for the DLM and NEM. The TLM continuum model was quite successful at modeling Co(II) sorption data on α-Al$_2$O$_3$ over wide ranges in ionic strength and pH.

This chapter has also illustrated that independent confirmation of surface species by XAS is essential to identifying the relevant surface processes and reactions. Without this information, it is impossible to pinpoint the failure of the different SCMs or to select additional appropriate

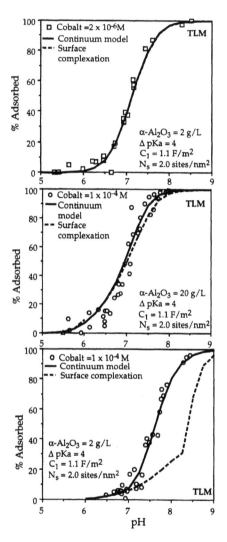

Figure 30
Continuum model predictions of cobalt sorption edges α-Al₂O₃.

reactions or the best SCM to utilize for a given set of conditions. Much still remains to be accomplished, however, before SCMs can be more widely applied even for divalent cation sorption. For example, while this study focused on moderate to high surface coverage, modeling of lower coverage data still needs to be performed. Based on our own past experience and that of others, in order to extend SCMs to lower coverage, a two-site model having an additional higher energy site at relatively low concentration is required.[4,22] The second site at low concentration has been found necessary to predict the shift of the position of the pH sorption

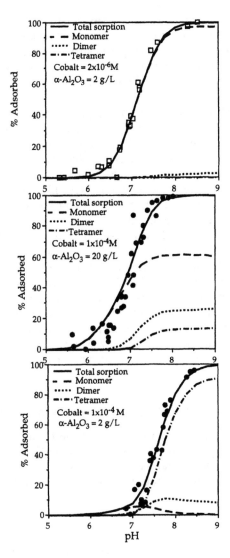

Figure 31
Continuum model species distribution predictions for cobalt sorption edges α-Al$_2$O$_3$.

edge at very low coverages where no shift should result if only one type of site existed.

Additional work is also still needed to improve our ability to model the sorption of more weakly sorbing cations, like the alkaline earth ions, with metal oxides. Past attempts to model the effect of changing ionic strength on the position of the sorption edge by assuming only outer-sphere complexes have met with limited success.[10] More-recent studies with weakly sorbing divalent cations have shown that a combination of both inner- and outer-sphere surface complexes allows for more successful modeling.[59] However, to date no spectroscopic data have been generated supporting the selection of both types of surface complexes. Ongoing XAS studies of Sr sorption on metal oxides and clays by Hayes and co-workers[36] may soon shed more light on this aspect of alkaline earth sorption.

Perhaps the biggest obstacle for applying SCMs that remains is to find improved methods of characterizing more-complex mineral systems in terms of SCM parameters. Significant experimental difficulties still exist for obtaining SCM parameters for all but the simplest metal oxide surfaces. Before we can extend the use of SCMs to more-complex minerals like aluminosilicate clays or mineral assemblages like soils, techniques need to be developed to more accurately represent and quantify the number and acid–base properties of the surface sites present.

While there is still much work needed in these areas, the potential for prediction of metal ion sorption at field scale using surface complexation models is evident. The development of data bases of SCM parameters for metal oxides and other mineral and soil components is ongoing, and the importance of utilizing spectroscopy for selecting model reactions is receiving widespread recognition among the modeling community. As the example presented here illustrates, improved prediction is possible over a wide range of conditions if relevant SCM reactions can be identified and proper modeling protocols followed, such as those outlined in this chapter.

VII. APPENDIX. MATERIALS AND EXPERIMENTAL METHODS

This appendix provides the details of the procedures and materials used for collecting the sorption and XAS data, XAS analysis, and the determinaton of model parameters for SCMs used in this chapter. For still further details, the reader should consult the references cited below.

A. Materials

The metal oxide selected for this work was corundum (α-Al_2O_3, Buehler LTD, Evanston, IL). It was selected because it is commercially available, has been subjected to extensive SCM surface characterization, is representative of metal oxides which form strong complexes with transition-metal and heavy metal divalent cations, and has similar sorption properties to γ-Al_2O_3 for which extensive XAS studies of divalent cation sorption have been carried out (see XAS review above, and Table 5).[22,40,63] Various physical properties of this material provided by the manufacturer or independently measured are presented in Table A-1.

Corundum was washed to remove surface impurities using a stirred cell ultrafiltration system (Model M2000, Amicon Corporation, Danvers,

TABLE A-1.

Properties of Corundum

Property	Value
Density	3.94 g/cc[a]
Particle size	0.3 μm[a]
Surface area	12.6 m^2/g[b]

[a] Reported by Beuhler Alumina (Evanston, IL).
[b] Determined in this work.

IL) modified to allow automatic refilling of the 2-L reactor. A 150-mm-diameter XM300 membrane was used to retain particles greater than 300,000 MWU. The washing procedure consisted of placing 50 to 100 g alumina in the cell and passing 10 L carbonate-free 0.001 M NaOH followed by 10 L 0.0001 M HNO$_3$ through the system. Following the acid–base washing procedure, the solids were rinsed by continuously pumping boiled, CO$_2$-free ultrapure water (deionized water which was passed through an additional set of ion-exchange resins and an activated carbon filter) through the ultrafiltration system. The solid suspension was rinsed until the specific conductance of the effluent water fell below 0.1 μS/cm. The BET N$_2$ surface area was determined to be 12.6 m^2/g; the N$_2$ gas isotherm data indicated that a slight degree of microporosity may be present within the samples, ranging between 8 and 11% of the total surface area.

Cobalt (II) was selected as the sorbate because it is a toxic and common radioactive contaminant and its sorption behavior is representative of many other divalent transition-metal and heavy metal cations of environmental concern.[72] It is also particularly well suited for *in situ* XAS studies, having sufficiently high K-shell fluorescence energies that it can be studied in the presence of corundum and water. In addition, because it has sufficiently different backscattering properties compared with Al and O, it is possible to distinguish among Co–Co, Co–O and Co–Al next-nearest neighbors, an important prerequisite for distinguishing among inner-sphere, outer-sphere, and multinuclear Co surface complexes using XAS. Co(II) has also been well studied in previous XAS and sorption studies on aluminum oxide sorbents which reduced the effort needed to select proper experimental conditions and facilitated analysis and interpretation of XAS data.[27,28]

Cobalt concentrations were determined using atomic absorption spectrometry (Perkin Elmer Model 1100B, Perkin Elmer Corp., Norwalk, CT). Both flame and graphite furnace methods were used depending on the sample concentration. Flame atomic absorption was used for samples containing concentrations above 10^{-5}M. Sodium nitrate was selected to serve as the background electrolyte in this work based on its previous use in studies of cobalt sorption to aluminum oxide. Relevant solution reactions

TABLE A-2.

Thermodynamic Constants for Aqueous Cobalt[a]

Reaction	log K
$Co^{2+} + H_2O = CoOH^+ + H^+$	−9.7
$Co^{2+} + 2H_2O = Co(OH)_2^0 + 2H^+$	−18.8
$Co^{2+} + 3H_2O = Co(OH)_3^- + 3H^+$	−31.5
$Co^{2+} + 4H_2O = Co(OH)_4^{2-} + 4H^+$	−46.3
$2Co^{2+} + H_2O = Co_2OH^{3+} + H^+$	−11.2
$4Co^{2+} + 4H_2O = Co_4(OH)_4^{4+} + 4H^+$	−30.5
$Co^{2+} + 2H_2O = Co(OH)_{2(s)} + 2H^+$	−13.1
$Co^{2+} + NO_3^- = CoNO_3^+$	0.2
$Co^{2+} + 2NO_3^- = Co(NO_3)_2^0$	−0.3 (I.S. = 0.5)

[a] Values obtained from References 70 and 71.

and thermodynamic constants for calculating Co(II) solution speciation and solubility are summarized in Table A-2. Except as noted below, experimental solution conditions were adjusted to avoid precipitation of cobalt oxide, carbonate, or hydroxide phases. In addition to the complexation and precipitation reactions shown, cobalt also undergoes oxidation reactions at high pH and high solute concentrations. Therefore, caution was also exercised to avoid reaching these conditions in the experiments conducted in this work.

B. Titration Experiments

Titration data for corundum used for SCM surface parameter evaluation were taken from Hayes et al.[40] The potentiometric titration experiments for α-Al_2O_3 were carried out as follows. Dried solid samples were suspended in deionized, argon-purged water for at least 2 days prior to the titrations. The aged suspensions were then placed in water-jacketed reactors to maintain the temperature at 25°C and stirred to prevent settling, while maintaining an argon atmosphere at all times. Electrolyte was added to adjust the ionic strength to the lowest level to be studied. Acid was added gradually to lower the pH to an arbitrary starting value at least 2 pH units away from the point of zero salt effect (PZSE), the crossover point of a family of constant ionic strength titration curves.[9] Then, the titration was carried out by adding incremental volumes of base. Between each incremental addition, 2 to 10 min were allowed for pH equilibration. The final pH of the base titration never exceeded 11 in order to minimize dissolution. A reverse acid titration was performed to return the suspension to the original starting pH. Additional electrolyte was then added to adjust the ionic strength to the next higher level desired, and the titration was repeated following the same protocol. Titrations for three ionic strength conditions, varying over two orders of magnitude, were performed. Based on extensive titration studies, no significant hysteresis was observed between acid and base titration "legs" using the above approach. In addition, comparison of results for titration experiments conducted over longer periods of time (24 to 48 h) to these data showed no significant differences between the techniques.[63]

C. Sorption Experiments

For sorption experiments, 10-ml high-density polyethylene (HDPE) centrifuge tubes were used as reactors. Each HDPE tube containing the solid suspensions was mixed by end-over-end rotation at 8 rpm in a constant temperature room maintained at 25°C. Mixing times were normally fixed at 48 h to allow sufficient time to achieve equilibrium conditions. pH was measured at 25°C using a Ross combination semi-microelectrode and an Orion Model 720 pH meter (Orion Research Inc., Boston, MA). The amount of metal ion sorbed was determined by analyzing the supernatant for solute following solid/liquid separation. Solid/liquid separation was accomplished by centrifugation at 22,000 RCF for 1 hour at 25°C using a high-speed centrifuge. Several control samples which did not contain solid were carried through each procedure to monitor system losses, and, in several cases, control samples which did not contain solute were also monitored.

Batch equilibrium data for sorption of Co(II) by α-Al$_2$O$_3$ were obtained for a range of surface coverages by conducting experiments over a range of pH values for different initial solute concentrations and varying solid concentrations (Table 13). For the purposes of calculating surface coverage, a coverage classification scheme assuming a hydrated radius (H.R.) of Co(II) of 4 Å was utilized in this work. The surface coverage values were calculated for the 50% adsorption point on the pH edge generated for the conditions shown. Each pH edge yielded a range of surface coverage extending from 0% to twice the value shown in Table 13.

D. FITEQL and the Goodness-of-Fit

FITEQL, a least-squares, fitting program, was used to find optimum SCM parameter values from sets of titration or metal ion sorption data.[63] This program optimizes the values of adjustable parameters by changing their values until the sum-of-the-squares of the residuals between the measured sorption data and FITEQL-calculated values is minimized. Optimization can be performed using single or multiple sets of data.

For estimation of the appropriate SCM surface parameters, combined titration data covering over two orders of magnitude in ionic strength were used (see Figure 9). In the case of metal ion sorption at moderate coverage, FITEQL was used to obtain the sorption constants using a single set of pH-edge data which fell completely within the moderate coverage range and each of the valid set of SCM parameters generated from the titration data. In turn, each of these surface complexation constants was then tested for its ability to predict other moderate coverage data. These data included pH edges collected at higher moderate surface coverage, pH edges collected at different ionic strengths, and sorption isotherm data collected at constant pH. In the case of the model predictions, the goodness-of-fit was based on a visual assessment of the ability of model predictions to match the data at other conditions. Under high coverage conditions, visual assessment of fitting rather than FITEQL was also used to select the number of surface complex formation reactions and the best values for multinuclear complex formation constants during the model development phase of the model.

In FITEQL, the relative goodness-of-fit for the different parameter sets is based on the F-statistic, defined as:

$$F = \frac{SOS}{DF} = \frac{\sum \left[\dfrac{Y_r}{S_r} \right]^2}{N_p \times N_c - N_u} \qquad (A\text{-}1)$$

where Y_r is the mass balance residual calculated from the deviation between the calculated and experimental mass balance for component r and S_r is the error calculated by propagating the error for total and free concentrations for component r. SOS is the sum-of-the-squares of weighted residuals for all components for which the total and free concentrations are known; DF is the number of degrees of freedom and is equal to the product of the number of data points, N_p, and the number of components for which both the total and the free concentration are known, N_c, minus the number of adjustable parameters, N_u. While it is apparent that the lower values of F indicate better fits to the data, the value of F is highly dependent on the error estimates assumed. Lower values of F result when higher error estimates are incorporated into the routine. The corollary of this is that the standard deviation of FITEQL-generated parameter values increases by increasing the error estimates. Caution must be exercised in selecting error estimates which are consistent with the data. In general, for a given set of error estimates, lower values of F are indicative of better agreement between the model fit and the data. Values of F less than 30 are usually considered to be indicative of good fits to the data. In arriving at valid sets of SCM constants, when using combined ionic strength data sets, F values were often found to be significantly greater than 30. For these cases, the fit to the titration data was considered satisfactory and the set of SCM constants valid if convergence was achieved and within 50% of the best value that was found.

For the analysis of titration data, the absolute error in pH was 0.05 and the relative and absolute errors in total hydronium ion concentration were 0.01 and 2.0×10^{-8} M, respectively. In the case of cobalt sorption data, the absolute error in pH and total cobalt concentration were 0.05 and

5.5×10^{-8} M, respectively, while the relative error in hydronium ion concentration was taken to be 0.1. A higher relative error selected for hydronium ion concentration in the case of sorption data was based on the reproducibility of the cobalt sorption edge of ±0.1 pH units. These error estimates are consistent with those used by others.[4,40]

E. XAS Data Collection and Analysis

Sorption samples were prepared for XAS at different coverages using the same procedures described above for generating the pH edges and isotherm data. Table A-3 summarizes surface coverage conditions for which XAS data were collected. All samples were equilibrated for a minimum of 48 h in HDPE centrifuge tubes. Following centrifugation, the pH was measured and an aliquot of supernatant removed and stored in acid prior to subsequent analysis for cobalt content. Most of the remaining supernatant cobalt solution was then removed with a Pasteur pipette leaving a slight residual to avoid resuspending the sample. To ensure that no excess supernatant remained, several careful washes of the supernatant remaining on the top of the solid plug in the centrifuge tube were performed. This entailed slowly pipetting ultrapure water into the centrifuge tube, carefully removing most of the excess water with the pipette, and finally using an absorbent tissue to remove the last few drops. Care was taken during the washing process to avoid removing any of the liquid entrapped in the pore spaces of the centrifuged samples to ensure that most of the wet paste would be undisturbed by the washing. This procedure was instituted to make the paste less watery, more uniform, and easier to transfer to XAS sample holders. XAS sample analysis indicated that this washing procedure had no effect on the spectra between washed and unwashed samples. Wet pastes for XAS were placed into aluminum sample holders, sealed with Kapton tape windows, and then placed in liquid nitrogen to maintain sample integrity and prevent oxidation.

TABLE A-3.

Surface Coverages for XAS Samples

Solid conc. (g/L)	Cobalt conc. (mol/L)	Coverage range	Coverage $\mu moles/m^2$ (%)	Coverage for H.R. = 4 Å (%)	Coverage for N_s = 3 sites/nm^2 (%)
20	1×10^{-4}	High	0.38	11.7	7.2
2	1×10^{-4}	High	1.8	52	34
2	1×10^{-4}	High	3.1	90	58
20	1×10^{-3}	High	3.2	106	60

XAS data were also collected for two reference samples, a crystalline solid, $Co(OH)_2(s)$, and a 12-mM aqueous cobalt nitrate solution, acidified to avoid hydrolysis. These samples were used for comparison to the unknown coordination environment of the sorption samples. Using a mortar and pestle, finely powdered boron nitride was mixed with the $Co(OH)_2(s)$ solid sample in a ratio sufficient to yield approximately 30% absorption of the incoming X-ray beam at the cobalt absorption edge. This sample was then transferred with a spatula into an aluminum sample holder and sealed with Kapton windows. The aqueous cobalt sample was pipetted into an aluminum sample holder that was sealed with Kapton tape. Both samples were then placed in liquid nitrogen and stored until scanned.

Data collection was performed at the Stanford Synchrotron Radiation Laboratory (SSRL) and at the National Synchrotron Light Source (NSLS). XAS spectra were collected at SSRL on wiggler magnet beamline 7-3, while at NSLS spectra were collected using a bending magnet beamline, either X-10C or X-11A. At SSRL, the beamline operated at a ring energy of 3 GeV, a current between 20 and 70 mA, and a wiggler field of 18 kG. At NSLS, the ring energy was 2.5 GeV with a current of 100 to 180 mA. Either Si (220) or Si (111) monochromator crystals were used with unfocused beam. The energy resolution using these crystals was 2 to 3 eV at the Co K-edge. In all cases, the monochromater was detuned 50% at the highest scanning energy to minimize contributions from higher harmonics in the incident beam. All samples were collected in fluorescence mode in either a liquid helium or nitrogen cryostat with the sample holder oriented at 45° to the incident beam. In the case of the reference samples, spectra were collected in transmission mode. Most of the sorption samples were collected using a 13-element Ge array detector with a Be window. However, some of the more concentrated samples were collected using a Stern–Heald-type detector with Soller slits and an Fe filter to reduce background scatter and fluorescence.[73] Multiple scans were collected for each sample and averaged to increase the signal-to-noise ratio. The number of scans ranged from 2 to 24 depending on the concentration of the sample.

F. XAS Data Reduction and Analysis

In XAS, the sample prepared as described above is exposed to a monochromatic beam of X-rays and scanned through a range of energies based on the central atom to be studied and the binding energy of a particular core electron (i.e., K1, L1, etc.). Typically, the sample is scanned from about 10 to 100 eV less than to 500 to 1000 eV greater than the binding energy of the core electron. In the region greater than 50 eV above the binding energy (the EXAFS region), the spectra exhibit oscillations due to interaction of an ejected electron from the central absorbing atom with its nearest neighbors (backscatterers). Quantitative data analysis techniques for the EXAFS region allow determination of the absorber–backscatterer distance and the number of backscatterers at a particular distance. The development and use of XAS data for acquiring structural information for neighboring atoms surrounding a central atom is well documented.[74,75] The application of this technique to the examination of water/metal oxide systems is also well documented.[38,67] The procedure used for data reduction essentially followed that described by others.[76] Specifically, the procedure involved the following steps:

1. Because of the potential drift in beamline energy during the course of data collection, an energy calibration was performed for each scan. This was accomplished by normalizing all spectra to a cobalt foil spectrum collected simultaneously with the spectra of the sample. A value of 7709.25 was assigned to the first inflection point in the K-edge of a cobalt foil spectrum for this adjustment.[77] In most cases the drift was less than 1 eV.

2. Fluorescence samples were collected using either a 13-element Ge detector or a Lytle-type detector. For samples collected using the 13-element Ge detector, all 13 channels were used and averaged to reduce the signal-to-noise ratio. In all cases multiple scans were collected, the number of which depended on the cobalt concentration, and averaged to give as low a signal-to-noise ratio as was feasible, based on time and sample concentration constraints.

3. Background absorption below the K-edge was removed from each averaged spectrum by subtracting a linear fit to the pre-edge region (7400–7700) from the entire spectrum.

4. Subtraction of background absorption above the K-edge for the averaged samples was then performed using a three-point spline fit from 7725 to 8375 eV and tabulated McMaster coefficients.[78] The EXAFS data were converted from photon E(eV) to $k(Å^{-1})$, the photoelectron wave vector, using

$$k = \left[\frac{2m}{\hbar^2 (E - E_o)} \right]^{1/2} \tag{A-2}$$

and weighted by k^3 to account for damping of oscillations at higher k. E_o is the threshold energy of the photoelectron at k = 0, E is the kinetic energy of the photoelectron, $\hbar$ is Planck's constant divided by 2π, and m is the mass of the electron.

5. The normalized EXAFS spectra were then Fourier transformed (FT) to produce radial structure functions (RSFs) using a Gaussian window ranging from k = 3.0 to 11.5 $Å^{-1}$. The FT peaks in the RSFs correspond to coordination shells about the central absorber and backscattering atoms, uncorrected for phase shift. Spurious FT peaks due to noise and termination effects were identified by noting variations in their amplitude and peak position with changes in Gaussian window limits of the transform.

6. Individual FT peaks in the RSFs were back transformed to produce filtered EXAFS spectra resulting from individual or dual contributions of particular absorber–backscatter pairs.

7. Nonlinear least-squares curve fitting of the filtered spectra to reference phase shift and amplitude functions (generated empirically or theoretically) provided coordination numbers and distances for the sorption samples. Empirical phase-shift and amplitude functions from dry, crystalline $Co(OH)_2(s)$ were used to obtain reference phase-shift and amplitude functions for Co–O and Co–Co absorber–backscatter pairs in filtered spectra from the sorption samples. Theoretical phase shift and amplitude functions were generated using *ab initio* single-scattering XAFS calculations for Co–Al.[79]

8. As a final check of parameters, the sample coordination and bonding distances obtained for the filtered EXAFS were used to generate an unfiltered EXAFS spectrum and compared with the normalized EXAFS spectrum generated with no Fourier filtering.

ACKNOWLEDGMENTS

The research presented in this chapter was funded by National Science Foundation Research Grants BCS-9210869 (LEK) and BCS-8958407 (KFH). We thank the staff at SSRL (supported by the Department of Energy and the National Institutes of Health). The authors would like to thank Britt Hedman and James Penner-Hahn for their assistance at SSRL; Garrison Sposito at the University of California at Berkeley and Chin-Fu Tseng at Lawrence Berkeley Laboratories for their sabbatical support of the senior author while preparing this paper; and Chia-Chen Chen, Justin Hildreth, and Eric Wight for their assistance in data collection.

REFERENCES

1. James, R.O.; Parks, G.A. Characterization of aqueous colloids by their electrical double-layer and intrinsic surface chemical properties, in *Surfaces and Colloid Science, Vol. 12*; E. Matijevic, Ed. Plenum Press: New York, 1982; pp. 119–216.

2. Hayes, K.F. Equilibrium, Spectroscopic, and Kinetic Studies of Ion Adsorption at the Oxide/Aqueous Interface; Ph.D. Thesis. Stanford University: Stanford, CA., 1987.

3. Dzombak, D.A. *J. Hydraulic Engr.* 1987, *113*, 430–446.

4. Dzombak, D.A.; Morel, F.M.M. *Surface Complexation Modeling: Hydrous Ferric Oxide.* John Wiley & Sons, Inc.: New York, 1990.

5. Schindler, P.W.; Stumm, W., The surface chemistry of oxides, hydroxides, and oxide minerals, in *Aquatic Surface Chemistry*; W. Stumm, Ed. Wiley-Interscience: New York, 1987; pp. 83–110.

6. Davis, J.A.; Kent, D.B. Surface complexation modeling in aqueous geochemistry, in *Mineral Water Interface Geochemistry, Vol. 23*; M.F. Hochella; A.F. White, Eds. *Mineralogical Society of America*: Washington, D.C., 1990; pp. 177–260.

7. Stumm, W. Chemistry of the solid/water interface processes at the mineral water interface and particle water interface, in *Natural Systems*. John Wiley & Sons, Inc.: New York, 1992.

8. McBride, M.B. *Environmental Chemistry of Soils*. Oxford University Press: New York, 1994; 405 pp.

9. Sposito, G. *The Surface Chemistry of Soils*. Oxford University Press: New York, 1984; 234 pp.

10. Hayes, K.F.; Leckie, J.O. *J. Coll. Interf. Sci.* 1987, *115*, 564–572.

11. Williams, A.F. *A Theoretical Approach to Inorganic Chemistry*. Springer: Berlin, 1979; 316 pp.

12. Morel, F.M.M.; Hering, J.G. *Principles and Applications of Aquatic Chemistry*. John Wiley & Sons: New York, 1993; 588 pp.

13. Kinniburgh, D.G.; Jackson, M.L.; Syers, J.K. *Soil Sci. Soc. Am.* 1976, *40*, 796–799.

14. McKenzie, R.M. *Aust. J. Soil Res.* 1980, *18*, 61–73.

15. Sposito, G. *The Chemistry of Soils*. Oxford University Press: New York, 1987.

16. Huang, C.P.; Stumm, W. *J. Coll. Interf. Sci.* 1973, *43*, 409–420.

17. Kinniburgh, D.G. *Environ. Sci. and Tech.* 1986, *20*, 895–904.

18. Honeyman, B.D. Cation and Anion Adsorption at the Oxide/Solution Interface in Systems Containing Binary Mixtures of Adsorbents: An Investigation of the Concept on Adsorptive Additivity; Ph.D. Thesis. Stanford University: Stanford, CA, 1984.

19. Farley, K.J.; Dzombak, D.A.; Morel, F.M.M. *J.Coll. Interf. Sci.* 1985, *106*, 226–242.

20. Dzombak, D.A.; Morel, F.M.M. *J. Coll. Interf. Sci.* 1986, *112*, 588–598.

21. Charlet, L; Manceau, A. *J. Coll. Interf. Sci.* 1992, *148*, 443–458.

22. Katz, L.E.; Hayes, K.F. *J. Coll. Interf. Sci.* 1995, *170*, 477–490.

23. Katz, L.E.; Hayes, K.F. *J. Coll. Interf. Sci.* 1995, *170*, 491–501.

24. Brown, G.E., Jr. Spectroscopic studies of chemisorption reaction mechanisms at oxide-water interfaces, in *Mineral Water Interface Geochemistry, Vol. 23*; M.F. Hochella; A.F. White, Eds. Mineralogical Society of America: Washington, D.C., 1990; pp. 309–363.

25. Hayes, K.F.; Roe, L.A.; Brown, G.E., Jr.; Hodgson, K.O.; Leckie, J.O.; Parks, G.A. *Science.* 1987, *238*, 783–786.

26. Chisholm-Brause, C.; Roe, A.L.; Hayes, K.F.; Brown, G.E., Jr.; Parks, G.A.; Leckie, J.O. *Geochim. Cosmochim. Acta.* 1990, *54*, 1897–1909.

27. Chisholm-Brause, C.J.; O'Day, P.A.; Brown, G.E., Jr. *Nature.* 1990, *348(6301)*, 528–530.

28. Chisholm-Brause, C.J. Spectroscopic and Equilibrium Study of Cobalt (II), Sorption Complexes at Oxide/Water Interfaces, Doctoral Thesis. Stanford University: Stanford, CA, 1991; 118 pp.

29. Roe, L.A.; Hayes, K.F.; Chisholm, C.; Brown, G.E., Jr.; Hodgson, K.O.; Parks, C.A.; Leckie, J.O. *Langmuir.* 1990, *7*, 367–373.

30. Dent, A.J.; Ramsay, J.D.F.; Swanton, S.W. *J. Coll. Interf. Sci.* 1992, *150(1)*, 45–60.

31. Xu, N. Spectroscopic and Solution Chemistry Studies of Cobalt (II) Sorption Mechanisms at the Calcite/Water Interface, Doctoral Thesis. Stanford University: Stanford, CA, 1993.

32. Combes, J.M.; Chisholm-Brause, C.J.; Brown, G.E., Jr.; Parks, G.A.; Conradson, S.D.; Eller, P.G.; Triay, I.R.; Hobart, D.E.; Meijer, A. *Environ. Sci. Tech.* 1992, *26(2)*, 376–342.

33. Papelis, C.; Brown, G.E., Jr.; Parks, G.A.; Leckie, J.O. *Langmuir.* 1995, *11*, 2041–2048.

34. Waychunas, G.A.; Rea, B.A.; Fuller, C.C. *Geochim. Cosmochim. Acta.* 1993, *57(10)*, 2251.

35. Fendorf, S.E.; Gamble, G.M.; Stapleton, M.G.; Kelley, M.J.; Sparks, D.L. *Environ. Sci. Tech.* 1994, *28*, 284–289.

36. Papelis, C.; Katz, L.E.; Hayes, K.F. unpublished data.

37. O'Day, P.A.; Brown, G.E., Jr.; Parks, G.A. EXAFS study of aqueous Co(II) sorption complexes on kaolinite and quartz surfaces, in *X-ray Absorption Fine Structure*; S.S. Hasnain, Ed. Ellis Horwood Ltd.: London, 1991; pp. 260–262.

38. O'Day, P.A.; Brown, G.E., Jr.; Parks, G.A. *J. Coll. Interf. Sci.* 1994, *165*, 269–289.

39. Papelis, C.; Hayes, K.F. *Colloids and Surfaces*, in press.

40. Hayes, K.F.; Redden, G.; Ela, W.; Leckie, J.O. *J. Coll. Interf. Sci.* 1991, *142*, 448–469.

41. Westall, J.C.; Zachary, J.L.; Morel, F.M.M. Technical Note No. 18, Department of Civil Engineering, MIT, Cambridge, MA, 1976.
42. Papeilis, C.; Hayes, K.F.; Leckie, J.O. Technical Report No. 306, Department of Civil Engineering, Stanford University, Stanford, CA, 1988.
43. Schecher, W.D.; McAvoy, D.C. Environ. Urban Systems. 1992, 16(1), 65–76.
44. Schindler, P.W.; Stumm, W. The surface chemistry of oxides, hydroxides, and oxide minerals, in Aquatic Surface Chemistry; W. Stumm, Ed. Wiley-Interscience: New York, 1987; pp. 83–110.
45. Davis, J.A.; James, R.O.; Leckie, J.O. J. Coll. Interf. Sci. 1978, 63, 480–499.
46. Hayes, K.F; Papelis, C.; Leckie, J.O. J. Coll. Interf. Sci. 1988, 125, 717–726.
47. Morel, F.M.M. Principles of Aquatic Chemistry. John Wiley & Sons, Inc.: New York, 1983; 446 pp.
48. Pitzer, K.S. J. Phys. Chem. 1973, 77, 268.
49. Chan, D.; Perram, J.W.; White, L.R.; Healy, T.W. J. Chem. Soc. Faraday Trans. I. 1975, 71, 1046–1057.
50. Hiemstra, T.; De Wit, J.C.M.; Van Riemsdijk, W.H. J. Coll. Interf. Sci. 1989, 133, 105–117.
51. Hiemstra, T.; Van Riemsdijk, W.H.; Bolt, G.H. J. Coll. Interf. Sci. 1989, 133, 91–104.
52. Sverjensky, D.A. Geochim. Cosmochim. Acta. 1994, 58, 3123–3129.
53. Schindler, P.W.; Kamber, H.R. Helv. Chim. Acta. 1968, 51, 1781–1786.
54. Hohl, H.; Stumm, W. J. Coll. Interf. Sci. 1976, 55, 281–288.
55. Stumm, W.; Huang, C.P.; Jenkins, S.R. Croat. Chem. Acta. 1970, 42, 223–244.
56. Huang, C.P.; Stumm, W. J. Coll. Interf. Sci. 1973, 22, 231–259.
57. Davis, J.A.; Leckie, J.O. J. Coll. Interf. Sci. 1978, 67, 90–107.
58. Hsi, C.D.; Langmuir, D. Geochim. Cosmochim. Acta. 1985, 49, 1931–1941.
59. Cowan, C.E.; Zachara, J.M.; Resch, C.T. Environ. Sci. Tech. 1991, 25, 437–446.
60. Manceau, A.; Charlet, L. J. Coll. Interf. Sci. 1973, 22, 231–259.
61. Westall, J.C.; Hohl, H. Adv. Coll. Interf. Sci. 1980, 12, 265–294.
62. Katz, L.E. Surface Complexation Modeling of Cobalt Ion Sorption at the α-Al_2O_3–Water Interface: Monomer, Polymer and Precipitation Reactions; Ph.D. Thesis. University of Michigan: Ann Arbor, MI, 1993.
63. Kent, D.B.; Tripathi, V.S.; Ball, N.B.; Leckie, J.O. Progress Report, Contract #SNL-25-1891, Sandia National Laboratory, 1986.
64. Westall, J.C. Report 82-01, Department of Chemistry, Oregon State University, Corvallis, 1982.
65. Brown, G.; Parks, G.A.; O'Day, P. Sorption at mineral/water interfaces: macroscopic and microscopic properties in Mineral Surfaces, Mineral Society Series; D.J. Vaughan; R.A.D. Pattrick, Eds. Chapman and Hall, 1995.
66. Benjamin, M.M.; Leckie, J.O. J. Coll. Interf. Sci. 1981, 79, 209–221.
67. Brown, Jr.; G.E.; Calas, G.; Waychunas, G.A.; Petiau, J. X-ray absorption spectroscopy and its applications in mineralogy and geochemistry, in Spectroscopic Methods in Mineralogy and Geology, Vol. 18; F.C. Hawthorne, Ed. Mineralogical Society of America: Washington, D.C., 1988; pp. 431–512.
68. Lotmar, W.; Feiknecht, W. Z. Krist. 1936, A93, 368–378.
69. Bol, W.; Gerrits, G.F.A.; van Panthaleon van Eck, C.L. J. Appl. Cryst. 1970, 3, 486–492.
70. Baes, C.F., Jr.; Mesmer, R.E. The Hydrolysis of Cations. John Wiley & Sons, Inc.: New York, 1976; 489 pp.
71. Smith, R.M.; Martell, A.E. Critical Stability Constants, Vol. 4, Inorganic Ligands. Plenum Press: New York, 1976.
72. Riley, R.G.; Zachara, J.M.; Wobber, F.J. Report DOE/ER-0547T, U.S. Department of Energy, Office of Energy Research, Washington, D.C., 1992.
73. Lytle, F.W.; Greegor, R.B.; Sandstrom, D.R.; Marques, E.C.; Wong, J.; Spiro, C.L.; Huffman, G.P.; Huggins, F.E. Nuclear Instr. Methods Phys. Res. 1984, 226, 542–548.
74. Teo, B.K. EXAFS: Basic Principles and Data Analysis. Inorganic Chemistry Concepts 9. Springer-Verlag: Berlin, 1986; 349 pp.

75. Koningsberger, D.C.; Prins, R. *X-ray Absorption: Principles, Applications, Techniques of EXAFS, SEXAFS, and XANES, Chemical Analysis Vol. 92*, John Wiley & Sons, Inc.: New York, 1988; 673 pp.
76. Cramer, S.P.; Hodgson, K.O. *Prog. Inorg. Chem.* 1979, *25*, 1–39.
77. Beardon, J.A.; Burr, A.F. *Rev. of Mod. Physics*, 1967, *39*(1), 125–142.
78. McMaster, W.H.; Del Grande, N.K.; Mallet, J.J.; Hubbell, J.H. Compilation of X-ray cross sections III: U.S. Atomic Energy Commission; UCRL-50174; 1969.
79. Rehr, J.J.; de Leon, J.M.; Zabinsky, S.I.; Albers, R.C. *J. Am. Chem. Soc.* 1991, *113*(4), 5135–5140.

Chapter 4

SURFACE-CONTROLLED DISSOLUTION AND GROWTH OF MINERALS

Patrick V. Brady and William A. House

CONTENTS

0-8493-8351-X/96/$0.00+$.50
© 1996 by CRC Press, Inc.

I. INTRODUCTION

The ability to quantitatively predict rates of mineral growth and dissolution is one of the most fundamental goals of geochemistry and materials science. A reasonable approach to the problem is first to examine
the macroscopic factors (temperature, pH, ionic strength, solution composition, etc.) which affect reaction rates. If these observations can be
rationalized in terms of likely reaction mechanisms at the atomic level,
fundamental laws may be established whereby rate dependencies can be
predicted. The ultimate goal is to use molecular chemical kinetics to
predict the reaction paths and fates of macroscopic systems.

The objectives of this chapter are twofold:

1. Outline the observed link between mineral surface chemistry and macroscopic rates of growth and dissolution and
2. Examine the probable microscopic reactions which explain this link.

The connection between surface chemistry and mineral dissolution and growth in aqueous fluids has historically focused on model metal (hydr)oxides. The following treatment will therefore dwell initially on reactions on SiO_2, $FeOOH$, and Al_2O_3 surfaces. The general approach to (hydr)oxide surface kinetics can, with a few modifications, then be applied to mixed (hydr)oxide silicate minerals and glasses. The link between surface chemistry and the dissolution of carbonates and sulfides will then be explored. The literature describing sulfide and carbonate corrosion is extensive, in part because of the implications for, respectively, acid mine drainage and the chemical behavior of the ocean. The elevated reactivity of sulfides and carbonates makes experimental surface chemical measurements somewhat problematic, so much so that it has only recently been possible to predict the likely stoichiometries of species present at the respective surfaces.

The connection between mineral surface chemistry and mineral growth rates is far less clear than is the link between surface chemistry and mineral dissolution. As a result, the treatment below will lean heavily on dissolution rates.

II. RATE LAWS FOR SINGLE (HYDR)OXIDE DISSOLUTION

A. Overview

The dissolution rates of minerals are commonly believed to be controlled by the chemical species present at the surface. Several such species are depicted in Figure 1. The dissolution rates of a number of single (hydr)oxides appear to be simple functions of the number of charged sites (protonated + deprotonated) at their respective surfaces.[1,2] Thus, the pH dependence of mineral dissolution arises largely from the fact that the number and type of surface species also depend on the pH of the fluid. Adsorbing species, such as oxalate and other organic acids, can accelerate mineral corrosion. Generally, ligands which bind strongly to a given metal in solution have been found to accelerate the dissolution of the same metal (hydr)oxide.[3] A number of adsorbents are seen to retard or "poison" the dissolution (as well as growth) of minerals. In other words, rates are, respectively, lower or higher than rates measured in the absence of the adsorbent. Measured rates in multicomponent systems, therefore, reflect the competition between adsorbents for surface sites.[4,5] For example, the adsorption of many trace substances onto calcite affects the precipitation and dissolution rates, see, for example, the review by Compton and Pritchard.[6] Adsorbed sulfate slows the dissolution of calcite in marine systems, thereby retarding the exchange of Mg for Ca in limestones. It has therefore been argued to slow the growth of dolomite.[7] Adsorbed Mg (and Pb) retards corrosion of silica glass.[8] In surface waters hydrophobic organic macromolecules block potentially reactive surface sites, in the process stabilizing mineral surfaces against hydrolysis. The adsorbed molecules also influence the electrical surface properties of the minerals and their interaction with such microorganic compounds as pesticides.[9]

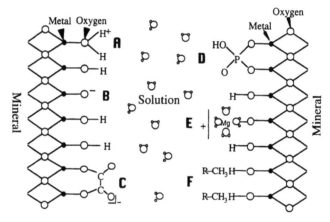

Figure 1
Schematic of metal oxide surfaces and potentially rate-determining adsorbents. A, Protonated surface site; B, Deprotonated surface site; C, Adsorbed oxalate; D, Dissolution inhibition by adsorbed phosphate; E, Adsorbed magnesium; F, Adsorbed hydrophobic organic (e.g., humics, hydrophobic moieties of fatty acids, etc.).

B. (Hydr)oxide Surface Chemistry at 25°C

The dissolution rate of a single (hydr)oxide ultimately depends on the relative proportion of adsorbed species which promote or retard atomic detachment from a surface site. For rate law analysis, site distributions of adsorbed species are calculated by first considering the acidity of the mineral surface. The acid–base properties of metal (hydr)oxide surfaces can be simply explained with a three-site model (see Chapter 3 and references therein for more details). This model postulates that three surface complexes of differing charge exist in dilute solutions free of adsorbents other than H^+ and OH^-: $>M-O-H_2^+$, $>M-O-H$, $>M-O^-$. $>M-$ denotes a surface metal site which is bound to the solid through one, two, or three surface oxygens. (See Figure 1.) Proton transfer reactions between the three can be written as:

$$>M-O-H_2^+ \leftrightarrow >M-O-H + H^+ \quad Ka_1 = \frac{[>M-O-H]a_{H^+}}{[>M-O-H_2^+]\exp\left(\dfrac{F\psi}{RT}\right)} \quad (1)$$

$$>M-O-H \leftrightarrow >M-O^- + H^+ \quad Ka_2 = \frac{[>M-O^-]\exp\left(\dfrac{-F\psi}{RT}\right)a_{H^+}}{[>M-O-H]} \quad (2)$$

In Equations 1 and 2 bracketed terms are site concentrations in mol cm^{-2}; Ka_1 and Ka_2 are thermodynamic equilibrium constants which are specific to individual oxides; a_{H^+} is the activity of protons in solution; F is Faraday's constant (96,485 C mol^{-1}); and ψ is the electrochemical potential of the surface (in volts). The exponential term models the work entailed in the formation of the respective charged surface species. R is the gas constant (8.314 J mol^{-1} K^{-1}) and T is temperature (K). The total number of surface exchange sites, $[>M-O-H^+_2] + [>M-O-H] + [>M-O^-]$, is generally close to the value calculated from the mineral surface area and crystallographic constraints. These range from 1 to 6 sites nm^{-2} (1.7 to 10×10^{-6} mol m^{-2}). At the pH of zero charge, pH_{zpc}, where pH = $(pKa_1 + pKa_2)/2$, the surface has zero net charge (pKa_1 and pKa_2 represent the negative logarithm of the equilibrium constants of Equations 1 and 2, respectively). Representative pH_{zpc} values for a number of solids are shown in Table 1. The pH_{zpc} for calcite is variable, depending on the levels of the lattice ions Ca^{2+} and CO_3^{2-} in solution[13] (see below).

TABLE 1.

25°C pH_{zpc} values[a]

Oxide	pH_{zpc}
SiO_2	2.2
TiO_2	4.7
BeO	10.2
β-MnO_2	7.0
α-Fe_2O_3	8.5
α-FeOOH	7.3
a-Al_2O_3	9.5
MgO	12.4

[a] The BeO value is from Furrer and Stumm.[11] All other values are from Parks.[12]

Surface charge in dilute solutions free of other multivalent adsorbing species is positive at pH < pH_{zpc} and negative at pH > pH_{zpc}. Specific adsorption of multivalent cations onto anionic surface sites causes an upward shift in the observed pH_{zpc}. Adsorption of multivalent anions to cationic surface sites causes an analogous downward shift in the pH_{zpc}. In solutions containing additional species, adsorption of rate-determining adsorbates such as oxalate and Mg can be written as ion association reactions. Combining Equations 1 and 2 with adsorbate–surface association reactions and expressions for charge and mass balance provides sufficient information to calculate the distribution of adsorbed species at the mineral–solution interface. A number of programs such as MINEQL[14] and HYDRAQL[15] exist to perform such calculations if the necessary surface equilibrium constants and total site densities are available (See Chapter 3.) The pH, T, and ionic strength dependencies of site occupancies can then be compared with the empirically determined rate dependencies to determine which species control dissolution and growth.

C. Experimental Methods

Before examining the systematics of adsorbent-promoted dissolution, a few caveats must be mentioned. The general controls on dissolution have sometimes been obscured by the wide variety of experimental methods used to measure rates. Rates are often measured with batch reactors where the rate constants are determined by plotting increases in the dissolved mineral component with time. Generally, there is a short-term accelerated and nonlinear release of components from high-energy surface sites produced by particle grinding and incomplete removal of ultrafine particles.[16,17] By pre-etching the mineral with a strong acid or by calculating rates only from the subsequent linear release region, the effects of particle grinding can be avoided.[18] Early dissolution of high-energy sites caused by grinding may also be avoided by simply storing the ground mineral for long periods of time before measuring rates[19] (although this may change the surface chemistry through reaction with atmospheric gases).

It is possible that as mineral components accumulate in solution other, secondary phases can form and the dissolving mineral may approach equilibrium with the solution. Near-equilibrium dissolution rates are much slower than those measured in dilute solutions, reflecting appreciable back reaction (growth) and the change in the affinity to dissolve. It is very difficult to measure rate dependencies near equilibrium because of analytical difficulties and because the number of possible reactions occurring becomes much greater.[20] Flow-through reactors[21-23] prevent both the effects of transient dissolution of high-energy sites, as well as the buildup of components in solution, although short-term surface adjustments to pH changes in such systems are still not understood. Recent work using isotopic doping and flow-through reactors[24] may ultimately give the most precise measurements of dissolution near equilibrium.

Chemical buffers are routinely used to control solution pH. This presents a problem if the buffer components adsorb and catalyze the reaction being measured. Many of the buffers which are routinely used are weak organic acids that are capable of accelerating dissolution rates.[25,26] It is therefore often necessary to carefully separate out the independent, but routinely parallel, effects of pH and organic acid concentration.

D. Dissolving Surface Areas

Because rates are generally assumed to be proportional to the amount of mineral exposed to solution, the method for determining the surface area of the dissolving mineral is important. Surface areas calculated from the geometry of crystals are generally three to five times smaller than the same areas measured with BET[27] (Brunauer-Emmett-Teller — surface-area measurement technique) analysis of gas adsorption isotherms.[28] The difference is thought to arise from microscopic rough features at the mineral surface which are not immediately visible. Holdren and Speyer[29] have shown that for alkali feldspars with surface areas greater than ~50 cm² g⁻¹ dissolution rates are not directly proportional to the amount of mineral surface area exposed to solution. They suggested that defects associated with exsolution lamella are the primary reaction sites and that the lack of proportionality between rates and surface area came from differential (reduced) exposure of the lamellae when particles were ground to fine size. Holdren and Speyer's data also suggest that specific surface areas for plagioclase minerals increase significantly with anorthite (Ca) content, which may arise from the differences in formation of the various plagioclase series minerals. Feldspar dissolution data from the literature suggest that BET-normalized dissolution rates decrease with mineral grinding.[30] At the same time, recent work indicates that BET-measured surface areas can vary significantly throughout dissolution experiments. Zhang et al.[31] measured BET surface areas of dissolving hornblende, a Ca- and Mg-rich double-chain silicate, at pH 4 and showed that surface areas increased by ~100% after being dissolved for 30 days. An even larger increase in BET surface area is seen for wollastonite ($CaSiO_3$) dissolution.[32] The surface area increase is thought to come from the opening of etch pits, hollow dissolution cores, and the formation of micropores.[31] Recent X-ray work has also shown that terrace areas on calcite roughen substantially during dissolution, increasing surface area as the dissolution reaction proceeds.[33]

An alternative method for measuring dissolution rates is to focus specifically on the corrosion of the mineral surface. Dissolution rates are calculated by monitoring ledge motion and etch pit opening over time by microscopic techniques.[34] Work in this vein has the added attraction of linking macroscopic rate dependencies and specific molecular interactions at the mineral surface.[35-37]

To glean some systematics from the existing measurements of mineral dissolution, the focus here will initially be on experiments conducted far from equilibrium. Here back reaction and saturation effects should be minimal. At the same time, only long-term rates measured after the dissolution of high-energy surface sites will be considered and, with few exceptions, only experiments in which mineral surface areas were measured by BET isotherms will be considered.

E. Rate Laws

Surface-controlled mineral dissolution, far from equilibrium, is generally described by a rate law of the form:[38]

$$R = k[>M-L]^n \tag{3}$$

R is the dissolution rate (in mol cm^{-2} s^{-1}); k is the experimentally determined rate constant; and [>M–L] is the concentration of the surface species controlling the process (in mol cm^{-2}); >M–L is generally an adsorbed proton, electronegative oxygen site, or other ligand, often an adsorbed organic molecule. n, the reaction rate order, is commonly observed to be an integer, at least for single (hydr)oxides. In Figure 2 are shown low pH (pH <pH$_{zpc}$) 25°C dissolution rates of BeO,[11] α-FeOOH,[39] and δ-Al$_2$O$_3$.[11] Note that dissolution rates are proportional to the number of protonated sites on the respective (hydr)oxide surface. Because rates increase with the concentration of adsorbed protons, it appears that adsorbed protons destabilize and weaken metal–oxygen bonds.

In Figure 3 are plotted high pH (pH > pH$_{zpc}$) 25°C dissolution rates of γ-Al$_2$O$_3$[40] and quartz.[41] γ-Al$_2$O$_3$ dissolution rates appear to be simple functions of the density of protonated surface sites; with Rate = k[>Al–O$^-$]^4 (Furrer and Stumm[11] have argued that the rate order should be 3). Quartz rates at pH > 8 are proportional to the concentration of deprotonated surface sites[41,42] with Rate = k[>Si–O$^-$].

Some generalizations can be made which apply to the data above, and indeed apparently to all single (hydr)oxides. In dilute solutions at pHs below the pH$_{zpc}$ of the (hydr)oxide, dissolution rates increase with decreasing pH. This is because the number of dissolution-promoting species (adsorbed protons) increases. At pHs above the pH$_{zpc}$, rates increase because the rate-promoting deprotonated sites increase in number.

It is difficult to discern any systematic reason for the observed rate orders to be what they are. Stumm and co-workers[11,39] argued from their work with BeO (n = 2), α-FeOOH (n = 3), and Al$_2$O$_3$ (n = 3) that n was equivalent to the valence of the metal center. This does not coincide with the rate order observed for SiO$_2$ (n = 1), or with the rate order observed at pH > pH$_{zpc}$ for Al$_2$O$_3$ (n = 1) by Carroll-Webb and Walther.[40] It is not

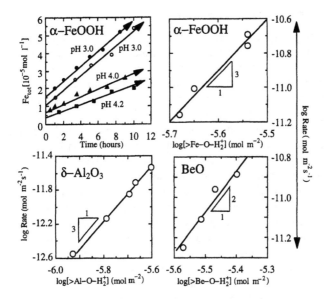

Figure 2
Dissolution rates of α-FeOOH,[39] δ-Al$_2$O$_3$,[11] and BeO[11] plotted as a function of the measured number of protonated sites at the mineral surface, [>M-O-H$_2^+$] (mol m^{-2}). The latter were calculated from potentiometric titrations. Dissolution rates were calculated from metal release rates measured with batch reactors (see the original papers for details).

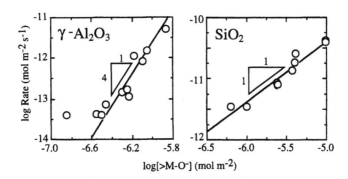

Figure 3
γ-Al$_2$O$_3$[40] and quartz[41] dissolution rates plotted as a function of the measured number of deprotonated sites, [>M-O$^-$]. The latter were measured by potentiometric titration[40] or taken from published values. Dissolution rates were calculated from metal release rates measured with batch reactors.

difficult to write activation reactions (see the transition state theory section below) involving n charged sites as reactants to give a rate law with n as the reaction order. However, considering crystallographic constraints, triply and quadruply protonated sites are likely rare.

F. Rate–Free Energy Correlations

At low pH (pH $\ll$ pH$_{zpc}$) (hydr)oxide surfaces are completely saturated with adsorbed protons. The number of protonated sites is similar from mineral to mineral at low pH. Any differences in dissolution rates at very low pH are therefore likely to reflect variations in bond stabilities at the respective mineral surfaces. Dissolution rates of alkaline earth oxides at pH = 1 and pH = 2 correlate with ionic radius of the respective metal cation.[43] Dissolution rates of transition-metal oxides correlate with the exchange rate of water into the inner coordination sphere of the respective aquated metal cation.[43] Both these observations suggest that surface species–normalized dissolution is controlled by the strength of metal–oxygen bonds. Sverjensky and Molling[44] have refined the correlation between dissolution and mineral by explicitly considering the effect of crystal structure (see Figure 4). Note that dissolution rates have been put into the context of transition state theory (see below). Specifically, measured dissolution rates were used to calculate $\Delta G_{f(surface\ complex)}$, the free energy of formation (kcal mol^{-1}) of the surface complex controlling dissolution. In Figure 4, β is an independently derived fit factor and r_{M+2} is the radius of the cation in the given oxide. ΔG_n is the free energy of formation of the given oxide. The vertical axis is a proxy for measured rates. The critical result is that rates (surface complex free energies) can be predicted from tabulated mineral free energies and the correlation in Figure 4.

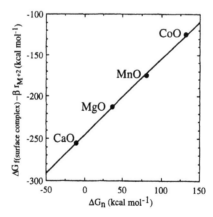

Figure 4
Correlation between pH 1 oxide dissolution rates and oxide free energies (see text for explanation). (From Sverjensky, D.; *Nature.* 1992, *358*, 310–313. With permission.)

According to Equation 3, R = k[>M–L]^n, rates should increase in response to any system changes which cause the reacting surface species concentration, [>M–L], or reaction order to increase. Given that the total number of surface sites is constant and that it is the charged sites which promote dissolution, Equations 1 and 2 suggest that surface charge, hence rates, are sensitive functions of the surface electrostatic potential, ψ. ψ increases with ionic strength (salt content of the adjacent solution). Because different models of the surface–solution interface predict different values for ψ, a precise relation between surface speciation and rate can be difficult to determine unambiguously. A good example of this is the dissolution rate law for quartz. pH-dependent quartz dissolution rates at pH >8 and 25°C are proportional to the number of deprotonated surface species at the quartz–solution interface[41,42,45,46] (see Figure 3). Note that each anionic surface site has associated with it a charge-balancing counterion (in most experimental methods at high pH this is Na^+ or K^+). It is a continuing question how close these counterions approach the mineral surface and how strongly bound they are to charged surface sites. In the one extreme, triple layer models[47] generally predict surface ion pairs of >Si–O$^-$ and Na^+. Other surface chemical formulations emphasize that one or more hydration layers separate the two charged species, with very weak interaction between the two.[48] The degree of interaction is important to an understanding of the factors which control detachment of Si sites. If counterion binding to the surface drives dissolution (as opposed to the simple presence of the deprotonated site), then rates should vary depending on the attraction of the given counterion for the surface.[49,50] Salts may facilitate surface charge-promoted dissolution by weakly coordinating with the surface and enhancing water dissociation within the quartz–solution interface.[49] In contrast, the results of House[51] demonstrate similar effects of NaCl and Na_2SO_4 salts on the dissolution rates of silica at 25°C with a greater than twofold increase in the rate constant for an ionic strength range of 0.01 to 1 M. These results can be explained by the effect of the electrolyte concentration on the ionization of the silanol surface and support the conclusion that the complexation of Na^+ with the surface of silica, i.e., by the formation of an inner-sphere complex, is not an important influence on the dissolution reaction at this temperature. Similar experiments with $CaCl_2$ solutions produced results consistent with the complexation of calcium with the silica surface to form an inner-sphere complex: >(SiO)$_2$Ca and dissolution by hydrolysis at the complexed site. Unlike the results with Na salts, the changes in the reaction rate could not be explained by the changes in the ionic strength of the solution alone and needed some specific interaction of the cation with the surface reaction site.

The key to settling this question probably lies in the measurement of *in situ* >Si–O$^-$–Na–nH$_2$O$^+$ bond distances. Independent verification may ultimately come from quantum mechanical calculations.[49,50,52-54] These studies have to this point been somewhat limited by their reliance on assumed reaction mechanisms and models of the near-surface potential field. The results, therefore, tend to be nonunique, even for a single mineral such as quartz. More computationally intensive models may, in the future, shed light on the role of counterion binding in mineral dissolution.

G. Effects of Temperature

Temperature affects rates most importantly by changing the value of the rate constant, k. This change is commonly quantified through the Arrhenius expression as:

$$\frac{k_{T_2}}{k_{T_1}} = \exp\left[\frac{E_a}{R}\left(\frac{1}{T_1} - \frac{1}{T_2}\right)\right] \tag{4}$$

where E_a is the activation energy of dissolution (kJ mol^{-1}). Surface equilibria, hence [>M–L], also can change with temperature. The Arrhenius equation is generally used in the following form by workers to calculate energies directly from dissolution rates;

$$\frac{R_{T_2}}{R_{T_1}} = \exp\left[\frac{E_a^{App.}}{R}\left(\frac{1}{T_1} - \frac{1}{T_2}\right)\right] \tag{5}$$

Equation 5 is commonly used with measured rates at two or more temperatures to determine E_a^{App}, an apparent activation energy. The apparent activation energy lumps together E_a and any effects of temperature on the concentration of the rate-controlling species at the mineral surface. The apparent activation energy can also be calculated from a linear regression of the logarithm of the rate constant or reaction rate with T^{-1} with the intercept related to the standard activation entropy.[55] The activation energies of mineral dissolution and growth generally fall in the range of 30 to 100 kJ mol^{-1} which is, for the most part, less than the energy involved in the breaking of bonds in solid state phase transformations. The latter are generally on the order of 80 to 320 kJ mol^{-1}. Representative activation energies are 53 to 92 kJ mol^{-1}, quartz dissolution;[34,41,49,53,54] 80 to 83 kJ mol^{-1}, silica dissolution;[56] 35 to 44 kJ mol^{-1}, calcite growth;[57-60] 46 to 59 kJ mol^{-1}, calcite dissolution.[61,62] Some of the variation in activation energies measured by various workers arises because of the fact that measured activation energies tend to vary as a function of pH because of temperature-dependent shifts in surface speciation (see below). Hence, activation energies measured on the same mineral at different pHs tend to differ.

Note that reactions which are controlled by the diffusion of material to and from the surface generally have activation energies at or below about 20 kJ mol^{-1}. Highly soluble alkali halide, hydroxycarbonate, and sulfate minerals (e.g., NaCl, $Na_2CO_3 \cdot 10H_2O$, $MgSO_4 \cdot 7H_2O$) generally possess low activation energies, and their dissolution is transport controlled.[63] Dissolution rates of soluble halides, hydroxycarbonates, and sulfates are somewhat less relevant to geochemical processes than their less soluble counterparts, as these minerals either reach equilibrium rapidly or are consumed entirely in the process.

To understand the energetics of reactions at the microscopic level, the effect of temperature on the rate constant must be separated from the independent effect of temperature on surface speciation. The pH_{zpc} values listed in Table 1 are routinely found to decrease with increasing temperature.[64,65] Also, at constant pH increased temperature leads to increased cation adsorption and decreased anion adsorption.[64-68]

Proton adsorption isotherms have the greatest pH dependency at pH > pH_{zpc}. At pHs much less than the pH_{zpc}, proton adsorption depends less on pH. This is because as [H+] increases all available proton acceptor sites become saturated and the slope of the proton adsorption isotherm flattens and approaches zero. A similar effect has been noted for organic chelate-promoted dissolution.[69] Because of the decrease in (hydr)oxide pH_{zpc} with increasing temperature, proton adsorption isotherms in the region below the original, but above the high T pH_{zpc}, can be expected to become more pH dependent.[70,71] This is important because any increase in the pH dependence of proton adsorption should cause rates to be more pH dependent. For example, Figure 5 shows the effect that temperature has on pH dependent alumina surface protonation at pH < pH_{zpc}.

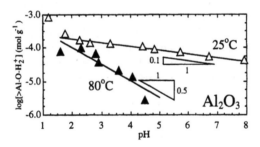

Figure 5
The effect of temperature on pH-dependent alumina surface protonation. (From Brady, P.V.; Walther, J.V. *Am. J. Sci.* 1992, *292*, 639–658. With permission.)

One of the important points here is that because activation energies give information about reaction dynamics, separating out temperature-dependent surface charging effects is an important step to understanding detailed rate mechanisms.[71,70] Specifically, the enthalpies of proton adsorption/desorption must be subtracted out of experimentally measured activation enthalpies to examine detailed bonding scenarios.

III. RATE LAWS FOR MULTIOXIDE SILICATE DISSOLUTION

Steady-state dissolution of multioxide silicates occurs after two early features are observed, one of which is the dissolution of high-energy sites and fine particles adhering to the surface (see above). While this accelerated early corrosion is going on, the alkali cations (Na, K, etc.) are exchanged rapidly for protons from the solution.[72] This leaves an alkali-depleted framework whose detachment ultimately determines the subsequent long-term steady-state release of mineral components. It is the latter dissolution regime which is most applicable to understanding mineral dissolution over geologic time. At very low pH (pH < 4) Si-rich leach layers form on feldspar surfaces because of the rapid detachment of Al as well as alkalis.[73,74] Leach layers form because Si sites are uncharged, hence unreactive, at low pH relative to the non-Si sites. At neutral and basic pH, where Si sites are considerably more reactive, silicate leach layers are seen to be much thinner. The dynamics of leach layer formation will be examined further below in the section on glass dissolution. Note, however, that leached layers apparently limit neither the charging of multioxide surfaces nor their steady-state dissolution except at extreme pHs (pH < 2).

On multioxide silicates, surface charge is often approximated as the summed contribution of adsorption reactions on each constituent (hydr)oxide.[75-77] Each (hydr)oxide at the mineral surface is treated as gaining or losing protons independently of its neighbors. This treatment assumes that the acidities of the constituent oxides are caused by short-range interactions and do not depend on the identity of their neighbors. Generally, the alkali elements, such as Na^+ and K^+ (and occasionally Ca^{++}),[32] are ignored in the calculation as they are rapidly exchanged with protons from solution. The assumption of summed acidities is an approximation which is semiqualitative.[78] Nevertheless, it is a useful tool which can be used to roughly predict which surface sites control silicate surface charge and dissolution.

A. Multioxide Surface Chemistry

The surfaces of complex silicates appear to behave similarly to zwitterions in that they possess two ionizable surface groups which gain and lose protons over different pH ranges. Silica groups donate protons to solution, becoming anionic at pH > 2. Non-silica surface groups on many silicates appear to acquire protons from solutions less than neutral, giving the surface a net positive charge. This conclusion is arrived at by examining in turn the pH-dependent surface charge on each of the component oxides (SiO_2, Al_2O_3, etc.) and comparing them with pH-dependent complex silicate surface charge. From pH = 5 to 2 at 25°C, surface charge becomes increasingly positive on multioxide silicates, such as albite,[79] anorthite,[80] and wollastonite,[32] whereas silica surface charge does not become noticeably positive until well below pH 2.[12] This suggests that surface charge on complex silicates resides primarily on the exposed non-Si network-forming (hydr)oxides (Al_2O_3, MgO, etc.) at low pH.

Above neutral pH at 25°C, silicate surface charge appears to be controlled by the deprotonation of silica tetrahedra. Figure 6 is a plot of the measured surface charge for silica, kaolinite, and albite from pH 8 to pH

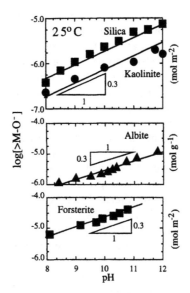

Figure 6
Measured surface charge[70] in 0.1 M NaCl solutions for silica (in mol m^{-2}), albite (in mol g^{-1}), and kaolinite (in mol m^{-2}) at 25°C. Forsterite surface charge[79] was measured in a dilute solution. (From Brady, P.V.; Walther, J.V. *Am. J. Sci.* 1992, 292, 639–658. With permission.)

12 at 25°C. Also shown is forsterite surface charge.[79] Note that SiO_2 is the only oxide common to each mineral surface and that for each mineral surface deprotonation is affected identically by increasing pH. The pH dependence of the log of the net negative surface charge (log[>M–O$^-$]) is modeled with a 0.3 slope. In fact, a slope of 0.33 at 25°C and ionic strength of 0.01 M are predicted from a simple surface speciation model[42] for quartz and silica dissolution. As surface charge on these complex silicates displays the same pH dependence as log[>Si–O$^-$] on SiO_2, the charge on the Si sites appears to dominate overall surface charge in near-neutral to basic pH solutions.

Taking account of albite dissolution rates[21] and albite surface charge measurements,[79] Brady and Walther[81] concluded that albite dissolution rates increase with decreasing pH below pH 5 because of increased protonation of tetrahedral aluminum sites exposed at the albite surface. This mechanism has since been echoed[82] and, subsequently, modeled with *ab initio* quantum mechanical techniques.[83] Carroll and Walther[84] argued that kaolinite rates increase with decreasing pH below pH 5 because of increased protonation on Al sites at the kaolinite surface, implying that dissolution rates are first-order functions of protonation on octahedral aluminum (hydr)oxide sites at the kaolinite structure. Similarly, Wieland and Stumm[85] concluded that proton-promoted dissolution of the gibbsite ($Al(OH)_3$) layer edges controlled kaolinite dissolution at low pH.

If surface charge on complex silicates is dominated at pH < 5 by protonation of the non-SiO_2 surface sites, the pH dependence of dissolution rates of complex silicates should themselves possess the same or similar pH dependence as the proton adsorption isotherms on the constituent non-SiO_2 oxides, assuming that acidities vary minimally (note that the latter is an assumption). For example, the pH dependence of kaolinite dissolution should possess the same pH dependence as the proton adsorption isotherm on Al_2O_3. Note that Al_2O_3 is being used here as a proxy for alumina sites at the kaolinite surface. 25°C and 80°C kaolinite dissolution rates are shown in Figure 7. At each temperature the low pH dissolution rates parallel the pH dependence of proton adsorption shown in Figure 6. This direct proportionality between rates and protonation is shown in Figure 8. Alternatively, based on the observation that the dissolution rates of albite, K-feldspar,[86] kyanite,[87] and kaolinite[88] are inversely related to the concentration of aqueous aluminum, Oelkers et al.[89] proposed that the dissolution of many aluminosilicate minerals is controlled by the concentration of an aluminum-deficient precursor complex formed by the exchange reaction, $M-Al + 3H^+ \leftrightarrow M-H_3 + Al^{3+}$, where M–Al denotes any aluminum on the mineral surface, and $M-H_3$ represents the site following exchange. It is assumed that this reaction is relatively rapid compared with the breaking of the silica tetrahedral framework and can be considered to be in local equilibrium with the adjacent solution. According to this approach, the pH dependence of dissolution at low pH arises from the pH dependence of the exchange reaction. Protonation of Al sites on kaolinite is experimentally observed, as is the exchange of Al for protons on the surface,[85] so both explanations for the rate-controlling surface species seem plausible.

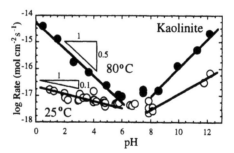

Figure 7
Kaolinite dissolution rates[84] (mol cm^{-2} s^{-1}) as a function of pH at 25°C and 80°C. The lines are simple linear fits to the data. (From Brady, P.V.; Walther, J.V. *Am. J. Sci.* 1992, 292, 639–658. With permission.)

The kaolinite results show three important features of long-term silicate dissolution: (1) multioxide silicates dissolve at low pH due to the presence of protons adsorbed onto, or exchanged for, non-Si sites; (2) the

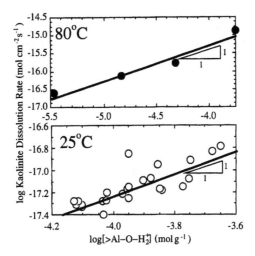

Figure 8
pH < 8 kaolinite dissolution rates[84] (mol cm^{-2} s^{-1}) as a function of measured Al$_2$O$_3$ surface charge[70] at 25°C and 80°C in I = 0.1 M NaCl (see Figure 6). (From Brady, P.V.; Walther, J.V. *Am. J. Sci.* 1992, *292*, 639–658. With permission.)

pH dependence of dissolution at low pH parallels the pH dependence of this proton adsorption/exchange; and (3) rates and proton adsorption/exchange appear to be affected similarly by temperature. To predict rates *a priori* requires an understanding of how rates vary from mineral to mineral at a given pH.

B. Rate–Free Energy Correlations for Silicates

Under highly acid conditions the non-Si surface sites for *all* silicates are likely to approach saturation with adsorbed protons. Silicate dissolution rates measured at pH 2 (and below), therefore, should give information about the contribution of mineral stability to the overall rate because the surface charge term should be approximately the same from mineral to mineral. Dissolution rates can thus be normalized to a single surface site density term by considering rates at very low pH. This isolates the crystal chemical, nonsurface component of dissolution. At pH 2 dissolution rates of olivines correlate well with ionic size, as well as with the rate of solvent exchange into the hydration sphere of the corresponding divalent cation.[43] The latter correlation is shown in Figure 9. The relationship is striking in that it successfully explains rates which vary from metal to metal by over six orders of magnitude. The correlation between dissolution rate and water exchange from a hydrated cation is not surprising as the two processes are mechanistically similar.[90,91] During proton-promoted dissolution, bridging metal–oxygen bonds are replaced by water molecules and, ultimately, a hydrated metal ion is released to solution.[43] The results in Figure 9 are exciting because they are a necessary first step toward actually predicting dissolution rate behavior of minerals.

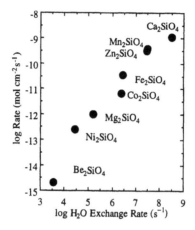

Figure 9

Dissolution rates for end-member orthosilicate minerals at pH = 2 plotted against the first-order rate coefficient (s^{-1}) for water exchange from the solvent into the hydration sphere of the corresponding dissolved cation. (From Casey, W.; Westrich, H. *Nature*. 1992, *355*, 157–159. With permission.)

To apply the pH 2 rates and correlations to predict rates at the more neutral pHs occurring in nature requires some technique for estimating the slope of the proton adsorption isotherm as a function of cation substitution, crystal structure, etc. By the same token, it is critical that the significance of n in (hydr)oxide dissolution rate laws, Equation 3, be pinned down. Ragnarsdóttir[92] showed that the pH dependency of aluminosilicate dissolution at 25°C and pH <7 decreases with increasing Si content in the mineral structure. Oxburgh et al.[93] have recently argued that the pH dependence of proton adsorption on feldspars is somewhat independent of composition (Ca content). Nevertheless dissolution rates of plagioclase series minerals at less than neutral pH become more pH dependent with An (Ca) content of the respective crystal, suggesting that the reaction order with respect to protonated species increases with An content. More work is needed in this direction to make clear the link between surface chemistry, dissolution rates, and mineral chemistry.

Dissolution rates of silicates correlate with their position in Bowen's reaction series.[94] Si-poor silicates, such as olivines, weather faster than chain silicates, which in turn weather faster than Si-rich, highly polymerized crystals such as amphiboles, sheet silicates, and quartz. To extend the acid pH correlation[43] to silicate structures more Si-rich than olivine requires some measure of bond strength that monitors the increasing stability of non-Si sites with polymerization of the silicate structure.

Conceptually, this can be thought of as an X_{Si} axis extending out perpendicular to the graph in Figure 9, where X_{Si} is the mole fraction of Si in the mineral formula. In Figure 10 is shown Casey and co-workers'[95] correlation between pH 2 feldspar dissolution rates and albite (Si) content. On a dissolving anorthite surface, silica tetrahedra are never linked to each other, existing rather as isolated sites. Albite, with a higher Si content, consists of linked silica tetrahedra. Obviously, increasing polymerization of the silicate structure makes it more resistant to hydrolysis.

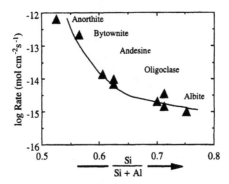

Figure 10
25°C dissolution rates of plagioclase feldspars as a function of An (Ca) content. (From Casey, W.H.; Westrich, H.R.; Holdren, G.R. *Am. Miner.* 1991, *76*, 211–217. With permission.)

At high pH (pH > 7) it is somewhat easier to model dissolution because the Si sites are those most highly charged (again, see Figure 6). The connection between surface charge and silicate dissolution at basic pH can be shown by plotting dissolution rates as a function of surface charge. In Figure 11 are shown basic pH 25°C dissolution rates of quartz, forsterite, albite, and kaolinite as a function of surface charge. Note again that quartz dissolution rates are first-order functions of [>Si–O⁻], indicating that the rate-limiting step in quartz dissolution involves a single deprotonated surface complex.[41] Thus, for quartz, as well as forsterite, kaolinite, and albite, there is a first-order dependence of dissolution rate on surface charge. Similarly, while the surface charge remains to be measured, the pH dependence of the dissolution of the above minerals, dlogRate/dpH ~ 0.3, is paralleled by that of nepheline,[96] andalusite,[97] bytownite, anorthite, and enstatite,[81] and chrysotile,[98] heulandite,[93] and sanidine.[99]

Silicate surface charge parallels surface charging on silica tetrahedra, and the pH dependency of silicate dissolution is identical to that of quartz dissolution. It therefore appears that silicate dissolution is controlled at basic pH by adsorption-mediated detachment of silica sites. That is, the rate-determining step for silicate dissolution is the surface charge controlled–detachment of Si sites. This explains why most silicate dissolution

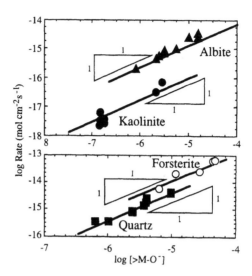

Figure 11
Measured 25°C dissolution rates of quartz (I = 0.1 M),[41] kaolinite,[40] forsterite,[79] and albite[21] as a function of the negative surface charge on the respective mineral.[70,79] Surface charge was calculated from the lines given in Figure 6. [>M–O⁻] is the concentration of negatively charged surface sites. *Note:* For quartz >M–O⁻ stands for >Si–O⁻. (From Brady, P.V.; Walther, J.V. *Am. J. Sci.* 1992, 292, 639–658. With permission.)

rates are affected identically by changes in pH above pH 8. Interestingly, Kubicki et al.[100] have recently used *ab initio* techniques to calculate that $Si(OH)_5^-$ sites at silicate surfaces may be important reaction intermediates for silicate dissolution at above neutral pH.

Although the pH dependence is the same, the absolute values of the rates are not. In other words, if all high pH dissolution rates are normalized to a given pH, hence a nearly common value of [>Si–O⁻], rates will vary by orders of magnitude, though their pH dependence will be the same. These surface charge–normalized rates can be correlated with energetic measures of surface structure similar to low pH silicate rates. Because Si detachment appears to control dissolution at pH > 8, absolute rates of silica detachment during stoichiometric dissolution should correlate with the energetics of the Si–O surface bonds. The measure of bond energy used below is the simple mean ionic site potential of oxygen bonds attached to Si atoms in a mineral. At constant pH and temperature, rates appear to increase with decreasing potential of the oxygen sites attaching the silica group to the surface. Ionic site potential increases with increased sharing of an oxygen by Si and is lowest for oxygen sites coordinated with only one silicon atom, such as in an olivine structure.[101] Site potential, therefore, generally increases in the order: island silicates, single-chain silicates, double-chain silicates, sheet silicates. Site potential also positively correlates with X_{Si}. Figure 12 shows the correlation between site energy and pH 8 dissolution rates.[81] The latter are simply the release rate of silica from a given mineral divided by the ratio of silica to non-alkali oxides in the mineral formula. This is done to account for differences in surface coverage by silica tetrahedra between different minerals. Above pH 8 at 25°C, all dissolution rates increase with pH such that dlogRate/dpH = 0.3, reflecting the increase in [>Si–O⁻] with pH.

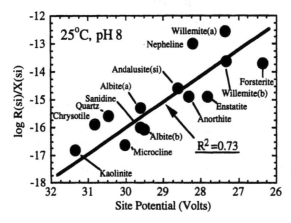

Figure 12
Correlation between normalized silica release rates at pH 8 (mol cm⁻² s⁻¹) of silicates at 25°C and mean potential of oxygen sites[101] bound to Si in the mineral structure. Rates are shown for kaolinite,[40] quartz,[41] chrysotile,[98] albite(a),[21] albite(b),[22] andalusite,[97] anorthite,[80,81] enstatite,[81] forsterite,[69,79] nepheline,[96] willemite(a),[35] and dissolution rates of ⟨2110⟩, Willemite(b). (From Brady, P.V.; Walther, J.V. *Am. J. Sci.* 1992, 292, 639–658. With permission.)

The same approach works at high temperature if the T-dependent shift in the SiO_2 deprotonation isotherm is first accounted for. The latter becomes slightly steeper with temperature:[67] dlog[>Si–O⁻]/dpH ~ 0.4 at 60°C vs. 0.3 at 25°C. Mineral dissolution rates at T ~ 60 to 70°C are first-order functions of negative surface charge[70] as they are at 25°C, suggesting that dissolution rates at high temperature are controlled by the same mechanisms that operate at 25°C. Specifically, surface charge–mediated detachment of Si sites controls multioxide silicate dissolution at near-neutral and basic pH. Rates become more pH dependent with increasing temperature because the slope of the isotherm becomes more positive with increasing temperature. As a result, at constant solution pH the temperature dependence of dissolution increases. This is best seen in the pH > 8 data for kaolinite shown in Figure 7.

To consider how the temperature dependence varies from mineral to mineral, it is reasonable to correlate the high temperature rates with site potentials, as done previously at 25°C. Silica release rates normalized to the mole fraction of surface silica sites at pH 7 are plotted against site potential in Figure 13.

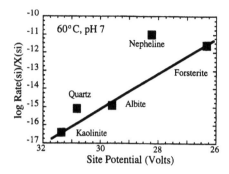

Figure 13
Correlation between normalized silica release rates at pH 7 (mol cm^{-2} s^{-1}) of silicates at 60°C and mean potential[101] of oxygen sites bound to Si in the mineral structure. Rates are shown for kaolinite,[40] quartz,[41] albite,[22] nepheline,[96] and forsterite.[69] The 60°C albite rates are inter-polations of Knauss and Wolery's[22] 25° and 70°C results using their measured activation energy. Because no pH 7 nepheline rates exist, pH 5 60°C nepheline rates of Tole et al.[96] are shown in Figure 12. Note that the nepheline rate was calculated from a geometrically measured surface area. As a result, the nepheline rate shown in Figures 12 and 13 is probably a factor of 3 or 4 too high. (From Brady, P.V.; Walther, J.V. *Am. J. Sci.* 1992, *292*, 639–658. With permission.)

C. Summary and Exceptions

It is useful to summarize the important features of silicate dissolution before proceeding further.

1. At pH < 5

1. Dissolution rates increase with decreasing pH, reflecting increased pro-tonation and detachment of non-Si surface sites.

2. At a given pH, rates can be correlated with the water exchange rate of the respective metal — at least for orthosilicates. Rates for other isos-tructural silicates probably do likewise. At the same time, rates decrease with increasing Si content.

2. At pH > 8

1. Dissolution increases with pH, reflecting increased deprotonation and detachment of Si surface sites.

2. At a given pH, rates can be correlated with the site potential of oxygen sites bound to Si (and possibly other measures of bond strength) — at least for most silicates.

Between pH 5 and 8 both mechanisms are presumably operative. At the same time, neutral surface species may control dissolution. Because ferric iron and aluminum are relatively insoluble at near-neutral pH, the dissolution of silicates containing these elements becomes incongruent.

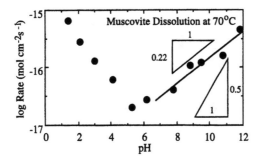

Figure 14
70°C dissolution rates of muscovite as a function of pH. (Adapted from Reference 102.)

This is due to the formation of amorphous (hydr)oxides and clays. It is therefore often difficult to determine the exact controls on silicate dissolution at neutral pH.

Some silicates apparently do not dissolve as the model outlined above would predict. 70°C dissolution rates of muscovite are shown in Figure 14. Rates increase with decreasing pH below pH 6 similar to other silicates. Like the other silicates, pH < 6 steady-state muscovite dissolution is also probably driven by the proton-promoted detachment of non-Si sites. However, muscovite dissolution rates increase only slightly with pH[102] above pH 7. The pH dependence of this, dlogRate/dpH = 0.22, is significantly less than the value of 0.5 to 0.4 which is commonly observed for the other silicates at 70°C. Recall that the pH dependence of high pH rates reflects the acidity of the Si surface sites. One explanation for the slight pH dependence of muscovite rates is that the acidity of an Si site on a muscovite edge is appreciably different from the same site on pure SiO_2. This may arise from the substitution of Al into the Si-tetrahedral sheet. At the same time Oelkers et al.[89] have argued that the lower pH–log Rate slope may be an artifact of a dependence of dissolution rates on dissolved Al. In any case, muscovite may not dissolve in the systematic fashion the other silicates do. Because very few pH > 8 rates exist, it is difficult to expand the muscovite results to speculate about site acidities and dissolution behavior of the other three-layer clays.

Diopside and prehnite are the two other notable exceptions to the rate law outlined above for silicates. Prehnite rates at 90°C are essentially independent of pH[103] above pH 8. Figure 15 presents steady-state diopside dissolution rates measured in flow-through reactors.[104] At basic pH, diopside and prehnite dissolve atypically for silicates. Rates are unaffected (or decrease), whereas other silicate rates roughly double (or more at higher T) for every unit increase in pH. The lack of pH dependence in diopside rates may indicate that site acidities do not sum for diopside and prehnite as they do for other silicates. This seems unlikely, as the site-summing approximation appears to work well for such structurally different Mg-silicates as forsterite, enstatite, and chrysotile. Adsorbed

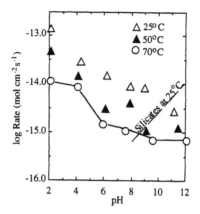

Figure 15

Diopside dissolution rates as a function of pH and temperature.[104] The dashed line gives the pH-dependence observed for other silicates at pH > 8 at 25°C.

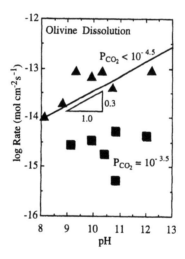

Figure 16

Mg-olivine dissolution rates[69,79] as a function of pH and CO_2 at 25°C. Rates are calculated from Mg release.

bicarbonate species can decrease dissolution rates at high pH, as they do with forsterite (see below); however, CO_2 effects were minimized in the diopside experiments. This suggests that another explanation is needed.

Figure 16 shows forsterite dissolution rates[26] measured as a function of P_{CO_2}. Note that, similar to most silicates, dissolution rates at pH > 8 in the absence of abundant CO_2 roughly double for every unit increase in pH. At high P_{CO_2}, though, rates are minimal and independent of pH. The inhibition of dissolution probably occurs because of the adsorption of

HCO_3^- or CO_3^{-2} at exposed positively charged Mg sites.[26] The process occurring here may be similar to the inhibition of hydroxyapatite (HAP) dissolution. There, the binding of ligands to positively charged Ca sites decreases the rate at which dissolution nuclei are formed (see below). Adsorption of CO_2 also substantially affects dissolution of wollastonite.[32]

IV. GLASS DISSOLUTION

A. Overview

The surface charge analysis for silicate minerals was based largely on the assumption that (hydr)oxide acidities can be summed up stoichiometrically. In essence, the surface chemistry of silicate crystals can be described by treating them as a mixture of their (hydr)oxide components. This approach may be more valid for random mixtures of (hydr)oxides like glasses, where long-range ordering is minimal. For this reason the macroscopic rate law for silicate dissolution should be applicable to the dissolution of noncrystalline glasses. Hence, a general rate law for glass should be somewhat similar to that for some crystals. Note, though, that the overall dissolution of a glass matrix may be more complex because a number of competing processes may occur in parallel (see below).

A macroscopic rate law describing glass dissolution would be useful in a large number of industrial and environmental processes. Corrosion of glass-based ceramics and glazes limits their usefulness. Dissolution of leaded glass (crystal) is an important source of lead toxicity.[105] Borosilicate glass may be one of the primary waste forms used for the isolation of nuclear waste. Some portion of the radioactive waste produced globally may be incorporated in borosilicate matrices before final storage, with the enclosed radionuclides remaining toxic for spans of 10,000 years or more.[106] Resistance of the glass matrix to corrosion, therefore, will be one of the most important controls on radionuclide release rates to the environment. From a scientific standpoint it is important that the mechanisms of the dissolution process be understood. From an engineering perspective it is critical that a rate law be sufficiently robust that it can predict glass corrosion far into the future in the face of potentially large-scale changes in temperature and groundwater composition (pH, redox potential, salinity, etc.).

Early work on glass dissolution suggested that the diffusion of non-Si components, network-modifying cations (Mg, Ca, etc.) through the silicate matrix in binary glasses determined the rate of dissolution over short time periods.[107] Under high fluid-to-rock ratios, early leaching of non-Si metals from the glass was thought to cause a reaction front, between glass and cation-poor glass, to advance away from the surface into the interior. What resulted was a nonstoichiometric, Si-rich coating at the glass surface. The rate of diffusion of modifying cations out of the glass (or hydrated surface layer) equals the detachment rate of Si from the leached layer over long-term steady state dissolution.[108] The Si accumulated at the surface limits neither movement of protons (or H_2O) to the unaltered glass nor the transport of newly detached modifying cations across the leach layer into solution. Apparently, the stepwise dissolution of the leached layer plays little role in the overall dissolution of silica glasses. Bunker et al.[109] suggested that the reactions which control steady-state dissolution occur at the leached layer–glass contact. A cross section of dissolving glass is shown schematically in Figure 17.

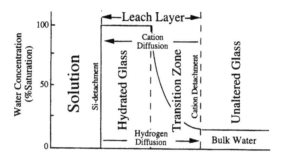

Figure 17
Proposed mechanism for steady-state glass dissolution. (From Bunker, B.; Arnold, G.W.; Beauchamp, E.K.; Day, D.E. *J. Non-Cryst. Sol.* 1983, *58*, 295–322. With permission.)

At low pH, what drives the detachment is the release of cations at the contact between the transition zone and unaltered glass. If Si detachment controlled steady-state dissolution rates of multioxide glass, then pH-dependent dissolution rates should be very similar to those of amorphous silica. The latter are pH independent at low T from pH 2 to 7 (see Figure 18[285]). Because, however, glass dissolution rates are strongly pH dependent in this pH region, detachment of the non-Si cations must control dissolution.

In other words, the analogy to crystalline multioxide silicates would suggest that the proton-promoted detachment of non-Si sites at the base of the transition zone controls the overall dissolution of multioxide glasses at low pH. The key point here is that the Si-rich layer caused by the nonstoichiometric leaching apparently does not affect the rate-determining steps for long-term, steady-state glass dissolution. Long term in this case means the length of most laboratory experiments (tens of days at most). The specific chemical factors which control glass dissolution are made clearer if the simple case of pure Si glass is discussed first. Afterward, using this simple case as a basis, multicomponent glasses containing B, Mg, Al, etc., as well as Si, can be considered.

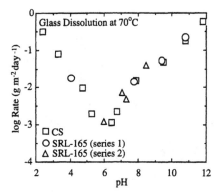

Figure 18
Long-term dissolution of CSG and SRL-165 waste glasses at 70°C.[285] Series 1 and 2 denote the use of different buffers.

B. Surface Chemistry and Rates

Although the surface of dissolving pure silica glass has no long-range surface structure, its macroscopic surface chemistry is routinely assumed to be roughly equivalent to that of quartz and silica gel.[8,42,45,56] Surface charge measured on extensively washed quartz[111] shows similar pH and ionic strength dependencies as surface charge measured on amorphous silica.[112] The assumption that silica polymorphs have similar acidity may be appropriate, as each surface will consist of similarly hydrated Si groups. Nevertheless, a noncrystalline solid has appreciably less durability than its crystalline equivalent. The identical charging behavior only means that, for a given surface area, charging, averaged over all the kinks, ledges, and planes that make up the surface of a quartz crystal, roughly matches up with the acidity of Si tetrahedra summed over the whole glass surface.

Figure 19 shows pH dependent glass and quartz dissolution rates far from equilibrium at 25°, 60°, and 65°C. Both quartz and glass rates are minimal near the pH_{zpc} ($pH_{zpc} \sim 2$ for both quartz and SiO_2 glass), increasing only at pH > 7. Above pH 8 at 25°C, dlogRate/dpH ~ 0.3 for quartz. At 60°C and 65°C, dlogRate/dpH ~ 0.4 for both quartz and Si glass. The steepening of the rate isotherm between 25° and 60°C reflects a slight steepening of the deprotonation isotherm with temperature.[67] (The small difference between the 60° and 65°C data probably causes very little additional change in the shape of the deprotonation isotherm.)

As argued earlier, the increase in quartz rates above pH 7 appears to reflect the pH-dependent increase in [>Si–O⁻], the rate-determining surface species. Glass dissolution rates likely increase above pH 8 for the same reason.[114] Wirth and Gieskes[8] concluded that glass dissolution rates

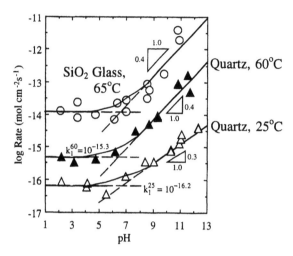

Figure 19

Dissolution rates of quartz[41] at 25° and 60°C and SiO$_2$ glass[113] at 65°C. The quartz rates are rates from 0.1 M NaCl solutions. At pH > 8 the solid lines describe a first-order dependence of rates on [>Si–O$^-$]. The dashed lines fitting the glass data are used to show that the pH dependency of rates of glass dissolution is equal to that of quartz.

increase with pH in response to pH-dependent increases in the number of deprotonated Si sites. Fleming[110] subsequently showed that rates were first order with respect to [>Si–O$^-$]. From pH 2 to 7 rates are roughly independent of pH, presumably reflecting a rate dependency on uncharged, hydrated surface sites. These observations can be summarized in a rate equation for quartz and SiO$_2$ glass dissolution, which can be written as:

$$\log \text{Rate} = \log k_1 + \log k_2[\text{>Si–O}^-] \qquad (6)$$

where k_1 is the rate constant (mol cm^{-2}s^{-1}) for dissolution which dominates from pH 2 to 7 and k_2 is the rate constant (s^{-1}) for surface charge–promoted dissolution at pH > 7. Remember that the [>Si–O$^-$] term in the rate law models the pH dependence of rates.

C. Multicomponent Glass Dissolution

The dissolution rate behavior of basaltic glass[115] is shown in Figure 20, along with the SiO$_2$ glass[113] and quartz[41] rates of the previous figure. At pH > 3, increases in pH affect basalt glass rates identically to those of quartz and silica glass. Rates are independent of pH until approximately pH 7, but increase by approximately 0.4 log units per unit increase in pH at pH > 7 at 65°C. Basalt glass rates increase below pH 3 with decreasing pH, whereas quartz and silica glass rates are unaffected by pH. It is

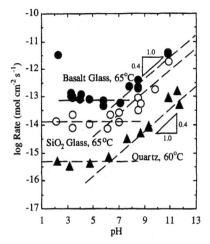

Figure 20

pH-dependent quartz[41] rates at 60°C and SiO₂[113] and basalt glass[115] dissolution at 65°C. At pH > 8 the dashed lines describe a first-order dependence of rates on [>Si–O⁻].

reasonable to ascribe the increase in basalt glass rates at acid pH to protonation/exchange and detachment of non-Si oxides such as MgO. This is analogous to the observed behavior of complex silicates; rates increase in acid pH with decreasing pH because of increased proton loading of non-Si surface sites.

The most important point to be gained from Figures 19 and 20 is that both silica and basalt glasses dissolve by a mechanism which is very similar (if not identical) to that of quartz at near-neutral to basic pH. It is therefore reasonable to conclude that >Si–O⁻-promoted detachment of Si sites controls the corrosion of mixed (hydr)oxide glasses at all but acid pHs. To extend this analysis and formulate a useful rate law for glass dissolution requires that rates be normalized to a common pH, that is, to a common value of (>Si–O⁻). Multicomponent glass dissolution rates can then be compared and potentially calibrated as a function of surface durability. Recall that this is the same procedure followed with crystalline silicates at low and high pH.

D. Rate–Bond Energy Correlations

There are no studies of glass dissolution which can be used by themselves to reliably calibrate a glass dissolution rate law. This is because a wide variety of glass compositions have been dissolved at a number of temperatures under widely varying experimental conditions. A number of workers have used Materials Characterization Center MCC-1 glass durability tests to test rate laws for glass dissolution.[116,117] In the MCC-1 tests, 4 cm² of geometrically measured mineral surface area is exposed to

40 cm^3 distilled H_2O at 90°C for 28 days. Silica release rates, $[Rate]^{Si}_{28-day}$ (g m^{-2}), are determined by measuring the amount of Si leached from a given glass and using the following formula,

$$[Rate]^{Si}_{28-day} = \frac{M_{Si}}{X_{Si}SA} \qquad (7)$$

M_{Si} is the mass of Si released into solution over 28 days; X_{Si} is the fraction of Si in the glass; and SA is the specimen surface area (m^2). The solution pH, which is measured at the end of the experiment, generally rises because of cation release and exchange during the experiment. Therefore, pH is not fixed, nor is the mineral surface area measured at the end of the experiment. Also, because the measured rate comes from one Si analysis at the end of the experiment, the early dissolution of fine-grained material at the mineral surface is included in the rate. Moreover, the faster dissolving glasses may experience appreciable back reaction to form new phases as solutions approach saturation. Keeping these uncertainties in mind, a portion of the MCC-1 compilation will be used to examine the effects that substituting various cations, such as Mg, Al, B, etc., into the glass matrix has on glass durability. Specifically, it will be assumed (as done by previous workers) that (1) the early dissolution of high-energy surface sites is minimal relative to long-term dissolution and (2) the pH measured at the end of the run is not far from the average pH the mineral surface reacted to during the 28-day experiment.

Figure 21 shows 137 pH > 7 MCC-1 results. This group includes Si-rich glasses, Pb- and alkali-rich glasses, medieval glasses, fritted glasses, as well as a variety of waste glasses.[118] At a given pH, rates vary by roughly two orders of magnitude. For comparison, pH-normalized dissolution rates of crystalline silicates vary by nearly six orders of magnitude (see Figure 9.) Some of the scatter in glass dissolution rates is almost certainly due to the experimental uncertainties noted previously. Most of the scatter arises, however, because of the wide range in glass compositions used in the tests. Despite the wide scatter in the data, one feature does stand out: Si release rates can be interpreted to increase by about 0.4 log units per unit increase in pH. This is very similar to the pH dependence seen for the dissolution of pure silica glass and quartz in Figure 19. It is therefore reasonable to assume that the same dissolution mechanism which gives rise to the pH dependence of quartz and simple Si glass rates does likewise for all multicomponent Si-bearing glasses: >Si–O$^-$-promoted detachment of Si sites apparently controls the rate of glass corrosion. Substitution of network-modifying cations into a silica glass apparently does not affect this, although their presence is certain to affect the stability of the Si–O bonds and, therefore, the absolute rates at any pH.[119,120]

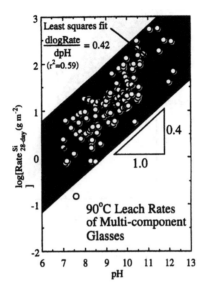

Figure 21
28-day silica release rates from multicomponent glasses[118] as a function of the final measured pH.

At this point it is useful to separate out the effect of surface charging (pH) to examine the role of glass composition in dissolution. This is done below by regressing rates according to the following Equation:

$$\log[\text{Rate}]^{\text{Si}}_{\text{28-day}} = \log\{[\text{Rate}]^{\text{Si}}_{\text{28-day}}\}_{\text{Normalized}} + 0.4\,\text{pH} \qquad (8)$$

The rates as a group seem to follow this relation. {[Rate]$^{\text{Si}}_{\text{28-day}}$}, the first term on the right-hand side, is calculated for each data point at pH 7 and shown in Figure 22. Rates are plotted against $G_{\text{hydration}}$, the calculated free energy of hydration (kcal mol^{-1}) for each glass. The latter is the free energy required to convert each of the glass components to its hydrated form.[121,122] These free energies also include two additional terms to model solubility changes in the Si and B components at high pH.[122] Free energies of hydration are useful only to the extent that they are linked to bond strength and, hence, can be used to correlate rates far from equilibrium. Their thermodynamic applicability near equilibrium is minimal as glasses are metastable and do not grow from low-temperature solutions. Nevertheless, $G_{\text{hydration}}$ correlates well with the fraction of SiO$_2$ in the glass, as well as with the number of nonbridging oxygen bonds in a given glass. Surface charge–normalized rates, therefore, correlate with each quantity.

The important point to be gained from this is that above neutral pH glass durability is a function of both surface chemistry and glass composition. Rates increase with pH in the basic region in response to

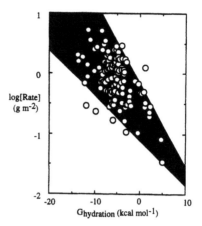

Figure 22
28-day release rates of multicomponent glasses[118] normalized with respect to pH (see above) as a function of the calculated free energy of hydration.

>Si–O⁻-promoted detachment of Si sites. At a given pH, glasses which are Si-rich and highly polymerized dissolve more slowly than glasses with large amounts of cation substitution.

To actually predict the long-term (time $> 10^3$ years) behavior of glasses, the back reactions between glasses and solutions must be understood. Under conditions where contact with groundwaters is intermittent, molecular diffusion of water into the glass may control dissolution and growth of secondary phases.[123] At high fluid-to-rock ratios it is important that the effect of saturation (affinity) on rates be understood. Secondary (hydr)oxide and clay phases eventually form from glass components leached into groundwaters. Newly formed minerals will have the effect of fixing the activities of their components in the aqueous phase (and may also sorb radionuclides). By controlling solution compositions and degree of saturation, the newly formed minerals may play an important role in the overall lifetime of glasses.

V. LIGAND-ENHANCED DISSOLUTION OF (HYDR)OXIDES AND SILICATES

Because coordinatively bound protons and negatively charged, deprotonated sites are seen to modify the reactivity of mineral surfaces in a reasonably predictable fashion, it is expected that other adsorbed ligands affect dissolution rates as well. Obvious ligands to consider are the organic acids found in soil weathering profiles,.as these are often more abundant than are either protons or hydroxyls. (Organic acid concentrations can approach 10^{-3} mol L^{-1} in soils.) Proton donors such as oxalic and acetic acid are critical to soil nutrient cycling. They accelerate the break-down of minerals and elevate the solubility of mineral components, making them more bioavailable.[124]

Stumm and co-workers[11] showed that organic ligand–promoted dissolution can be described by a macroscopic rate law of the form

$$\text{Rate} = k_L[\text{>Org}] \tag{9}$$

k_L is the ligand–promoted dissolution rate constant (s^{-1}) and [Org] is the number of ligands adsorbed at the mineral surface (mol cm^{-2}). In Figure 23 are shown aluminum oxide dissolution rates measured in a variety of organic acids.[11] Along with each rate isotherm is a proposed configuration for the respective adsorbed ligand. Note first of all that dissolution depends strongly on the identity of the adsorbing ligand. For dicarboxylic acids $k_{oxalate} > k_{malonate} > k_{succinate}$. The five-membered ring of oxalate is therefore more corrosive than the six and seven member rings of malonate and succinate, respectively. The Al_2O_3 dissolution–promoting ability of adsorbed ligands appears to be correlated with the affinity of their aqueous counterparts for dissolved Al.[11]

Another point to be gained from Figure 23 is that ligands adsorbed through two functional groups to a single site (bidentate adsorption) are much more corrosive than an organic ligand adsorbed through a single functional group (unidentate adsorption). One implication is that at low pH, where aqueous and adsorbed organic acids become protonated, ligand-promoted rates should become less important. This protonation and opening of surface complexes can be schematically shown as:[125,126]

$$\tag{10}$$

In natural waters containing organic acids, mineral surfaces will possess adsorbed protons as well as organic ligands and the actual rates of dissolution will reflect the contribution of each. Adsorbed bicarbonate may also play a role in the weathering of minerals (e.g., hematite) at alkaline pH.[127] Inorganic ions in natural waters are also expected to influence the electrical properties of the solid interface through complexation of cations with surface groups.[128] Although the effects of commonly occurring cations, such as Ca^{2+} and Na^+, on dissolution rates have been studied,[42,54] there has been little attention to inorganic anions, such as hydrogen phosphate and bicarbonate. Indeed, it is usually necessary to separate the effects of the metal cation in studies of the effects of organic acids added as their salts.

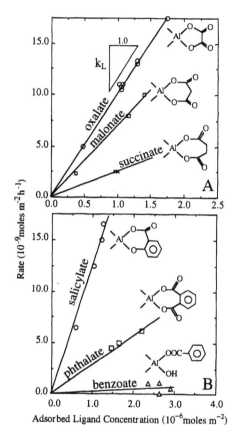

Figure 23

Dissolution rates of aluminum oxide[11] as a function of adsorbed organic ligand concentrations for (A) carboxylates and (B) phenolates. (From Furrer, G.; Stumm, W. *Geochim. Cosmochim. Acta.* 1986, 50, 1847–1860. With permission.)

Before discussing organic-mediated dissolution of multioxides, it is useful to mention the ability of organic ligands to reductively dissolve minerals which are more soluble in their reduced state. A number of workers have shown that adsorbed organic ligands can transfer electrons to Fe(III) and Mn(III) and Mn(IV) (hydr)oxides. This oxidizes the ligand and reduces the charge on the metal center. In reducing solutions, Mn(II) and Fe(II) are far more soluble than their oxidized counterparts. Hence, those sites receiving electrons from adsorbed organic ligands tend to detach more readily. Reductive (and oxidative) dissolution of minerals has been extensively reviewed elsewhere.[129-131]

A number of organic ligands accelerate the dissolution of quartz, particularly multifunctional ligands with closely spaced oxygen functional groups.[25] The accelerating effect proceeds in the order:

$$R_{Citrate} > R_{Oxalate} >> R_{Salicylate} > R_{Phthalate} > R_{Acetate} \qquad (11)$$

A similar result is seen for plagioclase dissolved in 0.1 M acid solutions.[132] Specifically, rates go from high to low in the order: citric > salicylic > aspartic > acetic. Multifunctional acids such as citrate and oxalate may accelerate dissolution by decreasing the activation energy of dissolution.[54] The organic acid–catalyzed dissolution of multioxide silicates is similar to the dissolution of the component oxides. Below neutral pH organic-mediated detachment of the non-Si sites is probably more important than detachment of Si sites. At pH 4.5 organic acids accelerated dissolution of forsterite in the order EDTA ~ citrate > oxalate > tannic acid >> succinate > phthalate > acetate.[133] Grandstaff[134] suggested that Mg–organic surface complexes controlled dissolution. More-recent olivine dissolution experiments with phthalate and ascorbate agree with this and show that at very low pH the organic effect is minimal both because the olivine surface is saturated with protons and because the organic ligands are themselves appreciably protonated.[26]

Generally, those ligands which complex strongly with a given metal in solution also bind strongly and accelerate the corrosion of the same metal (hydr)oxide.[90] There is probably a crystallochemical component as well which must be examined in further detail since a number of workers have found organic ligands to have less effect on dissolution rates of minerals than one might predict. 0.1 M citrate has no effect on bronzite dissolution at pH 3.[134] Likewise, 0.2 M phthalate has no effect on dissolution rates of a number of magnesium silicates,[18] nor do millimolar quantities of oxalate have an effect on dissolution rates of oligoclase and tremolite.[135] The correlation between metal coordination in solution and metal detachment from mineral surfaces is a rule of thumb which unfortunately does not always work. Taking account of the observation that dissolution rates of numerous aluminosilicate minerals decrease with increasing aqueous aluminum concentration, Oelkers et al.[89] proposed that the effect of organic acids on these rates is due to the formation of aluminum–organic acids complexes[136] which thus lower aqueous aluminum concentrations.

One of the solute species which apparently does not directly affect dissolution rates of silicates is dissolved CO_2 (H_2CO_3).[137] Earlier work[138,139] suggested that dissolution rates of silicates were amplified at soil pHs by dissolved carbon dioxide. This hypothesis has important implications for weathering because, due to microbial activity, soil carbon dioxide levels are 10 to 100 times that of the atmosphere. It would also suggest an additional link between the biosphere and weathering (see Chapter 5). In Figure 24 are shown pH 4 dissolution rates of augite and bytownite (An_{60}) as a function of temperature in CO_2-saturated solutions ($P_{CO_2} = 1$ atm) and in solutions in equilibrium with $P_{CO_2} = 10^{-3.5}$ atm. Note that augite and anorthite exposed to high-CO_2 solutions dissolve at essentially the same rate as in solutions relatively depleted in CO_2.

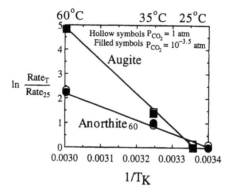

Figure 24

Augite (squares) and bytownite (circles) dissolution rates,[137] r, (mol Si cm^{-2} s^{-1}) as a function of reciprocal temperature (K) and P_{CO_2}. (From Brady, P.V.; Carroll, S.A. *Geochim. Cosmochim. Acta.* 1994, *58*, 1853–1856. With permission.)

VI. DISSOLUTION OF CARBONATE MINERALS

Calcite and aragonite ($CaCO_3$), dolomite ($CaMg(CO_3)_2$), and magnesite ($MgCO_3$) are four of the more prevalent carbonate minerals found in Earth's crust. Reactions at the carbonate mineral–solution interface control such macroscopic processes as:[140] the differential solubilities of calcite and aragonite and the saturation state of seawater with respect to each; inhibition of calcite and aragonite dissolution and growth; crystal morphology; and overall pathways of carbonate diagenesis. The dissolution of calcite by acidic waters is also linked with neutralization strategies for lakes.[141,142] When the pH of the water is less than approximately 5.6, the controlling reaction is that of H^+ with the mineral rather than H_2O. Various models of the dissolution process have been proposed and used in designing liming strategies. By dissolution, growth, and sorption, carbonate mineral surfaces may place important controls on the chemistry of soils, rivers, lakes, groundwaters, the ocean, and marine sediments.

It is not possible in this section to give an exhaustive coverage of the work done on carbonate surface reactivity, or even that done in just the past 20 years. Such an approach would lead into the extensive literature of sedimentary petrology and chemical oceanography. The approach here will instead be to look first at dissolution rate laws for carbonate minerals and then to briefly consider the chemistry of the respective surfaces, focusing on the link between the two. The mechanistic steps controlling calcite growth and inhibition will then be considered in a later section.

A. Calcite Dissolution

The majority of studies of calcite dissolution over the last 100 years have involved the use of stirred suspensions of powdered calcite (see, for example, the review of Compton and Unwin[143]) with fewer studies of single crystals. Most of the research with single crystals has utilized a rotating disk technique[61,62,144,145] to permit more control over the hydrodynamics at the crystal interface than is possible in stirred suspensions. This approach also permits the reproducible treatment of the crystal surface prior to dissolution. The disadvantages of the rotating disk technique are that, although the transport of solutes in the vicinity of the crystal surface is known, the transport at the detector in the bulk solution is not defined and the reaction products accumulate in the solution in the same way as in stirred suspensions. This usually limits the reaction to short times so that significant changes in the bulk composition are avoided. In response to the limitations of the rotating disk technique and to maintain the use of single crystals in well-defined hydrodynamic conditions, the channel electrode method was developed and applied to examining the dissolution kinetics of Iceland spar and blocks of natural limestones.[6] With this technique, the single crystal is mounted in a channel with the solution flowing over the exposed surface in laminar flow conditions. A short platinum detector electrode is located immediately downstream of the crystal, and this is held at a potential such that all the H^+ reaching it is reduced. The electrical current at this electrode then provides a measure of how much H^+ has reacted with the calcite surface. The well-defined flow at the calcite interface permits the modeling of convective/diffusion fluxes in the cell.

In spite of the limitations of the stirred cell method, the experimental results have provided important information about the reaction mechanism. In particular, the mechanistic model of Plummer et al.[146] has been successfully used to develop rate equations describing both the dissolution and precipitation of calcite in a range of solution conditions. The model was originally developed from data on the dissolution of semi-optical-grade crystals of Iceland spar as a function of pH, P_{CO_2}, and temperature. Plummer et al.[146] noted three regions in the rate of dissolution against pH diagram.

1. At low pH, the rate is dependent on pH and the stirring rate, but shows little dependence on the concentration of CO_2.[147-150]

2. At intermediate pH (3.5 to 5.5 at 1 atm CO_2), the rate depends on the concentration of CO_2 and pH.

3. At higher pH and CO_2 below 20 mm Hg, the rate is essentially independent of both pH and the concentration of CO_2.

The boundaries between these regions are a function of pH and the concentration of CO_2 with the division occurring at increasing pH as the concentration of CO_2 decreases. The Plummer et al. model may then be summarized as:

$$R = k_1[H^+] + k_2[H_2CO_3^*] + k_3[H_2O] - k_4[Ca^{2+}][HCO_3^-] \qquad (12)$$

where the square brackets indicate ion activities, R is the dissolution rate, and $H_2CO_3^*$ represents the sum of the concentrations of H_2CO_3 and CO_2 in solution. The last term on the right accounts for the backward reaction, i.e., calcite growth. Other workers[57,151] have argued for the growth term to be proportional, instead, to $[Ca^{++}][CO_3^{2-}]$.

In the Plummer et al. model, the elementary reactions for growth and dissolution are thought to occur in an immobile surface layer on a relatively small number of surface sites. The model assumes that reaction of surface $CaCO_3$ with H^+ is faster than reaction with $H_2CO_3^*$ and H_2O and that the activities of water and carbonic acid in the surface layer are equivalent to those in the bulk solution (in other words, the bulk diffusion of carbonic acid and water is assumed to be much faster than reaction in the surface layer). The pH in the surface layer will be determined by the bulk concentration of $H_2CO_3^*$ and the condition that the surface layer is in equilibrium with calcite. By detailed balancing of Ca^{2+}, HCO_3^-, CO_3^{2-}, H^+, OH^- and H_2O movement from the bulk to the surface layer, Plummer et al.[152] derived the following expression for the reverse rate constant:

$$k_4 = \frac{K_2}{K_C}\left\{k_1' + \frac{1}{a_{H^+}}\left[k_2 a_{H_2CO_3^*} + k_3 a_{H_2O}\right]\right\} \qquad (13)$$

where K_2 and K_c are the equilibrium constants for, respectively, bicarbonate dissociation and $CaCO_{3(calcite)} \leftrightarrow Ca^{2+} + CO_3^{2-}$. This rate law has subsequently been found to fit observed rates from a variety of workers.[60,153-156]

Results from experiments with Iceland spar in a rotating disk have confirmed the general form of Equation 12 with $k_3 = 1.4 \times 10^{-9}$ mol cm^{-2} s^{-1} (see Table 2) and k_1 and k_2 given by the rate of transport of H^+ and H_2CO_3, respectively, to the crystal surface.[145] This value of k_3 is slightly higher than that measured by Plummer et al.[146] of 1.2×10^{-10} mol cm^{-2} s^{-1}, probably as a result of the uncertainties in the value of the surface areas of the powder and exposed crystal. As shown in Table 2, these values are in excellent agreement with dissolution results[61] and with the crystallization measurements using calcite powder as a seed material[60] in suspension.

More recently, with the development of the channel electrode technique, it has been shown that under conditions of high mass transport, the reaction of H^+ with Iceland spar is controlled by a surface process

TABLE 2.

Comparison of the Rate Constant, k_3 (Equation 12), for Dissolution in Neutral to High pH and Zero Concentration of CO_2 Solutions

Method	Sample	$k_3/10^{-10}$ mol cm^{-2} s^{-1}	Ref.
Dissolution suspension	Crushed Iceland spar	1.2	152
Precipitation suspension	BDH calcite	1.35	60
Rotating disk	Polished Iceland spar	1.4	61
Rotating disk	Iceland spar	1.4	144
Fluidized bed	Powder	0.65	151
Channel flow cell	Iceland spar	0.95	6
Rotating disk	Shtall calcite	3.1	62

rather than by the rate of convection/diffusion in solution.[23,143] This leads to a rate equation: R (rate in mol cm^{-2} s^{-1}) = $k_1[H^+]_o$, where $[H^+]_o$ is the concentration of protons at the calcite–water interface. This so-called surface rate law contrasts with the first term in Equation 12, where a_{H^+} relates to the bulk activity and k_1 to the rate of convective diffusion. The first-order heterogeneous rate constant was found to be 0.043 ± 0.015 cm s^{-1} at 25°C with the assumptions that the thermodynamic dissociation constant of H_2CO_3 is 1.74×10^{-4} M and the dehydration rate constant of H_2CO_3 is 20 s^{-1}. This mechanism has also been tested at 25°C using a variety of calcium carbonate–containing rocks.[143] The channel electrode was used with dilute HCl, pH 3, with Carrara marble, Portland stone, Abbeytown limestone, freshwater limestone, chalk, and Connemara marble. Connemara marble contained a high proportion of acid insoluble material and was too heterogeneous for meaningful results to be obtained, whereas both freshwater limestone and chalk produced very rapid dissolution (k_1 > 0.5 cm s^{-1}) and gave good agreement with a transport-controlled reaction. The other rocks gave dissolution rates in agreement with the heterogeneous rate law with $k_1 = 0.099$ (Carrara marble), $k_1 = 0.039$ (Portland stone), and 4.5×10^{-3} (Abbeytown limestone), all in units of centimeters per second. Abbeytown limestone (Abbeytown, Eire) is a dolomitic limestone which probably accounts for the lower rate constant for this sample.

One consequence of these results is that in lake-liming conditions, the rate of calcite dissolution will be appreciable slower than expected from classic calculations for mass transfer to spheres falling at their terminal velocities in water. One estimate is that the mass transport to particles <100 μm in size is appreciably greater than the flux caused by the surface heterogeneous reaction.[157] In effect, the dissolution rates may be overestimated, depending on the nature of the source rock, if a transport-controlled mechanism is assumed. The discrepancy between field observations[158] of the dissolution of single crystals of Iceland spar in an acid stream draining a mining area and predictions from Equation 12 may also be partly the result of an overestimation of the dissolution rate using a transport control rather than the surface control mechanism proposed by Compton et al.[157] However, it is important to remember that dissolution inhibitors and promoters are likely to occur in natural waters and any comparison between field and laboratory rates must seek to identify the important surface active solutes (see below for further discussion).

Compton and Pritchard[6] have also applied the channel flow cell with pH and calcium potentiometric microelectrodes to investigate the dissolution of Iceland spar at pH > 7 and in the absence of CO_2. They proposed a variation of the Plummer et al.[146] rate equation of the form (high pH, $[H_2CO_3] = 0$):

$$R \text{ (rate in mol cm}^{-2} \text{ s}^{-1}) = k_3 - (k_3/K_c)[Ca^{2+}]_s[CO_3^{2-}]_s \qquad (14)$$

where the subscripts s refer to surface concentrations and K_c is the solubility product of calcite. The rate constant, k_3, was calculated as 0.95×10^{-10} mol cm^{-2} s^{-1}, which is slightly lower than the previous results from the rotating disk experiments noted in Table 2 and the results of Sjöberg and Rickard[61] of 1.4×10^{-10} mol cm^{-2} s^{-1}. This strongly suggests that the morphology of the crystals and method of surface pretreatment largely determine the rate constant and that large differences between samples of single crystals can be expected.

Various aspects of the effects of surface morphology on dissolution rates of single crystals have been investigated.[62,144] Freshly cleaved Iceland spar crystals have been found to be unreactive until etch pits have developed on the surface. Preparation procedures which produce rougher crystal surfaces will lead to higher rate constants, although the general agreement with the mechanistic model with changing solution conditions is still expected. Similar observations have been made for the precipitation reaction (rate constant, k_4) using different crystal seed preparations.

There is no doubt that the differences in the absolute rates, measured in the same conditions but on different mineral preparations, raise some concerns about the general applicability of the models to natural situations. These differences are thought to arise from variations in the densities of point defects, dislocations, microfractures, kinks, grain and twin boundaries, corners, edges, and ledges.[159] Although the rate constants in the mechanistic model are normalized with respect to the surface area of the mineral, this may not be linearly related to the active site densities for different samples. In general, it has been discovered that dissolution rates are relatively insensitive to the density of defects, e.g., rate enhancements of two to three times are expected for dislocation densities varying from 10^6 to 10^9 cm^{-2}, increasing at higher pH and lower temperature. There is a relatively low enhancement factor of 2.3 for dissolution of calcite[62] at 25°C and pH 8.6 for an increase in dislocation densities of 10^3 to 6×10^8 cm^{-2}. This insensitivity is likely to be caused by a relatively small flux from point defects and dislocations compared with that from receding edges and ledges on the crystal surface.[62] This implies that kinetic data from macroscopic crystals with low strain faces are more likely to be generally applicable. The high dissolution rates from mechanically polished surfaces of single crystals are attributed to enhanced dissolution of cracks and dislocation loops produced in the process of grinding. It is generally recommended that mechanically polished surfaces be treated with dilute acid, e.g., 1 mM HCl for 10 to 15 s, to clean the surface of contaminants and fully roughen the surface in a reproducible way.[6]

B. Dolomite Dissolution

Dolomite dissolution was studied extensively by Busenberg and Plummer[160] and by Chou et al.[151] In Figure 25 are shown dissolution rates of calcite, dolomite, magnesite, aragonite, and witherite.[151] Busenberg and Plummer suggested that defect densities control absolute dissolution rates and that the back reaction at high pH depends on the $FeCO_3$ content of the growing crystal. Initial rates are nonstoichiometric, $CaCO_3$ being released at a greater rate than $MgCO_3$. This causes an Mg-rich surface layer to form. Over longer periods dissolution is stoichiometric. Absolute dissolution rates depend primarily on the degree of crystal ordering, and compositional variations give a second-order effect.[160]

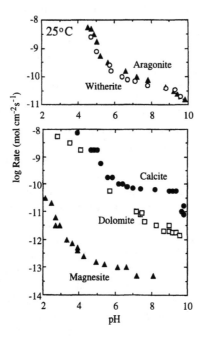

Figure 25

Far from equilibrium dissolution rates of witherite, aragonite, calcite, dolomite and magnesite as a function of pH at 25°C. (Adapted from Reference 151.)

The pH dependency of dolomite dissolution at low $[H_2CO_3]$ may be less than that for dissolution of the other carbonates. Busenberg and Plummer argued that at $P_{CO_2} = 0$, dlogRate/dpH ~ 0.5. Chou et al. measured a higher value, dlogRate/dpH ~ 0.75 (see Reference 151 for a discussion of this). Note that, with the possible exception of the dolomite, the dissolution behaviors of the various carbonate minerals are fairly similar in their pH dependence. Nevertheless, at a given pH, absolute rates vary significantly, calcite dissolving much faster than dolomite, which dissolves more rapidly than magnesite.

C. Rates, Surface Chemistry, and Bond Strength

The pH and P_{CO_2} dependencies of rates presumably reflect the changing concentrations of rate-controlling surface species. The surface chemistry of carbonate minerals has been examined by a number of authors.[13,140,161-165] Results from streaming potential measurements[13] on calcite with no gas phase have shown that the major surface ions are Ca^{2+} and CO_3^{2-}, with no evidence of either $CaOH^+$ or $CaHCO_3^+$ ions determining the surface charge. The pH dependence of the electrophoretic mobility arises from the effects of pH on the concentration of the lattice ions in solution. This conclusion is also supported by the results of Foxall et al.[166] Van Cappellen et al.[167] have recently attempted to explain rate dependencies using a theoretical model of metal carbonate surfaces which assumes that bicarbonate and metal cations determine surface charge. Carroll and Papenguth[165] explain surface-controlled calcite dissolution and growth near equilibrium by a simple exchange reaction between Ca^{2+} and HCO_3^- at the calcite surface. Because the origin of metal carbonate surface charge remains a matter of debate, it is hard to tie rate dependencies to specific surface equilibria. It is fairly clear, though, that the variables which control metal carbonate dissolution (pH, $[M^{2+}]$, and P_{CO_2}) also appear to affect metal carbonate surface charge. The microscopic origins of the specific dependencies remain to be worked out.

Absolute rates (i.e., the rate constants) vary by up to four orders of magnitude. This difference presumably reflects differential stabilities of the respective mineral surface sites. In Figure 26 are plotted first-order rate constants as a function of mean oxygen site potential. Rates tend to plot very roughly in the order one might expect from the earlier site potential results. Recall from above that silicates having a high mean O site potential tend to be more stable and resistant to hydrolysis. The witherite results are the only values which tend to fall off a simple linear fit.

Any comparison between laboratory experiments and measurements in field conditions or application of dissolution (or precipitation) models must consider the importance of dissolved inhibitors or promoters. These are substances that generally adsorb to the calcite surface, and in doing so, either "block" active dissolution sites or aid dissolution by forming

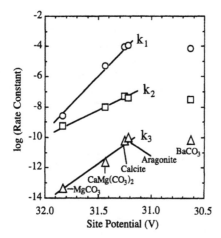

Figure 26
Rate constants for carbonate dissolution[151] as a function of mean site potential of oxygen for the respective minerals.[101]

surface complexes. Although the kinetics of the adsorption have not been extensively studied, the results for phosphate adsorption[168] and Cu adsorption[169] show that the kinetics are likely to be fast, i.e., a few seconds rather than minutes and in suspensions probably transport controlled. If isomorphous substitution of the adsorbate in the lattice is a possibility, then a slower kinetic phase is usually evident.

The phenomena and mechanisms of reaction inhibition have been extensively reviewed, most recently by Morse[150] and Zhang and Nancollas.[170] The literature dealing specifically with inhibitors of calcite dissolution and precipitation has been summarized by Compton and Pritchard.[6] Rate-affecting compounds include a number of organic molecules, phosphates, polyphosphates, glycerophosphates, phosphonates, sulfates, and metals such as Cu^{2+}, Pb^{2+}, La^{3+}, Y^{3+}, Sc^{3+}, Cd^{2+}, Au^{2+}, Zn^{2+}, Mn^{4+}, Ni^{2+}, Ba^{2+}, Mg^{2+}, Co^{2+}, Mg^{2+}, Am^{3+}, UO_2^{2+}. For some of these ions, e.g., Mg^{2+}, Zn^{2+}, and Sc^{3+}, inhibition is caused by ion exchange with lattice calcium ions or by direct competitive adsorption with surface calcium ions at growth sites. In the latter, the extent of the inhibition may be described by a Langmuirian adsorption model involving only the inhibitor concentration or by a two-component Langmuir model which accounts for competition between the inhibitor ion and calcium or carbonate. For example, the inhibition of calcite dissolution by Cu^{2+}, mesotartrate, and maleic acid[6,171,172] has been modeled successfully using the Langmuir site-blocking mechanism. Inhibition by organic adsorbates depends on the molecular interaction with the surface and lattice cation spacing expected for a particular crystal plane. Although maleic acid is an effective inhibitor of dissolution, fumaric acid is less so. The Ca^{2+}–Ca^{2+} separation on a calcite surface correlates closely with the carboxylate group separation in the maleate dianion, leading to a more pronounced reduction in the rate of dissolution (fumaric acid is the trans isomer and maleic acid is the cis isomer; both carboxylates are on the same side of the C=C bond). Two of the most notable inhibitors of calcite dissolution and growth are magnesium[155,173-175] and sulfate.[175] Other inhibitors, important in freshwater and marine environments, are phosphate,[176-178] Fe(II), Fe(III), and, in some situations, organic compounds.[179,180]

VII. DISSOLUTION OF METAL SULFIDES

Dissolution of pyrite (FeS_2) and other sulfides is critical to the supergene enrichment of ore deposits, acid mine drainage, and the biogeochemical cycling of Fe, S, acidity, free oxygen, and electrons. Seams of pyrite and marcasite commonly exist as crystalline agglomerates in coal mines. Both corrode rapidly when exposed to atmospheric oxygen, releasing acidity to the environment in the process. As a result, remediation of acid mine wastes relies on an understanding of the specific dissolution mechanisms involved.[181] At the same time, pyrite dissolution kinetics are important in chemical coal desulfurization and *in situ* leaching of copper and uranium ore.[182] In oxidizing, near-surface environments FeS_2 is highly soluble and is oxidatively corroded by electron acceptors such as Fe^{+++} and O_2,

$$FeS_2 + \frac{7}{2}O_2 + H_2O \rightarrow Fe^{2+} + 2SO_4^{-2} + 2H^+ \tag{15}$$

$$FeS_2 + 14Fe^{3+} + 8H_2O \rightarrow 15Fe^{2+} + 2SO_4^{-2} + 16H^+ \tag{16}$$

Note that, in addition to the obvious production of acidity, the reactions above involve the transfer of 14 to 16 electrons from sulfide to sulfate. This potential for electron transfer causes sulfide oxidation in nature to be largely catalyzed by the activity of microorganisms. *Thiobacillus ferrooxidans* accelerates the weathering of pyrite by catalyzing the oxidation of Fe^{++} to Fe^{+++}, using free oxygen as an electron acceptor.

The dynamic interaction between free oxygen, iron, and dissolving pyrite is schematically represented as:[76]

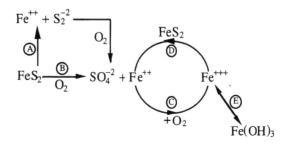

By reaction A, pyrite is dissolved and the sulfide subsequently oxidized to sulfate. Alternatively, by reaction B, the pyrite surface is oxidized directly by free oxygen and sulfate leached to solution. At step C, ferrous iron is oxidized and possibly precipitated as relatively insoluble iron hydroxide (step E). Alternatively, Fe^{+++} oxidatively dissolves pyrite in step D.[183,184]

McKibben and Barnes[185] envisioned pyrite oxidation to occur at acid pH through one of two mechanisms. Oxidation by dissolved oxygen involved adsorption of the latter onto the surface and subsequent oxidation by water to H_2O_2. Pyrite oxidation was then thought to be driven by adsorbed hydrogen peroxide. The second reaction pathway involves adsorption of Fe(III), electron transfer, dissociation of the surface complex, and desorption into solution. The last step was considered to be the rate-limiting one. Wiersma and Rimstidt[186] measured pyrite dissolution rates at pH 2 and emphasized the importance of Fe^{+++} control of oxidation. Moses et al.[187] have focused on the likely surface steps involved in the autocatalytic oxidation of pyrite by ferric iron (step D above). The initial oxidation step is the movement of a hydroxyl radical from a hydrated ferric iron to a sulfur site exposed at the mineral surface. By a subsequent series of hydroxyl transfers and dehydration steps a thiosulfate ion is produced. This reaction hypothesis predicts that rates should depend on the availability of Fe^{+++} and its affinity for sulfide surfaces. Luther[188] has applied molecular orbital theory to arrive at a very similar mechanism for pyrite dissolution. More-recent electrochemical studies of pyrite oxidation[189] point to an initial surface oxidation step, followed by the surface association of hydroxyls onto Fe^{2+} sites and subsequent oxidation of sulfide groups.

Rimstidt et al.[190] showed that Fe^{+++} could oxidatively dissolve galena, sphalerite, chalcopyrite, and arsenopyrite by donating electrons to the sulfide surface in acidic solutions. Rates are explained by the rate law:

$$Rate_{dissolution} = kA[Fe^{+++}]^n \qquad (17)$$

where A is the mineral surface area exposed to solution. k, the rate constant, decreases in the order of $k_{galena} > k_{arsenopyrite} \gg k_{sphalerite} > k_{chalcopyrite}$. n, the reaction order is approximately one for arsenopyrite and galena and roughly one half for sphalerite and chalcopyrite (n = 0.58 and 0.48, respectively). The rate-limiting step for dissolution at acid pH probably involves electron transfer from surface sulfide sites to adsorbed ferric iron.

Pyrrhotite (FeS) is also an important component of sulfide mine tailings and, therefore, another source of acid mine drainage. Rates are an order of magnitude or more greater than those of pyrite and are found to be independent of pH at 10°C (2 < pH < 6). At 33°C rates are minimal at pH 4 and increase at higher and lower pH.[191] The increase in pH dependence with temperature means that apparent activation energies at high and low pH are higher than they are at pH 4. Similar behavior is also seen for pyrite, at least on the low pH limb. The pH dependence of pyrite apparent activation energies is similar to what is observed for silicate dissolution (see above). Namely, rates at very high and very low pH are particularly sensitive to temperature, because of the temperature dependence of proton adsorption reactions.[70,71] The T and pH dependence of sulfide reaction rates, therefore, probably point to a T-dependent shift in rate-controlling adsorption of protons and/or electron acceptors at the respective mineral surfaces. This is a link which has yet to be explored in detail.[191]

Before surface-charging reactions can be used to mechanistically understand sulfide rate dependencies it will be important to first constrain the actual stoichiometries of charged groups at sulfide surfaces, something which has only recently received extensive attention.[192-195] Sulfide surface charge appears to vary as a function of solution metal and sulfide content, in addition to pH.[196,197] Potential functional groups include hydroxylated adsorbed metal cations[198] whose net charge depends on pH; metal–proton exchange complexes;[196] and anionic sulfide groups. Sulfide surface charge may arise solely from protonation reactions occurring on exposed thiol groups.[198] In other words, pH-dependent surface charge is thought to be caused primarily by the species: $>S–H_2^+$, $>S–H$, and $>S^-$. Presumably, it is the adsorption of electron acceptors (Fe^{+++}) to the latter sites which precedes oxidation and detachment of sulfide surface sites. ζ potential measurements of PbS point to surface oxidation and subsequent protonation of metal sites:[199]

$$PbS + 2OH^- \leftrightarrow Pb(OH)_2 + S + 2e^- \tag{18}$$

$$Pb(OH)_2 \leftrightarrow\ >Pb{\overset{\displaystyle OH}{\underset{\displaystyle O^-}{\diagup\!\!\diagdown}}} + H^+ \tag{19}$$

Although the pH dependence of surface charge was not greatly affected by light intensity, ζ potentials were observed to be more negative in sunlight, pointing to the importance of photooxidation in determining sulfide surface properties.

VIII. RATES NEAR EQUILIBRIUM

As the components of a dissolving mineral accumulate in solution, chemical kinetics predicts that the rate should slow down. At equilibrium the rate of forward reaction (dissolution) must exactly equal the rate of the reverse reaction. The reverse reaction in this case is mineral growth. Until this point only dissolution rates far from equilibrium have been considered. A truly general rate law must account for the change in reaction velocity with saturation, or thermodynamic affinity.[55,200-202] This is generally done by multiplying the far-from-equilibrium rate law by an experimentally measured scaling factor which is a function of the thermodynamic driving force for the reaction. The formalism for the latter can be derived from transition state theory (TST)[203] and is most simply illustrated using silica dissolution and growth as an example.

A. Transition State Theory and Silica–H$_2$O Kinetics

In the context of TST the overall macroscopic rate of a reaction is proportional to the concentration of the "activated complex" in an elementary rate-limiting reaction, the activated complex being in equilibrium with the reactants but irreversibly reacting to products once formed. That is:

$$\text{Rate} \propto [\dagger] = \frac{kT}{h}[\dagger] \qquad (20)$$

The cross denotes the activated complex; k is Boltzmann's constant; h is Planck's constant; and brackets denote concentrations. The rate-limiting activation reaction for silica and quartz dissolution above pH 8 can be written as:[41]

$$SiO_2 > Si–O^- + nH_2O \leftrightarrow (nH_2O)(SiO_2 > Si–O^-)^\dagger$$

$$K_{(21)} = \frac{\left\{(nH_2O)(SiO_2 > Si–O^-)^\dagger\right\}}{\left\{SiO_2 > Si–O^-\right\}\left\{H_2O\right\}^n} \qquad (21)$$

n denotes the number of water molecules in the activated complex. Brackets denote thermodynamic activities. Note that it is impossible at this point to state unambiguously the exact stoichiometry of the activated complex (e.g., number of water molecules, bridging oxygen bonds, etc.), hence, the somewhat general equation above. For example, the experimental data are such that the reaction can also be written to include adsorption of a hydroxyl onto an >Si–OH site.[34] By the same token, Hiemstra and Van Riemsdijk[45] have argued that the [>Si–O$^-$] measured in the laboratory actually represents a number of distinct deprotonated Si sites which differ in their connectivity to the bulk SiO$_2$ surface (i.e., through one, two or three O bonds). Hiemstra et al.[204] have advanced a model for separating out the specific sites responsible for surface charge and dissolution/growth by their coordination to the mineral surface. The

MUltiSIte Complexation (MUSIC) model,[45] applied to the SiO_2–H_2O system, suggests that basic pH dissolution of silica may be treatable as the weighted sum of detachment of activated complexes from each site.

The TST dissolution rate law is written for the reaction above as:

$$Rate_{dissolution} = \frac{kT}{h}\left[(nH_2O)(SiO_2 > Si\text{–}O^-)^\dagger\right] \qquad (22)$$

Substituting with the equilibrium constant for the formation of the activated complex, Equation 21 gives:

$$Rate_{dissolution} = \frac{kTK_{(21)}}{h}[> Si\text{–}O^-]a_{silica}a_{H_2O}^n\gamma \qquad (23)$$

a_{silica} denotes the activity of the pure silica solid; γ is the ratio of the activity coefficient of the activated complex to that of $>Si\text{–}O^-$. This will be assumed to be unity as both entities possess a single negative charge. Rates depend on pH and ionic strength because $[>Si\text{–}O^-]$ depends on each of these factors. Note again that specific interaction between $>Si\text{–}O^-$ sites and dissolved salts may also affect rates.[49,51,205]

Similar to dissolution rates, growth rates of silica from solution increase with pH at pH > 7 and with increasing salt content.[206] Fleming[110] explained these dependencies by showing that far-from-equilibrium precipitation rates of amorphous silica are proportional to the number of deprotonated sites at the surface and that rates change with pH and ionic strength due to the effect of the latter on the number of deprotonated sites. The precipitation rate law far from equilibrium is:[110]

$$Rate_{growth} = k[>Si\text{–}O^-][SiO_2^{aq}] \qquad (24)$$

The molecular deposition of SiO_2 from aqueous solution is the reverse of dissolution,[206] that is, that the rate-controlling steps for both dissolution and precipitation are symmetrical. This is similar to the results of Rimstidt and Barnes[207] and Gratz et al.,[34] who have likewise shown that quartz growth and dissolution rate laws are mirror images. This means that the activated complexes for growth and dissolution are the same (or energetically very similar). The activation reaction for silica growth is therefore written as:

$$>Si\text{–}O^- + SiO_2^{aq} + nH_2O \leftrightarrow (SiO_2 > Si\text{–}O^-)(nH_2O)^\dagger \qquad (25)$$

and the TST rate law as:

$$Rate_{dissolution} = \frac{kT}{h}\left[(nH_2O)(SiO_2 > Si\text{–}O^-)^\dagger\right] \qquad (26)$$

Substituting from the activation reaction gives:

$$\text{Rate}_{\text{dissolution}} = \frac{kTK_{(25)}}{h}\left[> \text{Si-O}^-\right]a_{\text{SiO}_2^{aq}}a_{\text{H}_2\text{O}}^n\gamma \tag{27}$$

$K_{(25)}$ is the equilibrium constant for Reaction 25, and γ is here the product of the activity coefficients of $>\text{Si-O}^-$ and aqueous silica divided by the activity coefficient of the activated complex. Again, γ is set equal to unity. Remember that the objective is to come up with a formalism for understanding mineral dissolution and growth near equilibrium. To do this Equations 23 and 27 are combined to calculate the total reaction due to simultaneous dissolution and growth, $\text{Rate}_{\text{total}}$:

$$\begin{aligned}
\text{Rate}_{\text{total}} &= \text{Rate}_{\text{dissolution}} - \text{Rate}_{\text{growth}} \\
&= \frac{kTK_{(21)}}{h}\left[> \text{Si-O}^-\right]a_{\text{silica}}a_{\text{H}_2\text{O}}^n - \frac{kTK_{(25)}}{h}\left[> \text{Si-O}^-\right]a_{\text{SiO}_2^{aq}}a_{\text{H}_2\text{O}}^n \\
&= \frac{kT\left[> \text{Si-O}^-\right]}{h}\left[K_{(21)}a_{\text{silica}}a_{\text{H}_2\text{O}}^n - K_{(25)}a_{\text{SiO}_2^{aq}}a_{\text{H}_2\text{O}}^n\right]
\end{aligned} \tag{28}$$

At equilibrium, $\text{Rate}_{\text{total}}$ goes to zero; hence, the two products in brackets must cancel and:

$$K_{(21)}a_{\text{silica}}a_{\text{H}_2\text{O}}^n = K_{(25)}a_{\text{SiO}_2}^{aq}a_{\text{H}_2\text{O}}^n \tag{29}$$

This points out the fact that at equilibrium the ratio of $K_{(21)}$ and $K_{(25)}$ must equal the equilibrium constant for the reaction of silica dissolving to aqueous silica. Therefore;

$$\frac{K_{(21)}}{K_{(25)}} = \frac{a_{\text{SiO}_2^{aq}}}{a_{\text{silica}}} = K_{\text{silica}\leftrightarrow\text{SiO}_2^{aq}} = K_{\text{sol}} \tag{30}$$

K_{sol} will be taken below to be the equilibrium constant for the general dissolution reaction for a given mineral and is strictly defined as:

$$K_{\text{sol}} = \prod_i a_{i(\text{eq})}^m \tag{31}$$

$a_{i(\text{eq})}^m$ is the activity of each species in the dissolution reaction at equilibrium raised to m, the reaction coefficient of the respective species. Q_{sol} is defined to be the activity product for a given dissolution reaction and is equal to:

$$Q_{sol} = \prod_i a_i^m \qquad (32)$$

For the case of silica dissolution $Q_{sol} = a_{SiO_2^{aq}}$. Substituting these expressions and letting the activities of water and pure silica equal unity give a complete rate law for silica growth and dissolution:

$$\text{Rate}_{total} = \frac{kT[> Si-O^-]}{h} \left[K_{(21)} - K_{(25)} a_{SiO_2^{aq}} \right]$$

$$= \frac{kT[> Si-O^-] K_{(21)}}{h} \left[1 - \frac{Q_{sol}}{K_{sol}} \right] \qquad (33)$$

The rate law is equivalently expressed in thermodynamic terms as:

$$\text{Rate}_{total} = \frac{kT[> Si-O^-] K_{(21)}}{h} \left[1 - \exp\left(\frac{\Delta G_r}{RT} \right) \right]$$

$$= \frac{kT[> Si-O^-] K_{(21)}}{h} \left[1 - \exp\left(\frac{-A}{RT} \right) \right] \qquad (34)$$

ΔG_r, the free energy of reaction, measures the distance of the silica–fluid system from equilibrium. At equilibrium, ΔG_r equals zero as does the overall rate. ΔG_r is large and positive when the solution is supersaturated with respect to silica. The dissolution rate is therefore negative, and silica grows. ΔG_r is negative when the solution is dilute and undersaturated with respect to silica. The overall rate is then positive and silica dissolves. Note that $\Delta G_r = -A$, where A is the chemical affinity of the reaction.

A general rate law for the net effect of mineral dissolution and precipitation is:[208]

$$\text{Rate}_{total} = k_{dissolution} a_{>M}^m \left[1 - \exp\left(\frac{n_e \Delta G_r}{RT} \right) \right] \qquad (35)$$

$a_{>M}^m$ is the activity of the rate-promoting surface species, and m is the number of species involved in the formation of the activated complex. n_e depends on the number of times each elementary reaction must occur to yield the overall reaction stoichiometry and upon the relative magnitudes of the rates at steady state.[208] Near-equilibrium rates are linear functions of free energy:

$$\text{Rate}_{\text{total}} = k_{\text{dissolution}} a_{>M}^m \left(\frac{n_e \Delta G_r}{RT} \right) \tag{36}$$

Experimental rate laws like the one above are useful for fitting experimental data because they can potentially be used to constrain elementary reaction mechanisms.[110]

The attraction of a rate law which covers both mineral growth and dissolution is that it might conceivably give some information about absolute rates near equilibrium, where many geologic processes occur and where it is very difficult to measure rates in the laboratory. Also, if rates behave according to the reasoning above, mineral growth rates can be predicted from measured dissolution rates[201] and vice versa. The underlying assumptions of the rate law are, however, formidable. First, it was assumed that the activated complex was the same for dissolution and growth. Although probably true for silica and quartz dissolution/growth, it is overly optimistic to expect similar behavior for other multioxide minerals. Multiple reaction pathways may exist, each limiting reaction rates over a set range of saturation state.[209] The only way to actually link rates to thermodynamic driving forces (and elementary rate laws) is to experimentally measure the dependence of rates on saturation state.

B. Dissolution and Growth of Single and Mixed-Oxide Silicate Minerals as a Function of Chemical Affinity

The application of dissolution rates measured in the laboratory, typically at far-from-equilibrium conditions, to near-equilibrium conditions typical of many natural processes requires detailed understanding of the variation of these rates as a function of chemical equilibrium. In accord with Equation 35, if the activity of the rate-promoting species involved in the formation of the activated complex is independent of chemical affinity, this variation should have a relatively simple functional form, as shown in Figure 27. Rates are independent of chemical affinity far from equilibrium ($\exp[n_e \Delta G_r /_{RT}] \ll 1$) and a near-linear function of chemical affinity near equilibrium. The part of the curve at far-from-equilibrium conditions, where the dissolution rate is independent of chemical affinity, has been referred to as the dissolution plateau.[210] A large number of studies have been performed over the past few years to determine the extent to which Equation 35 applies to (hydr)oxides, such as gibbsite and quartz, as well as sparingly soluble silicates.[20,86-89,205,207,211-219] Dissolution rates of a number of minerals are shown as a function of chemical affinity in Figure 28 . A comparison of Figures 27 and 28 illustrates that Equation 35 can be used to accurately describe the variation of quartz dissolution rates as a function of saturation state. Analcime dissolution and growth results from

Murphy and co-workers[212,213] can also be explained by Equation 36. In contrast, Figures 29 through 36 show that, except for possibly anorthite,[89] the rate behavior of the other minerals is more complex. In addition, from Figures 32 and 34 through 36 it appears that the dissolution rates of many aluminosilicates at constant saturation state tend to decrease with increasing aqueous aluminum concentration.

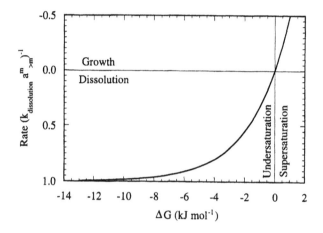

Figure 27
Variation of dissolution and precipitation rates as a function of ΔG, at 25°C according to Equation (35).

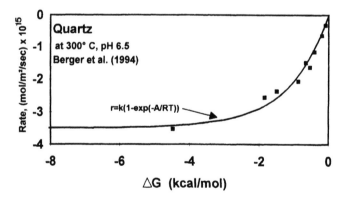

Figure 28
Variation of quartz dissolution rate[205] with saturation state. The curve was computed using Equation (35) and $n_e = 1$. (Courtesy of E.H. Oelkers, Université Paul Sabatier, Toulouse, France.)

Two explanations have been proposed for dependence of reaction rates on chemical affinity. Burch et al.[215] proposed an equation to describe their albite dissolution data based on the assumption that the overall rate

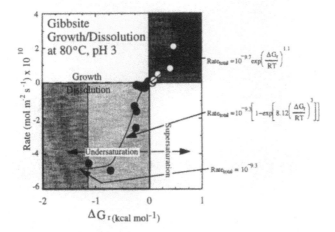

Figure 29
Dissolution and precipitation rates of gibbsite[20] as a function of ΔG_r. (See original reference for error bars and experimental set-up).

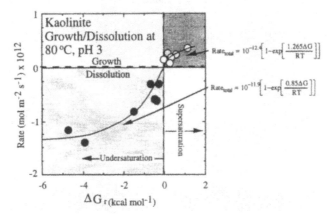

Figure 30
Kaolinite dissolution[211] and growth[219] rates. Within the error of the data n ~ 1.

is the sum of the rates of two parallel reactions. At extremely undersaturated conditions, dissolution rates are assumed to be dominated by reactions occurring at dissolution outcrops. Relatively close to equilibrium, the overall rate is dominated by detachment occurring at normal surface sites. The difference in saturation state between the two regimes corresponds to ΔG_r^{crit}, the critical free energy beyond which a microscopic hole will form a macroscopic etch pit at a dislocation outcrop. From independent calculations Burch et al.[215] determined ΔG_r^{crit} to be approximately 32 kJ mol⁻¹, which corresponds exactly to the degree of undersaturation where the sharp decrease in albite dissolution rates is observed. Smectite

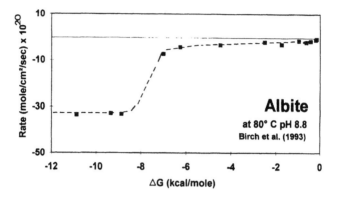

Figure 31
Variation of dissolution rates of albite as a function of saturation state at pH 8.8 and 80°C.
(Courtesy of E.H. Oelkers, Université Paul Sabatier, Toulouse, France.)

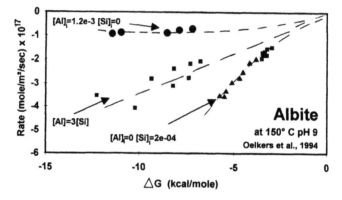

Figure 32
Variation of dissolution rates of albite as a function of saturation state at pH 9.0 and 150°C.[89]
The dashed curves serve only to group the different data sets for the reader. (Courtesy of
E.H. Oelkers, Université Paul Sabatier, Toulouse, France.)

dissolution results can be explained by a similar approach.[216] Schott and
Oelkers[218] have argued against the rate-affinity model above, noting that
(1) the same approach applied to quartz gives ΔG_r^{crit} = 5 kJ mol^{-1} at 300°C,
yet the predicted dramatic decrease in rate behavior is not found in the
300°C rates[205] and (2) the model does not explain the observed dependency of some aluminosilicate dissolution rates on the aqueous Al/Si ratio.
As noted in Section III.A Oelkers and Schott and co-workers instead argue
that exchange of protons for surface aluminum controls the dissolution
of many aluminosilicates, including kaolinite,[88] K-feldspar,[87] and kyanite.[217] Exchange of protons for aluminum is envisioned to break Al–O
bonds, in the process leading to the formation of an Si-rich precursor,
which subsequently detaches into solution. Note that in this model

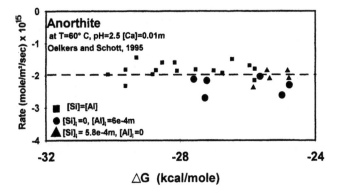

Figure 33
Variation of dissolution rates of anorthite as a function of saturation state at pH 2.5, [Ca] = 0.01M and 60°C. (Courtesy of E.H. Oelkers, Université Paul Sabatier, Toulouse, France.)

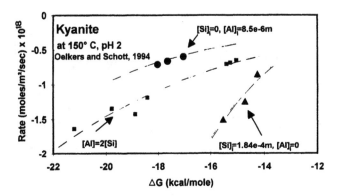

Figure 34
Variation of dissolution rates of kyanite as a function of saturation state at pH 2.0 and 150°C.[217] The dashed curves serve only to group the different data sets for the reader. (Courtesy of E.H. Oelkers, Université Paul Sabatier, Toulouse, France.)

absolute rates far from equilibrium will be proportional to the number of exchanged Al sites at the surface, which will be proportional to the ratio $a_{H^+}/a_{Al^{+3}}^{1/3}$. Because the activity of aqueous protons and aluminum shows up in the affinity term, as well as in the preexponential rate-promoting surface species term, if the terms are not resolved separately, dissolution rates will appear to be affected by chemical affinity over wide ranges of the chemical affinity. Clearly, determining the role of exchange reactions in multioxide mineral dissolution is a critical step to establishing rate controls near equilibrium, as well as far from equilibrium.

A number of important points can be drawn from the above observations. Although mineral dissolution far from equilibrium can often be explained by a single rate-determining elementary step, mineral growth

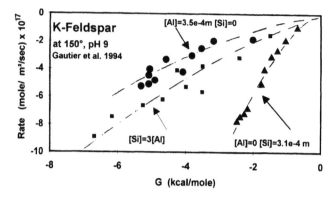

Figure 35

Variation of dissolution rates of k-feldspar as a function of saturation state at pH 2.0 and 150°C.[217] The dashed curves serve only to group the different data sets for the reader. (Courtesy of E.H. Oelkers, Université Paul Sabatier, Toulouse, France.)

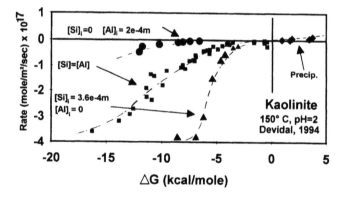

Figure 36

Variation of dissolution rates of kaolinite as a function of saturation state at pH 2.0 and 150°C.[217] The dashed curves serve only to group the different data sets for the reader. (Courtesy of E.H. Oelkers, Université Paul Sabatier, Toulouse, France.)

can only rarely be so treated. Note also that the rate constant in the growth/dissolution rate laws for gibbsite and kaolinite makes no explicit provision for surface speciation effects on rates. Gibbsite dissolution rates are related to the number of protons adsorbed at the surface.[220,221] Hiemstra et al.[204] have argued with the MUSIC model that this behavior arises specifically from doubly protonated Al sites. Similarly, kaolinite dissolution rates are first-order functions of surface protonation at low pH (see above). The rate constants in the gibbsite and kaolinite dissolution rate laws in Figures 29 and 30, therefore, must implicitly contain the respective rate-promoting concentration term.[219] By the same token, the growth laws for each of these minerals probably include yet undetermined surface speciation terms (see below).

C. Microscopic Reaction Steps

Up to this point mineral dissolution has been described by rate laws which account for the effects of solution chemistry (pH, T, organic ligands, saturation state) and mineralogy (bond strength) on macroscopic mineral surface chemistry. Clearly, observed rates of dissolution and growth integrate the reactivity of all the mineral faces, ledges, kinks, and etch pits found at the mineral surface. Because of variations in site density and bond strength, each of the surface microtopographic features are likely to dissolve or grow at different rates. For the same reasons, surface charging on the microscopic scale depends on the specific crystal faces exposed to solution.[75,204] Microtopography-controlled mineral dissolution has been studied for quartz,[223,224] goethite,[225] willemite,[35,36] and calcite.[62,159] The latter work has focused on the possible role of defects and, specifically, on the link between etch pit opening and measured dissolution rates.

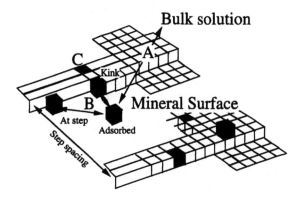

Figure 37
Kössel crystal[226] showing mechanistic steps involved in dissolution and growth.

In Figure 37 is shown a schematic view of mineral surface topography. Each cube (ion) in Figure 37 is coordinated to the surface by between one and five bonds. Each face represents a bond for such a "Kössel crystal".[226] Mineral dissolution is essentially a decoordination process. Five-coordinated sites (C) are uncovered when the overlying layer dissolves away. The detachment of four-, three-, and two-coordinated ledge sites (B) cause monomolecular layers to move across the surface. Presumably, the final step in this decoordination sequence is the detachment of a singly coordinated site (A) from the surface to the bulk solution. This final detachment step may involve a component of surface diffusion from kink sites to the final liftoff point. At steady state the progression of a site from C to B to A requires that the number of each type of site (kink, ledge, etc.) remain constant. The overall rate of dissolution then is proportional to the rate at which the slowest decoordination step occurs. Near equilibrium, as mineral components build up in solution, the likelihood of back reaction increases. At equilibrium, the net detachment of mineral components is exactly balanced out by their attachment from the bulk solution. From highly saturated solutions mineral growth far outweighs site detachment.

Only recently have technical advances allowed the *in situ* imaging of mineral surfaces. Scanning Electron Microscopy (SEM) and Atomic Force Microscopy (AFM) have most notably been used to image ledge motion, etch pit formation, and dissolution of screw dislocations. This work is important in part because it can be used to constrain which steps control overall rates of surface reaction. When combined with surface complexation information, they potentially promise general rate laws describing mineral dissolution and growth. Two case studies, for quartz and willemite dissolution, that illustrate the approach are outlined below. These two, along with the work of Dove and Hochella[227] and Gratz et al.,[228,229] described in the calcite growth section, highlight the results coming from microscopic observation of dissolving and growing crystals.

1. Quartz Dissolution

Gratz and Bird[37,46] used AFM imaging to examine >Si–O⁻-promoted dissolution of quartz. Specifically, Gratz and Bird[37] considered α-quartz as a Kossel crystal, noting that quartz differed from the ideal in that surface units are made up of at most three-coordinated tetrahedral groups, not five-coordinated cubes. Ledge and kink sites then are bound through two shared oxygens to the surface and are indistinguishable from each other. Gratz et al.[34] used an AFM to image unit prisms ($10\bar{1}0$) and rhombohedra ($10\bar{1}1$) from dissolving quartz crystals and found that dissolution occurs by the opening of holes, termed *negative crystals*, at the quartz surface. The slopes produced by the ledges are on the order of 2 to 3°. Ledges nucleate on the slopes at the convex intersections of the walls with the surrounding faces. Crystal defects and detachment of triply bound tetrahedra nucleate ledges at the base of the negative crystal. Step motion was observed to be the primary mechanism of quartz dissolution. For surfaces with widely spaced steps, step interaction is minimal. Step interaction can dominate overall dissolution when steps approach five lattice units. At smaller step separation dissolution is dominated by detachment directly from the steps.

Gratz and Bird[37] used their negative crystal observations to propose a microtopographic model for quartz dissolution over a wide range of temperatures (25 to 300°C) and solution compositions (0.003 M < ionic strength < 0.1 M). The dissolution rate, R (M^3/s) of a negative crystal face is: $R = av_{ledge}/l$; where a is the step height; v_{ledge} is its velocity (M/s); and l is step spacing (M^{-1}). Rates were found to be proportional to the concentration of negatively charged surface sites, >Si–O⁻. Two of the important results were that step spacing was independent of temperature, pH, and ionic strength and that dissolution and growth rates of minerals should depend on the number of exposed steps, not simply the mineral surface area.

2. Dissolution of Zn_2SiO_4

Lin and Shen[35,36] focused on the pH-dependent directional dissolution of willemite (α-Zn_2SiO_4). Dissolution of the willemite basal plane (0001) in pH = 0 solutions generally results in the formation of etch pits and screw axis–controlled hillocks. The latter were pyramidal, extending parallel to the c-axis. Lin and Shen[36] used an SEM to monitor the etch pit and hillock formation process. Etch pits are observed to form at dislocation outcrops and depend on the dislocation strain energy, the surface energy of the crystal, impurities, and the degree of disequilibrium between crystal and solution.[231] Etch hillocks formed at the willemite surface because of the presence of dissolution-resistant impurities (e.g., Fe^{++}, Mg^{++}, and Mn^{++} substituting for Zn^{++} in the willemite lattice).

Lin and Shen calculated rates from the dissolution of the [0001] face and from the widening of etch pit openings on the basal plane for the $\langle 10\bar{1}0 \rangle$ and $\langle 2\bar{1}\bar{1}0 \rangle$ directions. At steady state, far from equilibrium, the dissolution rates for each of the directions were observed to increase with decreasing pH (from pH = 3 to pH = 1), presumably reflecting increasing protonation and proton-promoted detachment of Zn surface sites at low pH. Rates (in mol m^{-2} h^{-1}) for the [0001], $\langle 2\bar{1}\bar{1}0 \rangle$, and $\langle 10\bar{1}0 \rangle$ directions were fit to Rate = $1.51a_{H^+}^{0.88}$, $0.2a_{H^+}^{0.5}$, and $0.32a_{H^+}^{0.62}$, respectively. At a given pH from pH = 0 to pH = 2, dissolution rates decreased among the various crystallographic directions in the order of [0001] > $\langle 10\bar{1}0 \rangle$ ≥ $\langle 2\bar{1}\bar{1}0 \rangle$. From pH = 2 to pH = 4, rates decreased in the order $\langle 2\bar{1}\bar{1}0 \rangle$ ≥ $\langle 10\bar{1}0 \rangle$ > [0001]. To explain the differences in rate and reaction order between the various directions Lin and Shen suggested that protonation of Zn sites within the basal layer may have been favored by the presence of hexagonal hollow tubes parallel to the c-axis in willemite. Lin and Shen also noted that the pH dependence of directional dissolution was probably directly related to the protonation of the particular planes and that the charging of each of latter was likely to vary because of differences in site acidity. Lin and Shen subsequently showed that Zn detachment occurred along alternate screw axes through a closest-neighbor path. The work of Lin and Shen was one of the first to emphasize the need for understanding multioxide protonation on individual crystal faces.

IX. CRYSTAL GROWTH

A. Reaction Mechanisms

Crystal growth is generally treated as being mechanistically the converse of dissolution. Referring again to Figure 37, crystals are presumed to grow by the attachment of mineral components to kink sites at steps on the mineral surface. Steps form as a result of the presence of screw dislocations or becuase of the appearance of surface nuclei. In general, crystal growth theories have minerals growing through the movement of layers of unit cell or more across a planar surface. The layer steps move forward by adding material at coordinatively unsaturated kink sites. Steps form either from lattice imperfections or from surface nuclei made up of a critical number of crystal unit cells.

Growth rates have generally been described by a rate law of the form:

$$\text{Rate} = k_{\text{growth}} (S - 1)^n \tag{37}$$

$$S = \left(\frac{Q_{\text{sol}}}{K_{\text{sol}}} \right)^{\frac{1}{v}} \tag{37a}$$

S is the saturation ratio; v is the total number of formula units in the given mineral (e.g., for SiO_2, $v = 1$; for $CaSO_4$, $v = 2$; etc.). n, the reaction order, is an experimental fitting factor. The rate order is important as it gives information about the actual mechanisms occurring at the mineral surface. Note the similarity between the growth rate law above and the TST total rate law derived in the previous section (Equation 37 is the negative of Equation 34, as it monitors growth whereas the latter monitors dissolution).

Crystal growth requires (1) the diffusion of mineral components to the growing surface; (2) adsorption; (3) diffusion across the surface; (4) reaction at the mineral surface (kink site); and (5) incorporation into the crystal lattice. The first step (bulk diffusion) depends on the concentration gradient of mineral component from solution to the mineral–water interface. The second, adsorption step, depends upon the chemistry of the mineral surface, that is, upon surface potential, charge, etc. The latter terms are generally lumped together using any one of a number of adsorption isotherms (e.g., Langmuir, Freundlich, Temkin, etc.). Note that characteristic times for adsorption are generally very rapid, on the order of seconds.[231] Steps 4 and 5, reaction and integration, will depend upon the local chemical environment at the kink site.

Figure 38 shows concentration profiles of mineral components in solution for three types of rate control. When diffusion in the bulk solution

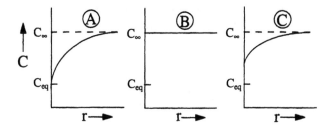

Figure 38
Concentration gradients at a growing crystal surface. C_{eq} is the concentration of the mineral component in solution at equilibrium (saturation). C_∞ is the concentration of the component in the bulk solution. r is distance from the mineral surface. A. Mineral growth control by diffusion in the bulk fluid. B. Control by surface reaction. C. Mixed control of growth. See text. (From Berner, R.A. *Early Diagenesis: A Theoretical Approach*, Princeton University, Princeton, NJ, 1980. With permission.)

is slower than surface reaction, hence, controlling the overall reaction rate, concentration gradients are non-zero (A). When diffusion is much more rapid than surface reaction, concentration gradients are quickly erased (B). When the flux of mineral component to a growing surface is roughly matched by the flux of material being incorporated into the mineral lattice, there is mixed control of the overall rate and concentration gradients are somewhat smeared out (C). For extensive, recent discussions of the coupling between diffusion and reaction see References 233 and 234.

When steps 2 to 5 are relatively rapid, bulk diffusion controls mineral growth rates. Diffusion of components from the bulk solution to the surface obeys Fick's law as:

$$\text{Rate} \propto D \frac{dC_i}{dx} = \frac{k}{\delta}\left(C_{i,b} - C_{i,s}\right) \tag{38}$$

dC_i/dx is the concentration gradient of mineral component i, and D is the diffusion coefficient of the component in aqueous solution. Note that multicomponent diffusion of electrolytes cannot be broken up into a single component problem without inducing potentially large numerical uncertainties;[234] however, to illustrate the role of aqueous diffusion in crystal growth this assumption will be adopted in the present discussion. D for ions is generally on the order of 10^{-5} cm^2 s^{-1} at 25°C in bulk water[235]; k is the growth rate constant; $C_{i,b}$ is the concentration of i in the bulk fluid; and $C_{i,s}$ is the concentration of i just outside of the crystal and any adsorption layer. δ is the distance across which i must diffuse. Values of δ calculated from experimental data with the rate law above are generally on the order of 10 to 100 mm. Values of δ calculated theoretically[236] are generally smaller, at $10 < \delta < 20$ mm. Note that the rate law reduces to a first order expression for mineral growth:

$$\text{Rate} = k_{growth}$$

$$\text{where } k_{growth} = \frac{k}{\delta C_{i,s}} \tag{39}$$

First-order growth rates can also arise because of nondiffusive control of growth (see below). Diffusion control can be roughly checked by calculating the rate predicted from the concentration and diffusion coefficient of the respective mineral component and from a theoretically estimated diffusion thickness. Rates can be estimated within an order of magnitude by the latter approach. If the measured rate is significantly less than the estimated rate, bulk diffusion is probably not an important factor. Recall also that solute diffusion-controlled rates are characterized by low activation energies (20 kJ mol^{-1} and less), reflecting the weak temperature dependence of diffusion coefficients. By measuring rates at two temperatures the possible role of diffusion can therefore be checked. A more rigorous test is that diffusion-controlled growth rates should be found to depend on the stirring rates of the experiments, as the latter affects the magnitude of δ, the diffusion layer thickness.

In the case where adsorption limits growth rates $C_{i,ad}$, the concentration of i adsorbed at the surface is set equal to $K_{ads}C_{i,eq}$. K_{ads} is the equilibrium constant for adsorption of i onto the surface. $C_{i,eq}$ is the concentration of i in solution at equilibrium with the respective crystal. It is assumed that this is the concentration of i at the immediate crystal–fluid interface. Adsorption proceeds at the rate of $r_{ads} = k_{ads}C_{i,b}$. The rate of desorption from the mineral surface is described by $r_{des} = k_{des}C_{i,ad} = k_{des}K_{ads}C_{i,eq}$. The absolute rate of adsorption is therefore:

$$r_{total} = r_{ads} - r_{des} = k_{ads}C_{i,b} - k_{des}K_{ads}C_{i,s} \tag{40}$$

Note that by detailed balancing $k_{ads} = k_{des}K_{ads}$. Adsorption rate constants and equilibrium constants ultimately depend on the ΔH and ΔS of bond formation between adsorbed species and the mineral surface. This in turn will depend on such variables as the charge of the mineral surface and the charge and hydration state of adsorbed ions. K_{ads} can be roughly estimated[237] by considering the magnitude of the ion-pair constant between the crystal components.

Because rates are proportional to adsorption rates in adsorption-limited growth the rate law is written as:

$$r_{total} = r_{ads} - r_{des} = k_{ads}C_{i,b} - k_{des}K_{ads}C_{i,s} \tag{41}$$

$$\text{Rate} \propto r_{total} = k\left(k_{ads}C_{i,b} - k_{des}K_{ads}C_{i,s}\right)$$

$$= k_{growth}(S-1)$$

$$\text{where } k_{growth} = \frac{kk_{ads}}{C_{i,s}}$$

(42)

This means that adsorption-limited rates are also first order.

If diffusion from the site of adsorption to the site of reaction/integration is rate limiting, crystal growth rates are found to obey the following rate law:[170,237,238]

$$\text{Rate} = k_{ad}\frac{(S-1)\ln S}{\sigma}\tanh\frac{\sigma}{\ln S}$$

$$\text{where } k_{ad} = v_{ad}aC_{i,eq} \text{ and } \sigma = \frac{19\gamma_e a}{2kTx_s}$$

(43)

v_{ad} is the adsorption frequency, a is the size of a growth unit, and γ_e is the edge free energy. Edge free energies cannot be measured independently but must be estimated from rate data. Values of γ_e for crystals are found to correlate with their solubilities.[239] The rate expression above comes from the classic theory of Burton, Cabrera, and Frank[238] who proposed that crystals grow by the addition of material to spiral steps emanating from screw and edge dislocations. The steps were predicted to move by material addition, rotating across interterrace areas in a perpetual spiral. When $\ln S > s$, the rate law is seen to be first order: Rate = $k_{ad}(S-1)$. This is the case where surface diffusion is rapid and growth is adsorption limited. On the other hand, when $\ln S \ll s$, the rate can be approximated by Rate = $k_{sd}(S-1)\ln S$, where k_{sd} is the surface diffusion constant. When $\ln S \sim (S-1)$, adsorption is rapid and diffusion of components to sites is rate limiting.

If all of the above steps are relatively fast, then reaction/integration will determine the overall growth rate. This leads to a rate law of the form:

$$\text{Rate} = k_{in}(S-1)S^{0.5}\ln S$$

$$\text{where } k_{in} = \frac{4v_{in}aK_{ads}C_{i,eq}}{19(\gamma_e/T)\exp(\gamma_e/kT)}$$

(44)

v_{in} is the integration frequency. Integration frequencies may be related to the ease with which hydrated ions lose their water molecules.[240] Note that dehydration is necessary for incorporation of crystal components into their respective lattice positions. Over most values of (S − 1), Equation 44 can be approximated by:[237]

$$\text{Rate} = k_{in}(S-1)^2 \qquad (45)$$

Polynuclear growth occurs when two-dimensional growth sites nucleate across the surface so rapidly as to control the addition of material to the surface. (See Figure 39.) In other words, they grow faster than ledges move. In such a case rates are described by:[170]

$$\text{Rate} = k_e S^{7/6}(S-1)^{2/3}(\ln S)^{1/6}\exp(-K_e/\ln S)$$

$$\text{where } k_e \approx 7 a v_{in}\left(K_{ads}C_{i,eq}\right)^{4/3} V_m^{1/3}\exp(-\gamma_e/kT)$$

$$\text{and } K_e = 4/3(\gamma_e/kT)^2 \qquad (46)$$

V_m is the molar volume of the crystal.

Figure 39
Polynucleation control of crystal growth.

Growth mechanisms are likely to change in response to variations in solution saturation state. At the same time crystal growth is routinely observed to be inhibited by the presence of additives which adsorb to and block kink sites. For surface diffusion–controlled growth, an increase in the interfacial free energy, γ_e, translates into a decrease in step spacing and absolute step velocity. For polynucleation an increase in γ_e means an increase in the size of the critical surface nuclei and corresponding drops in absolute rates of nucleation and overall growth.[241] An alternative explanation is that poisons adsorb strongly to kink sites, forcing steps to advance around. The net effect of ledges bending around sites is to increase ledge curvature (and surface free energy), ultimately causing rates to diminish as a result. At the same time additives can accelerate growth by lowering the surface free energy and by serving as new sites for nucleation.[238]

TABLE 3.

Growth Rate and Order

n	Mechanism
1	Bulk diffusion; adsorption; adsorption + volume diffusion
1–2	Adsorption + surface diffusion or integration; volume diffusion + integration or surface diffusion or polynucleation
2	Surface diffusion; integration; surface diffusion + integration
>2	Polynuclear growth

From Zhang, J.W.; Nancollas, G.W. *Rev. Min.* 1990, *23*, 365–393. With permission.

TABLE 4.

Selected Mineral Growth Rate
Dependencies

Mineral	n	Ref.
$CaCO_3$	1–3.7	228
$Ca_5(PO_4)_3F$	1.8–4.9	242
$NaAlSi_2O_6 \cdot H_2O$	1	213
SiO_2	1	110
SiO_2 (Quartz)	1	207
$Al(OH)_3$	1[a]	20
$Al_2Si_2O_5(OH)_4$	1[a]	219
$Mg_2Si_3O_{7.5}(OH)3H_2O$	3	243
CaF_2	1	244
$BaSO_4$	1–4	245
$PbSO_4$	2	246
$MgC_2O_42H_2O$	2	246
$CaC_2O_4H_2O$	2	247
$CaSO_42H_2O$	2	248

[a] $n \sim 1$ within the error of the data —
see text.

B. Reaction Orders and Rate Mechanisms

The rate orders for the various rate laws are described in Table 3. In Table 4 are listed observed growth rate dependencies for a number of crystals. A majority of mineral growth rate laws are first or second order, suggesting, first of all, that no single growth mechanism and rate law can describe the growth of minerals in general. Second, it should be reemphasized that the rate orders listed in Table 4 were each measured over limited saturation state ranges and that they probably do not describe growth over all composition states. Note also the rate laws in Table 4 do not always explicitly account for the role of surface-charging reactions on growth rates. Charged surface sites should affect (if not control) three of the five steps for crystal growth: adsorption, surface diffusion, and reaction at kink sites. Surface-coordination reactions may also control the formation of critical nuclei during polynuclear growth.

It is fairly apparent that the mechanistic understanding of mineral growth lags behind that of mineral dissolution. Although we are approaching a predictive footing for the latter, it is impossible to predict mineral growth rates *a priori*. The following sections attempt to give a coverage of recent work on mineral growth and dissolution at low temperatures where surface chemical data are more likely to exist. The focus again is on carbonates, silicates, and phosphates.

C. Case Studies — Calcite Growth

Calcite growth is an important control on alkalinity and pH in the oceans, porosity development in deep basins, and the efficiency of municipal water pipes. Calcite growth rates at 25°C have been observed to be first order;[57,152,249] second order;[249-251] and higher (n = 3 to 3.9[253,254]). As a result, growth data have been interpreted to support calcite growth by adsorption, surface diffusion, as well as polynucleation (i.e., the birth and spread model).[241]

Gratz et al.[228] imaged calcite growth with an AFM focusing specifically on the motion of spiral steps at low values of S (0.76 < S < 4.2), where step motion is thought to control long-term growth. The Burton–Cabrera–Frank (BCF) theory predicts that steps will interact over separation distances less than $6(D_s\tau_s)^{1/2}$, where D_s is the surface diffusion coefficient and τ_s is the mean residence time of an adsorbed molecule.[255] The mean displacement of an atom[228] is $2(D_s\tau_s)^{1/2}$; hence, steps interact when they are less than four diffusion lengths apart. This was observed to be on the order of ~2.5 nm, much less than the typical separation between steps. This suggests that surface diffusion plays a minor role in the step motion and that addition of material to kink sites at steps depends more on direct adsorption from solution.

Gratz et al.[228] proposed that the presence of charged surface sites (adsorbed Ca^{++} and CO_3^{-2}) hindered the diffusion of ions across the calcite surface. In fact, this should be true for a number of minerals. Specifically, on zwitterionic multioxide silicates, carbonates, and phosphate surfaces diffusion should be greatly diminished by the presence of both cationic and anionic surface sites except under conditions where the charge is minimal. For minerals such as quartz and TiO_2, whose ΔpK is large ($\Delta pK = pK_{a_2} - pK_{a_1}$), surface charging is low over large pH ranges. Hence, diffusing ions are not likely to be electrostatically hindered during dissolution or growth. (Hydr)oxides such as $Al(OH)_3$ and $Fe(OH)_3$, for which the ΔpK is relatively small, will always possess a relatively high absolute number of charged sites. On these surfaces diffusion will be minimal. If surface diffusion plays any role in multioxide silicate growth or dissolution, it will be at near neutral pH where surface charging is minimal. Chiang et al.[256] have proposed that surface diffusion is minimal during the growth of other salts from ionic solutions as well.

Dove and Hochella[227] imaged calcite growth under a wide range of S through scanning force microscopy (SFM) to show that calcite growth occurs by a variety of mechanisms. At very high values of S, calcite was observed to grow by polynucleation. Over short periods of time surface nuclei were observed to form, grow, and coalesce. Only over long periods of time was calcite observed to grow by spiral step motion. At S > 2, surface nuclei form, raising the effective surface area of the growing crystal and leading to initially high rates of growth. Long-term growth was dominated by step motion. Near equilibrium (S < 2), surface nucleation was found to be less important and monolayer growth controlled the reaction. Dove and Hochella showed that this transition in reaction mechanism with saturation state was consistent with the observed decrease in reaction order with decreasing saturation state. Hillner et al.[229] using AFM also observed growth at high pH 9.5 and S = 5 by the progression of steps across terraces; nucleation of islands on the terraces was not observed in this study.

The interaction between phosphate and calcite is thought to be an important mechanism in reducing phosphate concentrations in hardwater lakes[257,258] and a control over the variable morphology of the precipitates formed in natural conditions.[260] It has been observed that inorganic phosphate is removed from solution during the precipitation of calcite and that the rate of the coprecipitation is related to the rate of calcite growth.[168,258] Phosphate is also an inhibitor of calcite growth[177,260] with the degree of the inhibition depending on the saturation state of the solution with respect to calcite formation.[178] There is no direct evidence for the formation of a distinct calcium phosphate phase such as octacalcium phosphate or, less likely, hydroxyapatite (HAP). The coprecipitation has been modeled assuming the incorporation of phosphorus in the calcite lattice with the surface concentration at kink sites determined from the solution speciation of phosphorus and the general form of an adsorption isotherm.[168,259] Dove and Hochella[227] showed that the phosphate inhibition mechanism changed with saturation state to mirror the change in dominant reaction mechanism. At high values of S, phosphate adsorbed to and roughened surface nuclei blocking the formation of calcium carbonate. At low values of S, phosphate adsorbed to kink sites, inhibiting layer growth in the process. This is in agreement with the results of House and Donaldson[168] who showed that, within limits, the inhibition by phosphate can be overcome by increasing the solution saturation to induce calcite growth. The high resolution NMR results of Hinedi et al.[261] demonstrate that at low surface concentration of phosphate, e.g., 0.79 μmol P sorbed

g^{-1} $CaCO_3$, the adsorbed phosphate is not in the form of HAP or amorphous calcium phosphate minerals. Only at higher concentrations of surface
phosphate, e.g., 3.3 to 36.7 μmol P sorbed g^{-1} $CaCO_3$, was the formation
of carbonated, apatitic-like phase suggested from the NMR data. Adsorption data of phosphate on BDH calcite powders show that surface concentrations <0.5 μmol P sorbed g^{-1} $CaCO_3$, are likely when the phosphate
concentrations are <12 μM.[168]

House[154] tested several kinetic equations describing growth and concluded that the mechanistic model of Plummer et al.[152] best described
growth data from experiments in $Ca(HCO_3)_2$, i.e., in conditions similar to
natural hard waters. Other models such as the Davies and Jones[262] equation were less successful over a wide range of supersaturations. Subsequently, Cassford et al.[60] extended the Plummer model to include interfacial reactions at OH^- sites (leading to the kinetic equation of Plummer
et al.), CO_3^{2-} sites, or HCO_3^- sites on the surface. In effect, the rate-limiting
step is the dehydration of the ions and incorporation into the surface layer
(e.g., formation of inner-sphere complexes). The three mechanisms were
tested by adjusting the rate constant to optimize the agreement between
the CO_2 concentrations measured in solution and the calculated CO_2 concentrations at the surface, during free-drift precipitation experiments. This
is equivalent to a test of the assumption that the CO_2 concentration in the
bulk and surface are equal. The carbonate site model was found to produce significantly better agreement between the carbon dioxide concentrations. The surface CO_2 concentrations generated from the growth data
are in good agreement with the bulk CO_2 concentrations calculated from
the solution speciation, i.e. from experimental measurements of $[Ca^{2+}]$,
pH, $[HCO_3^-]$, and temperature. The rate constant, k_3, is in good agreement
with values calculated from dissolution data (see Table 2). The mechanistic
model has subsequently been used to investigate the effects of magnesium
and phosphate ions on precipitation kinetics.[155,178] The same approach
lends itself to modeling precipitation under field conditions using experimental streams.[263]

The pH-stat method has also been used to investigate the growth
kinetics between pH 8.3 and 10.0 with very low concentrations of CO_2.[264]
The data were successfully interpreted by a spiral growth mechanism
which assumed an equilibrium between the solution and adsorbed layer,
and that the rate limiting step for growth is the dehydration of the cation
and incorporation at the kink site. Other mechanisms, such as transport
control from the bulk solution and polynuclear and mononuclear growth,
were discounted because the values of the surface free energy needed to
produce agreement with the nucleation mechanism were unrealistically
small. Further research is needed to test the application of this model to
a wider range of solution conditions with a range of CO_2 concentrations
and to compare the results with the mechanistic model described above
for dissolution and precipitation.

D. Case Studies — Calcium Phosphate Dissolution and Growth

Phosphate is a critical nutrient on which life depends (see Chapter 5), and apatite is one of the most important phosphate-containing minerals at the surface of Earth. Growth and dissolution of apatite (and HAP) exert a strong control over phosphate availiability and biologic productivity in marine environments. At the same time, calcium phosphate minerals are the primary inorganic components of bones and teeth. HAP is an important soil mineral and plays a role in water purification processes. In metabolic processes HAP often buffers pH, Ca, and phosphate levels.[265]

Many freshwaters are supersaturated with respect to HAP. However, even when the concentration of phosphate is >120 μM, [Ca] = 3 mM at pH of approximately 8.5, HAP does not precipitate immediately.[266] Instead, an amorphous tricalcium phosphate, $Ca_3(PO_4)_2$ is formed. Most freshwaters have dissolved phosphate concentrations <20 μM, and, therefore, the direct precipitation of HAP is unlikely. Even with elevated concentrations of phosphate, the formation of a precursor phase such as octacalcium phosphate is more likely[267,268] because of the faster kinetics compared with HAP formation.

Dissolution of HAP has been extensively measured and found, at near neutral pH, to be controlled by a polynuclear mechanism at low levels of disequilibrium (0.1 < S < 0.7).[269] In the context of dissolution a polynuclear mechanism involves the opening of small holes at the mineral surface. Christoffersen and Christoffersen[270] showed that dissolution rates, normalized to a common value of S, are pH dependent. The pH dependence of the rate constant was interpreted to arise from the pH-dependent concentration of reactive protonated phosphate groups at the mineral surface. Protonated surface phosphate groups apparently promote the exchange of phosphate between the mineral surface and the bulk solution. Because of the connection between rates and surface speciation, the composition of the bulk solution is seen to affect rates in two ways: by controlling the saturation state of the solution and by determining the overall surface coverage by reaction-promoting adsorbed species.[270] For example, both Ca^{++} and H^+ can associate with surface phosphate, whereas only the latter promotes dissolution. High concentrations of Ca in solution, therefore, slow down the rate by displacing protons from the surface. Because the formation constant for protonated surface phosphate groups depends on the electric potential of the double layer between the crystal and the bulk solution, the latter may also indirectly affect rates through its effect on the work required to bring a proton to the mineral surface.

A number of biologically important ligands inhibit HAP dissolution. Inhibition of HAP dissolution is important for the modeling of tooth decay mechanisms. Christoffersen and Christoffersen[265] found that a number of ligands, in particular pyrophosphoric acid (PP), 1-hydroxyethylene-1,1-diphosphonic acid (EHDP), and methylene diphosphonic acid (MDP) inhibit HAP dissolution by adsorbing to cationic surface Ca sites. Binding of anionic ligands to Ca sites is thought to greatly decrease the rate at which holes (nuclei) appear and to slow their subsequent spread as well. Inhibition by ligands is thought to be favored by strong multidentate binding and by relatively short chain length. Conversely, bulky unidentate ligands were shown to be less effective inhibitors of dissolution.

A range of reaction orders have been measured for HAP, $Ca_5(PO_4)_3OH$, and fluorapatite (FAP) and $Ca_5(PO_4)_3F$. Van Cappellen and Berner[242] have suggested that some of the differences in rate order at low values of S are attributable to differences in the defect densities of the different seed crystals and that apatite growth is characterized by two different regimes of growth. At high values of S, where n > 2 (S > 37 for FAP), growth is by polynucleation. At low values of S (S < 37), growth is by spiral growth in accord with BCF theory (n ~ 2). The rate transition for FAP growth is shown in Figure 40. This is the sequence which is seen for calcite growth (see above), as well as for the growth of fluorite[251] and barium sulfate.[271]

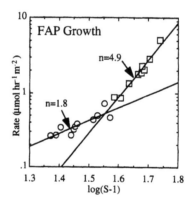

Figure 40

Fluorapatite growth rates as a function of saturation state.[242] Temperature = 25°C, pH = 8.00 and P_{CO_2} = 0. (From Van Cappellen, P.; Berner, R.A. *Geochim. Cosmochim. Acta.* 1991, *55*, 1219–1234. With permission.)

Mg^{++} inhibits the formation of apatite in marine environments.[274] In Figure 41 FAP growth inhibition by Mg^{++} is shown. Growth retardation can be explained by fitting Mg adsorption to the apatite surface with a Langmuir adsorption isotherm.[241] Rates normalized to the same degree of saturation depend on pH as well. This is shown in Figure 42 where saturation-normalized rates are plotted as a function of pH. Clearly hydrogen atoms catalyze the incorporation of mineral components into the growing FAP crystal. Because the reaction is surface controlled (i.e., bulk diffusion is not an important factor), it is probably adsorbed protons which control growth. Zawacki et al.[273] showed that HAP growth rates depended on the background electrolyte, being most rapid in NaCl and

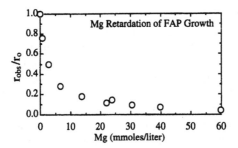

Figure 41
Retardation of fluorapatite growth rates[242] by Mg^{++}. (From Van Cappellen, P.; Berner, R.A. *Geochim. Cosmochim. Acta.* 1991, 55, 1219–1234. With permission.)

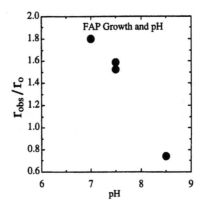

Figure 42
Fluorapatite growth rates[242] as a function of pH. (From Van Cappellen, P.; Berner, R.A. *Geochim. Cosmochim. Acta.* 1991, 55, 1219–1234. With permission.)

slowest in KNO_3. Under the conditions of their experiments, measured surface charge was most negative in NaCl and least so in KNO_3, suggesting a link between charging and crystallization. Chloride ion may accelerate phosphate mineral growth, whereas strontium and magnesium were observed to inhibit growth. The pH dependence is observed over saturation ranges where polynucleation controls growth (the region where n = 4.9). This is important because it suggests that surface species may control the formation of surface nuclei. Christoffersen and Christoffersen[274] have shown that the rate constant for dissolution of HAP by a polynuclear mechanism was likewise pH dependent (see above). The pH dependence of rates is thought to reflect the pH dependence of the concentration of HPO_4^{2-} adsorbed at the mineral surface. Specifically, protonation of adsorbed phosphate groups is thought to promote the formation of dissolution nuclei at the HAP surface. When solutions are supersaturated, the same surface species may accelerate growth as well.[242]

E. Silica Diagenesis

Amorphous silica is a commonly observed scale in geothermal operations and is associated with a number of diagenetic processes. SiO_2 is precipitated from many natural waters largely by microorganisms (e.g., diatoms and radiolaria) as amorphous silica (Opal-A) tests. Siliceous tests accumulate as a siliceous ooze on the ocean floor, and they undergo dissolution–reprecipitation reactions, transforming from Opal-A to Opal-CT and ultimately to the thermodynamically most stable silica polymorph quartz. Opal-CT is a disordered cristobalite–tridymite mixture which is intermediate in solubility between Opal-A and quartz. The Opal-A phase forms initially by the association of dense colloids in highly saturated alkaline solutions. Natural solutions become saturated with respect to silica by cooling of Si-saturated hydrothermal fluids, evaporation, and/or by sudden drops in fluid pH (silica is relatively insoluble below pH ~ 10 at 25°C).

Ernst and Calvert[275] suggested early on that the crystallization sequence occurs by a zero-order solid state transformation. Stein and Kirkpatrick,[276] using SEM and TEM reanalyzed Ernst and Calvert's results and argued for a first-order nucleation and growth mechanism. Williams et al.[277] showed that much of silica diagenesis in natural waters could be explained by the Ostwald step rule, which describes the replacement of less stable minerals by a series of progressively more stable minerals. The likely sequence for silica diagenesis is then.[277]

1. Opal-A is formed by the polymerization of colloidal silica from supersaturated alkali solutions, the latter caused by a drop in pH, temperature, or H_2O activity.

2. Opal-A dissolves in saline pore waters to form ordered polymers similar in structure to Opal-CT. Positively charged hydroxyl complexes (e.g., Mg or Ca) flocculate Si polymers.

3. The freshly deposited polymer increases in crystallinity, microdomains of tridymite being preferentially dissolved and microdomains of cristobalite being preserved, gaining in size through fresh growth.

4. Opal-CT redissolves at some point in the reordering, whereupon quartz grows.

The whole sequence occurs against the backdrop of the increasing temperature associated with burial. The last step in the sequence, Opal-CT transformation to quartz, may depend on both temperature and, indirectly, the presence of clay minerals.[277]

Condensation polymerization of silicic acid onto silica seeds depends largely on the surface chemistry of the latter and, particularly, on the degree of surface ionization.[110] Silica growth rates are proportional to calculated silica anionic surface charge. At high pH and elevated solution ionic strength, where $>Si-O^-$ surface sites are most abundant, growth rates are most rapid. At a given surface, charge rates increase with the concentration of silicic acid in solution (i.e., with increasing supersaturation). Freshly deposited silica appears to have an elevated pseudoequilibrium solubility, roughly twice the equilibrium level.[110] As a result, dissolution and growth rates depend on the amount of silica adsorbed at the surface. These results make the critical point that mineral dissolution and growth rate laws probably possess more than one surface species concentration term and that it is necessary to separate the combined effects of surface chemistry and saturation state on rates near equilibrium to arrive at a truly valid rate law.

F. Case Studies — Multioxide Silicate Growth

Precipitation rates of multioxide silicates have seldom been measured in the laboratory at low temperatures (T ~ 25°C), as rates are very slow. Moreover, unwanted polymorphs and hydroxide phases tend to form because geologically unreasonable levels of supersaturation are needed to produce measurable reaction progress. This obscures precipitation mechanisms and reaction stoichiometries as well. Low temperature (25°C) multioxide silicate growth experiments have focused primarily on the growth of clay minerals. Kaolinite and sepiolite ($Mg_2Si_3O_{7.5}(OH)3H_2O$) are two minerals which have probably received the most attention; kaolinite because it is important during diagenesis and sepiolite because it is easy to grow and potentially important for ocean chemistry.

Gastuche et al.[278] showed that kaolinite, despite being a common soil clay, was very difficult to grow in the laboratory because of the tendency of Al to remain in 4-fold coordination. Homogeneous kaolinite growth at low temperatures depends on Al being kept in 6-fold coordination in solution by organic acids.[279] Siffert[280] showed that oxalic, citric, tartaric and salicylic acids all form 6-coordinated complexes with Al. The binding between Al and organic acids is critical because it has the effect of dehydrating Al, making it easier to integrate into the kaolinite lattice. The mechanism for homogeneous precipitation of kaolinite at 25°C is thought to begin with the slow precipitation of colloidal aluminum hydroxide, followed by the adsorption of silica and the integration of the unit into the kaolinite structure. n ~ 1 for kaolinite growth at 80°C.[220] From Table 3 this suggests that the kaolinite growth mechanism is either bulk diffusion, or more likely adsorption (surface reaction).

Early interest in sepiolite growth derived from it being a potentially important sink for Mg in the oceans. Before the discovery of mid-ocean hydrothermal systems, the low temperature growth of Mg silicate clays was argued to be a potential mechanism for balancing the riverine input of Mg into the oceans. Sepiolite is apparently the only multioxide silicate which grows rapidly in the laboratory at 25°C, stoichiometrically and in a crystalline form. Sepiolite precipitates homogeneously from low temperature solutions at high pH.[281-283] Rates of homogeneous growth increase with pH. Measurements of seeded sepiolite growth[284] suggest that heterogenous growth is also pH dependent. Early work by Siffert[280] indicated that Mg-silicate growth might be limited by the adsorption of aqueous silica to Mg octahedra. Recent work suggests that growth may depend more strongly on the adsorption and dehydration of Mg.[243]

X. CONCLUSION

The emerging picture of reacting minerals is one of ensembles of crystal planes protonating and dissolving, or growing, semiindependently of each other. The formalism for most of the individual reaction rate laws is reasonably well established, as are the basic links between reaction order, surface complexation, and activated complex theory. This is particularly true for mineral dissolution. Recent advances in SEM and AFM imaging of dissolution and growth pathways have the advantage of revealing reaction modes on the individual planes on a pseudomolecular scale. To relate the latter results to the large number of bulk dissolution and growth experiments in the literature will ultimately require a rigorous approach for quantifying the surface chemistry of the microtopographic features of mineral lattices. To model overall reaction behavior in nature also requires a means for integrating microscopic reaction mechanisms and adsorption parameters over an evolving macroscopic mineral surface.

ACKNOWLEDGMENTS

PVB thanks the U.S. Department of Energy Office of Basic Energy Sciences/Geosciences, the U.S. National Science Foundation, and the American Chemical Society for funding. WAH thanks the Natural Environment Research Council, U.K., for funding. Many thanks to Madeleine Biber, Jim Mazer, Carl Moses, Eric Oelkers, John Walther, and Roy Wogelius for proofreading all or part of this chapter. Thanks also to Preston Poulter for help with references.

REFERENCES

1. Warren, I.H.; Devuyst, E. *International Symposium on Hydrometallurgy.* 1971; Chapter 11.
2. Terry, B. *Hydrometallurgy.* 1983, *11*, 315–344.
3. Clay, J.P.; Thomas, A.W. *J. Am. Chem. Soc.* 1938, *60*, 2384–2390.

4. Wieland, E.; Wehrli, B.; Stumm, W. *Geochim. Cosmochim. Acta.* 1988, 50, 1847–1860.
5. Biber, M.V. Ph.D. Thesis. Swiss Federal Inst. Tech: Zürich, Switzerland, 1992.
6. Compton, R.G.; Pritchard, K.L. *Phil. Trans. R. Soc. Lond.* 1990, A330, 47–70.
7. Baker, P.; Kastner, M. *Science.* 1981, 213, 214–216.
8. Wirth, G.S.; Gieskes, J.M. *J. Coll. Interf. Sci.* 1979, 68, 492–500.
9. Day, G.McD.; Hart, B.T.; McKelvie, I.D.; Beckett, R. *Coll. Surf.* 1994, 89, 1–13.
10. Schindler, P.W.; Stumm, W. *Aquatic Surface Chemistry: Chemical Processes at the Particle-Water Interface;* W. Stumm; Ed. John Wiley & Sons: New York, 1987; pp. 83–110.
11. Furrer, G.; Stumm, W. *Geochim. Cosmochim. Acta.* 1986, 50, 1847–1860.
12. Parks, G.A. *Chem. Rev.* 1965, 65, 177–198.
13. Thompson, D.W.; Pownall, P.G. *J. Coll. Interf. Sci.* 1989, 131, 74–82.
14. Westall, J.C.; Hohl, H. *Adv. Coll. Interf. Sci.* 1980, 12, 265–294.
15. Papelis, C.; Hayes, K.F.; Leckie, J.O. HYDRAQL: A program for the complexation of chemical equilibrium composition of aqueous batch systems including surface-complexation modeling of ion adsorption at the oxide/solution interface. Stanford University Technical Report No. 306, 1988.
16. Petrovich, R. *Geochim. Cosmochim. Acta.* 1981, 45, 1665–1674.
17. Petrovich, R. *Geochim. Cosmochim. Acta.* 1981, 45, 1675–1686.
18. Schott, J.; Berner, R.A.; Sjöberg, E.L. *Geochim. Cosmochim. Acta.* 1981, 45, 2123–2135.
19. Eggleston, C.M.; Hochell, M.F., Jr.; Parks, G.A. *Geochim. Cosmochim. Acta.* 1989, 53, 797–804.
20. Nagy, K.L.; Lasaga, A.C. *Geochim. Cosmochim. Acta.* 1992, 56, 3093–3111.
21. Chou, L.; Wollast, R. *Am. J. Sci.* 1985, 285, 963–993.
22. Knauss, K.; Wolery, T.J. *Geochim. Cosmochim. Acta.* 1986, 50, 2481–2497.
23. Compton, R.G.; Pritchard, K.L.; Unwin, P.R. *J. Chem. Soc. Chem. Commun.* 1989, 4, 249–251.
24. Beck, W.; Seyfried, W., Jr.; Berndt, M. *Geol. Soc. Am. Mtg. Abs.* 1993, 23, 319.
25. Bennett, P.C.; Melcer, M.; Siegel, D.; Hassett, J. *Geochim. Cosmochim. Acta.* 1988, 52, 1521–1530.
26. Wogelius, R.A.; Walther, J.V. *Geochim. Cosmochim. Acta.* 1991, 55, 943–954.
27. Brunauer, S.; Emmett, P.H.; Teller, E. *J. Am. Chem. Soc.* 1938, 60, 309–319.
28. White, A.F., Peterson, M.L. *Chemical Modeling in Aqueous Systems II;* D.C. Melchior; R.L. Bassett; Eds. ACS Symposium Series 416; American Chemical Society: Washington, D.C., 1990, Chapter 35.
29. Holdren, G.; Speyer, P. *Geochim. Cosmochim. Acta.* 1985, 49, 675–683.
30. Anbeek, C. *Geochim. Cosmochim. Acta.* 1992, 56, 1461–1470.
31. Zhang, H.; Bloom, P.R.; Nater, E.A. *Geochim. Cosmochim. Acta.* 1993, 57, 1681–1690.
32. Xie, Z.; Walther, J.V. *Geochim. Cosmochim. Acta.* 1994, 58, 2587–2589.
33. Chiarello, R.; Sturchio, N.; Wogelius, R. *Geochim. Cosmochim. Acta.* 1993, 57, 4103–4110.
34. Gratz, A.; Bird, P.; Quiro, G.B. *Geochim. Cosmochim. Acta.* 1990, 54, 2911–2922.
35. Lin, C.-C.; Shen, P. *Geochim. Cosmochim. Acta.* 1993, 57, 27–35.
36. Lin, C.-C.; Shen, P. *Geochim. Cosmochim. Acta.* 1993, 57, 1649–1655.
37. Gratz, A.; Bird, P. *Geochim. Cosmochim. Acta.* 1993, 57, 965–976.
38. Stumm, W.; Furrer, G.; and Kunz, B. *Croat. Chem. Acta.* 1983, 46, 593–611.
39. Zinder, B.; Furrer, G.; Stumm, W. *Geochim. Cosmochim. Acta.* 1986, 50, 1861–1870.
40. Carroll-Webb, S.A.; Walther, J.V. *Geochim. Cosmochim. Acta.* 1988, 52, 2609–2623.
41. Brady, P.V.; Walther, J.V. *Chem. Geol.* 1990, 82, 253–264.
42. House, W.A.; Orr, D.R. *J. Chem. Soc. Faraday Trans.* 1992, 88, 233–241.
43. Casey, W.; Westrich, H. *Nature.* 1992, 355, 157–159.
44. Sverjensky, D.; Molling, P.A. *Nature.* 1992, 358, 310–313.
45. Hiemstra, T.; Van Riemsdijk, W.H. *J. Coll. Interf. Sci.* 1990, 136, 132–150.
46. Gratz, A.; Bird, P. *Geochim. Cosmochim. Acta.* 1993, 57, 977–990.
47. Yates, D.; Healy, T.W. *J. Coll. Interf. Sci.* 1976, 55, 9–20.
48. Marinsky, J.A. *Aquatic Surface Chemistry.* W. Stumm, Ed. Wiley-Interscience: New York, 1987; pp. 49–81.

49. Dove, P.M.; Crerar, D.A. *Geochim. Cosmochim. Acta.* 1990, *54*, 955–969.
50. Dove, P.M. *Am. J. Sci.* 1994, *294*, 665–712.
51. House, W.A. *J. Coll. Interf. Sci.* 1994, *163*, 379–390.
52. Lasaga, A.C.; Gibbs, G.V. *Am. J. Sci.* 1990, *290*, 263–295.
53. Casey, W.H.; Lasaga, A.C.; Gibbs, G.V. *Geochim. Cosmochim. Acta.* 1991, *54*, 3369–3378.
54. Bennett, P.C. *Geochim. Cosmochim. Acta.* 1991, *55*, 1781–1798.
55. Murphy, W.M.; Helgeson, H.C. *Am. J. Sci.* 1989, *289*, 17–101.
56. House, W.A.; Hickinbotham, L.A. *J. Chem. Soc. Faraday Trans.* 1992, *88*, 2021–2026.
57. Nancollas, G.G.; Reddy, M.M. *J. Coll. Interf. Sci.* 1971, *37*, 824–830.
58. Wiechers, H.N.S.; Sturrock, P.; Marais, G.V.R. *Water Res.* 1975, *9*, 835.
59. Kazmierczak, T.F.; Tomson, M.B.; Nancollas, G.H. *J. Phys. Chem.* 1982, *86*, 103.
60. Cassford, G.E.; House, W.A.; Pethybridge, A.D. *J. Chem. Soc. Faraday Trans.* 1983, *79*, 1617–1632.
61. Sjöberg, E.L.; Rickard, D. *Geochim. Cosmochim. Acta.* 1983, *47*, 2281–2286.
62. MacInnis, I.N.; Brantley, S.L. *Geochim. Cosmochim. Acta.* 1992, *56*, 1113–1126.
63. Berner, R.A. *Am. J. Sci.* 1978, *278*, 1235–1252.
64. Machesky, M.L. *Chemical Modeling in Aqueous Systems II.* D.C. Melchior; R.L. Bassett; Eds. ACS Symposium Series 416. American Chemical Society: Washington, D.C., 1990; pp. 262–274.
65. Fokkink, L.G.J.; deKeizer, A.; Lyklema, J. *J. Coll. Interf. Sci.* 1989, *127*, 115–131.
66. Johnson, B.B. *Environ. Sci. Tech.* 1990, *24*, 112–118.
67. Brady, P.V. *Geochim. Cosmochim. Acta.* 1992, *56*, 2941–2946.
68. Brady, P.V. *Geochim. Cosmochim. Acta.* 1994, *58*, 1213–1217.
69. Wogelius, R.A.; Walther, J.V. *Chem. Geol.* 1992, *97*, 101–112.
70. Brady, P.V.; Walther, J.V. *Am. J. Sci.* 1992, *292*, 639–658.
71. Casey, W.H.; Sposito, G. *Geochim. Cosmochim. Acta.* 1992, *56*, 3825–3830.
72. Tamm, O. *Chem. Erde.* 1930, *4*, 429–440.
73. Casey, W.H.; Bunker, B. *Rev. Min.* 1990, *23*, 397–340.
74. Hellmann, R.; Eggleston, C.M.; Hochella, M.F., Jr.; Crerar, D.A. *Geochim. Cosmochim. Acta.* 1990, *54*, 1267–1281.
75. Parks, G. *Equilibrium Concepts in Natural Water Systems.* W. Stumm; Ed. Advances in Chemistry Series — American Chemical Society. 1967, *67*, 67–121.
76. Stumm, W.; Morgan, J.J. *Aquatic Chemistry;* 2nd Ed. John Wiley & Sons: New York, 1981.
77. Carré, A.; Roger, F.; Varinot, C. *J. Coll. Interf. Sci.* 1992, *154*, 174–183.
78. Sverjensky, D. *Geochim. Cosmochim. Acta.* 1994, *58*, 3123–3230.
79. Blum, A.E.; Lasaga, A.C. *Nature.* 1988, *331*, 431–433.
80. Amrhein, C.; Suarez, D. *Geochim. Cosmochim. Acta.* 1988, *52*, 2785–2794.
81. Brady, P.V.; Walther, J.V. *Geochim. Cosmochim. Acta.* 1989, *53*, 2823–2830.
82. Blum, A.E.; Lasaga, A.C. *Geochim. Cosmochim. Acta.* 1991, *55*, 2193–2202.
83. Xiao, Y.; Lasaga, A.C. *Geochim. Cosmochim. Acta.* 1994, *58*, 5379–5400.
84. Carroll, S.A.; Walther, J.V. *Am. J. Sci.* 1990, *290*, 797–810.
85. Wieland, E.; Stumm, W. *Geochim. Cosmochim. Acta.* 1992, *53*, 3339–3356.
86. Gautier, J.-M.; Oelkers, E.H.; Schott, J. *Geochim. Cosmochim. Acta.* 1994, *58*, 4549–4560.
87. Oelkers, E.H.; Schott, J. *Geochim. Cosmochim. Acta.* Submitted.
88. Devidal, J.L. Ph.D. Thesis. University Paul Sabatier: Toulouse, France, 1994.
89. Oelkers, E.H.; Schott, J.; Devidal, J.L. *Geochim. Cosmochim. Acta.* 1994, *58*, 2011–2024.
90. Stumm, W.; Wollast, R. *Rev. Geophys.* 1990, *28*, 53–69.
91. Casey, W.H. *J. Coll. Interf. Sci.* 1991, *146*, 586–589.
92. Ragnarsdóttir, K.V. *Geochim. Cosmochim. Acta.* 1993, *57*, 661–669.
93. Oxburgh, R.; Drever, J.I.; Sun, Y.T. *Geochim. Cosmochim. Acta.* 1994, *58*, 2439–2449.
94. Goldich, S. *J. Geol.* 1938, *46*, 17–58.
95. Casey, W.H.; Westrich, H.R.; Holdren, G.R. *Am. Miner.* 1991, *76*, 211–217.
96. Tolé, M.P.; Lasaga, A.C.; Pantano, C.; White, W. *Geochim. Cosmochim. Acta.* 1986, *49*, 1969–1981.

97. Carroll, S.A. Ph.D. Thesis, Northwestern University: Evanston, IL, 1989.
98. Bales, R.C.; Morgan, J.J. *Geochim. Cosmochim. Acta.* 1986, *49*, 2281–2288.
99. Schweda, P. *Water-Rock Interaction WRI-6.* A.A. Balkema, Rotterdam. 1989; pp. 609–612.
100. Kubicki, J.D.; Xiao, Y.; Lasaga, A.C. *Geochim. Cosmochim. Acta.* 1993, *57*, 3847–3853.
101. Smyth, J. *Geochim. Cosmochim. Acta.* 1989, *53*, 1101–1110.
102. Knauss, K.; Wolery, T.J. *Geochim. Cosmochim. Acta.* 1989, *53*, 1493–1501.
103. Rose, N.M. *Geochim. Cosmochim. Acta.* 1991, *55*, 3273–3286.
104. Knauss, K.G.; Nguyen, S.N.; Weed, H.C. *Geochim. Cosmochim. Acta.* 1993, *57*, 285–294.
105. Williams, Q.; Williams, K.J. *Lancet.* 1991, *337*, 791.
106. Ewing, R.C. *Scientific Basis for Nuclear Waste Management.* Plenum Press: New York, 1979; pp. 57–68.
107. Rana, M.A.; Douglas, R.W. *Phys. Chem. Glasses.* 1961, 2, 196–205.
108. Bunker, B. *Mat. Res. Soc. Symp. Proc.* 1987, *84*, 493–507.
109. Bunker, B.; Arnold, G.W.; Beauchamp, E.K.; Day, D.E. *J. Non-Cryst. Sol.* 1983, *58*, 295–322.
110. Fleming, B. *J. Coll. Interf. Sci.* 1986, *110*, 40–64.
111. Li, H.C.; deBruyn, P.L. *Surf. Sci.* 1966, *5*, 203–220.
112. Bolt, G. *J. Phys. Chem.* 1958, *61*, 1166.
113. Mazer, J.J.; Walther, J.V. *J. Non-Cryst. Sol.* 1994, *170*, 32–45.
114. Carroll, S.A.; Bourcier, W.L.; Phillips, L. *Mat. Res. Soc. Symp. Proc.* 1994, *333*, 533–540.
115. Mazer, J.J. Ph.D. Thesis, Northwestern University: Evanston, IL, 1987.
116. Jantzen, C.M.; Plodinec, M.J. *J. Non-Cryst. Sol.* 1984, *67*, 207–223.
117. Jantzen, C.M. Paper presented at the Conference on Advances in the Fusion of Glass, June 14–17, 1988, Alfred, NY, 1988; p. 18.
118. Abrajano, T.A.; Bates, J.K.; Bohlke, J.K. *Mat. Res. Soc. Symp. Proc.* 1988, *125*, 383–392.
119. Paul, A. *J. Mat. Sci.* 1977, *12*, 2246–2268.
120. Grambow, B. *Mat. Res. Soc. Symp. Proc.* 1985, *44*, 16–27.
121. Paul, A. *J. Mat. Sci.* 1977, *12*, 2246–2268.
122. Jantzen, C.M. *Advances in Ceramics. Nuclear Waste Management II;* D.E. Clark; W.B. White; A.J. Machiels; Eds. The American Ceramic Society: Westerville, OH, 1986; 20; pp. 703–712.
123. Mazer, J.J.; Bates, J.K.; Bradley, J.P.; Bradley, C.R.; Stevenson, C.M. *Nature.* 1992, *357*, 573–576.
124. Graustein, W.-C.; Cromack, K., Jr.; Sollins, P. *Science.* 1977, *198*, 1252–1254.
125. Biber, M.V.; Afonso, M.; Deng, Y.; Stumm, W. *Geochim. Cosmochim. Acta.* 1994, *58*, 1999–2010.
126. Biber, M.V. *Environ. Sci. Tech.* 1994, *28*, 763–768.
127. Bruno, J.; Wersin, P.; Stumm, W.; Brandberg, F. *Geochim. Cosmochim. Acta.* 1992, *56*, 1139–1148.
128. Dugger, D.L.; Stanton, J.H.; Irby, B.N.; McConnell, B.L.; Cummings, W.W.; Maatman, R.W. *J. Phys. Chem.* 1964, *68*, 757–760.
129. Stone, A.T.; Morgan, J.J. *Aquatic Surface Chemistry;* W. Stumm, Ed. Wiley-Interscience: New York, 1987.
130. Francis, A.J.; Dodge, C.J. *Appl. Env. Microbiol.* 1988, *54*, 1009–1014.
131. Hering, J.; Stumm, W. *Rev. Min.* 1990, *23*, 427–465.
132. Huang, W.H.; Kiang, W.C. *Am. Mineral.* 1972, *57*, 1849–1859.
133. Grandstaff, D.E. *Third International Syposium on Water–Rock Interactions, Proceedings.* Alberta Research Council: Edmonton, 1980; pp. 72–74.
134. Grandstaff, D.E. *Geochim. Cosmochim. Acta.* 1977, *41*, 1097–1103.
135. Mast, M.A.; Drever, J.I. *Geochim. Cosmochim. Acta.* 1987, *51*, 2259–2568.
136. Fein, J.B. *Geology.* 1991, *19*, 1037–1040.
137. Brady, P.V.; Carroll, S.A. *Geochim. Cosmochim. Acta.* 1994, *58*, 1853–1856.
138. Cawley, J.L.; Burruss, R.C.; Holland, H.D. *Science.* 1965, *165*, 391–392.
139. Lagache, M. *Geochim. Cosmochim. Acta.* 1976, *40*, 157–161.

140. Morse, J.W. *Mar. Chem.* 1986, *20*, 91–112.
141. Reuss, J.O.; Cosby, B.J.; Wright, R.F. *Nature*. 1987, *329*, 27–32.
142. Davison, W.; House, W.A. *Water Res.* 1988, *22*, 577–583.
143. Compton, R.G.; Unwin, P.R. *Phil. Trans. R. Soc. Lond.* 1990, *A330*, 1–45.
144. Compton, R.G.; Daly, P.J.; House, W.A. *J. Coll. Interf. Sci.* 1986, *113*, 12–20.
145. Compton, R.G.; Daly, P.J. *J. Col. Interf. Sci.* 1984, *101*, 159–166.
146. Plummer, L.N.; Parkhurst, D.L.; Wigley, T.M.L. *Chemical Modeling in Aqueous Systems*; ACS Symposium Series 93. American Chemical Society: Washington, D.C., 1979, *93*, 537.
147. Berner, R.A.; Morse, J.W. *Am. J. Sci.* 1974, *274*, 108–134.
148. Plummer, L.N.; Wigley, T.M.L. *Geochim. Cosmochim. Acta*. 1976, *40*, 191–202.
149. Rickard, D.; Sjöberg, E.L. *Am. J. Sci.* 1983, *283*, 815–830.
150. Morse, J.W. *Rev. Mineral.* 1983, *11*, 227–264.
151. Chou, L.; Garrels, R.M.; Wollast, R. *Chem. Geol.* 1989, *78*, 269–282.
152. Plummer, L.N.; Wigley, T.M.L.; Parkhurst, D.L. *Am. J. Sci.* 1978, *278*, 179–216.
153. Reddy, M.M.; Plummer, L.N.; Busenberg, E. *Geochim. Cosmochim. Acta*. 1981, *45*, 1281–1289.
154. House, W.A. *J. Chem. Soc. Faraday Trans. 1*. 1981, *77*, 341–359.
155. House, W.A.; Howson, M.R.; Pethybridge, A.D. *J. Chem. Soc. Faraday Trans*. 1988, *184*, 2723–2734.
156. Reddy, M.M. *Sci. Geol. Mem.* 1983, *71*, 109–117.
157. Compton, R.G.; Pritchard, K.L.; Unwin, P.R. *Freshwater Biol.* 1989, *22*, 285–288.
158. Wicks, C.M.; Groves, C.G. *J. Hydrol.* 1993, *146*, 13–27.
159. Schott, J.; Brantley, S.; Crerar, D.; Guy, C.; Borcsik, M.; Willaime, C. *Geochim. Cosmochim. Acta*. 1989, *53*, 373–382.
160. Busenberg, E.; Plummer, L.N. *Am. J. Sci.* 1982, *282*, 45–78.
161. Morse, J.W.; Mackenzie, F.T. *Geochemistry of Sedimentary Carbonates*. Elsevier: Amsterdam, 1990.
162. Zachara, J.M.; Cowan, C.E.; Resch, C.T. *Geochim. Cosmochim. Acta*. 1991, *55*, 1549–1562.
163. Stipp, S.L.; Hochella, M.F., Jr. *Geochim. Cosmochim. Acta*. 1991, *55*, 1723–1748.
164. Charlet, L.; Wersin, P.; Stumm, W. *Geochim. Cosmochim. Acta*. 1991, *54*, 2329–2336.
165. Carroll, S.A.; Papenguth, H.W. *Geochim. Cosmochim. Acta*. Submitted.
166. Foxall, T.; Peterson, G.C.; Rendall, H.M.; Smith, A.L. *J. Chem. Soc. Faraday Trans.* 1979, *175*, 1034.
167. Van Cappellen, P.; Charlet, L.; Stumm, W.; Wersin, P. *Geochim. Cosmochim. Acta*. 1993, *57*, 3505–3518.
168. House, W.A.; Donaldson, L. *J. Coll. Interf. Sci.* 1986, *112*, 309–324.
169. Compton, R.G.; Pritchard, K.L. *J. Chem. Soc. Faraday Trans.* 1990, *86*, 129–136.
170. Zhang, J.W.; Nancollas, G.H. *Rev. Min.* 1990, *23*, 365–393.
171. Barwise, A.J.; Compton R.G.; Unwin, P.R. *J. Chem. Soc. Faraday Trans.* 1990, *86*, 137–144.
172. Compton, R.G.; Pritchard, K.L.; Unwin, P.R.; Grigg, G.; Lees, M.; Silvester, P.; House, W.A. *J. Chem. Soc. Faraday Trans.* 1989, *185*, 4335–4366.
173. Weyl, P.K. *J. Sediment. Petrol.* 1960, *30*, 85–90.
174. Berner, R.A. *Am. J. Sci.* 1967, *267*, 45–70.
175. Sjöberg, E.L. *Acta Univ. Stockholm Contrib. Geol.* 1978, *332*, 1–92.
176. Morse, J.W. *Am. J. Sci.* 1974, *274*, 638–647.
177. Reddy, M.M. *J. Crystal Growth.* 1977, *41*, 287–295.
178. House, W.A. *J. Coll. Interf. Sci.* 1986, *119*, 505–511.
179. Suess, E. *Geochim. Cosmochim. Acta*. 1970, *34*, 157–168.
180. Berner, R.A. *Am. J. Sci.* 1978, *10*, 1475–1477.
181. Barton, P. *Sulfur in the Environment, Part II: Ecological Impacts*. Wiley-Interscience: New York, 1978; pp. 313–358.
182. Goddard, J.B.; Brosnahan, D.R. *Mining Eng.* 1982, *Nov.*, 1589–1596.
183. Singer, P.C.; Stumm, W. *Science*. 1970, *167*, 1121–1123.

184. Stumm-Zollinger, E. *Arch. Mikrob.* 1972, *83*, 110–119.
185. McKibben, M.A.; Barnes, H.L. *Geochim. Cosmochim. Acta.* 1986, *50*, 1509–1520.
186. Wiersma, C.L.; Rimstidt, J.D. *Geochim. Cosmochim. Acta.* 1984, *48*, 85–92.
187. Moses, C.O.; Nordstrom, D.K.; Herman, J.S.; Mills, A.L. *Geochim. Cosmochim. Acta.* 1987, *51*, 1561–1571.
188. Luther, G.W. III. *Geochim. Cosmochim. Acta.* 1987, *51*, 3193–3199.
189. Mishra, K.K.; Osseo-Asare, K. *J. Electrochem. Soc.* 1988, *135*, 2502–2509.
190. Rimstidt, J.D.; Chermak, J.A.; Gagen, P.M. *Environmental Geochemistry of Sulfide Oxidation;* C. N. Alpers; D. W. Blowes; Eds. ACS Symposium Series 550; American Chemical Society: Washington, D.C., 1994, pp. 3–13.
191. Nicholson, R.V.; Scharer, J.M. Environmental Geochemistry of Sulfide Oxidation; C.N. Alpers; D.W. Blowes; Eds. ACS Symposium Series 550; American Chemical Society: Washington, D.C., 1994, pp. 14–31.
192. Jean, G.E.; Bancroft, G.M. *Geochim. Cosmochim. Acta.* 1986, *50*, 1455–1463.
193. Morse, J.W.; Millero, F.; Cornwell, J.C.; Rickard, D. *Earth Sci. Rev.* 1987, *24*, 1–42.
194. Park, S.W.; Huang, C.P. *J. Coll. Interf. Sci.* 1987, *117*, 431–441.
195. Renders, P.J.; Seward, T.M. *Geochim. Cosmochim. Acta.* 1989, *53*, 255–267.
196. Rönngren, L.; Sjöberg, S.; Zhonxi, S.; Forsling W.; Schindler, P.W. *J. Coll. Interf. Sci.* 1991, *145*, 396–404.
197. Sun, Z.; Forsling, W.; Rönngren, L.; Sjöberg, S. *Intern. J. Min. Proc.* 1991, *33*, 83–93.
198. Dekkers, M.J.; Schoonen, M.A.A. *Geochim. Cosmochim. Acta.* 1994, *58*, 4147–4153.
199. Davis, A.P.; Huang, C.P. *Langmuir.* 1990, *7*, 803–808.
200. Aagaard, P.; Helgeson, H.C. *Am. J. Sci.* 1982, *282*, 237–285.
201. Lasaga, A.C. *J. Geophys. Res.* 1984, *89 B6*, 4009–4025.
202. Steefel, C.I.; Van Cappellen, P. *Geochim. Cosmochim. Acta.* 1990, *54*, 2657–2677.
203. Eyring, L. *J. Chem. Phys.* 1935, *3*, 107–115.
204. Hiemstra, T.; Van Riemsdijk, W.H.; Bolt, G.H. *J. Colloid Interf. Sci.* 1989, *133*, 105–117.
205. Berger, G.; Cadore, E.; Schott, J.; Dove, P. *Geochim. Cosmochim. Acta.* 1994, *58*, 541–551.
206. Iler, R.K. *The Chemistry of Silica.* Wiley-Interscience: New York, 1979.
207. Rimstidt, J.D.; Barnes, H.L. *Geochim. Cosmochim. Acta.* 1980, *44*, 1683–1699.
208. Morgan, J.J.; Stone, A.T. *Aquatic Surface Chemistry;* W. Stumm; Ed. John Wiley: New York, 1989; pp. 389–426.
209. Krupka, R.M.; Kaplan, H.; Laidler, K.J. *J. Chem. Soc. Trans. Faraday Soc.* 1966, *62*, 2754–2759.
210. Lasaga, A.C.; Soler, J.M.; Ganor, J.; Burch, T.E.; Nagy, K.L. *Geochim. Cosmochim. Acta.* 1990, *58*, 2361–2386.
211. Nagy, K.L.; Steefel, C.I.; Blum, A.E.; Lasaga, A.C. *AAPG Memoir.* 1990, *49*, 85–101.
212. Murphy, W.M.; Pabalan, R.T.; Prikryl, J.D.; Goulet, C.J.; *Water–Rock Interaction WRI-7;* Y. Kharaka; A. Maest; Eds., Balkema: Rotterdam, The Netherlands, 1992; pp. 107–110.
213. Murphy, W.M.; Pabalan, R.T.; Prikryl, J.D.; Goulet, J.C. *Am. J. Sci.* 1995. In press.
214. Devidal, J.L.; Dandurand, J.L.; Schott, J. *Water–Rock Interaction WRI-7;* Y. Kharaka; A. Maest; Eds. Balkema: Rotterdam, The Netherlands, 1992; pp. 93–96.
215. Burch, T.E.; Nagy, K.L.; Lasaga, A.C. *Chem. Geol.* 1993, *105*, 137–162.
216. Cama, J.; Ganor, J.; Lasaga, A.C. *Mineral. Mag.* 1994, *58A*, 1401.
217. Oelkers, E.H.; Schott, J. *Mineral. Mag.* 1994, *58A*, 559–560.
218. Schott, J.; Oelkers, E.H. *P. Appl. Chem.* 1995, *67*, 903–910.
219. Nagy, K.L.; Blum, A.E.; Lasaga, A.C. *Am. J. Sci.* 1991, *291*, 649–686.
220. Bloom, P.R. *Soil Sci. Soc. Am.* 1983, *47*, 164–168.
221. Bloom, P.R.; Erich, M.S. *Soil Sci. Soc. Am.* 1987, *51*, 1131–1136.
222. Hiemstra, T.; Van Riemsdijk, W.H. *J. Coll. Interf. Sci.* 1990, *136*, 132–150.
223. Ernsberger, F.M. *J. Phys. Chem. Sol.* 1960, *13*, 347–351.
224. Wegner, M.W.; Christie, J.M. *Phys. Chem. Minerals.* 1983, *9*, 621–632.
225. Cornell, R.M.; Posner, A.M.; Quirk, J.P. *J. Inorg. Nucl. Chem.* 1974, *36*, 1937–1946.

226. Kössel, W. *Nachr. Ges. Wiss. Göttingen Math. Phys. Kl.* 1927, 135–143.
227. Dove, P.M.; Hochella, M.F., Jr. *Geochim. Cosmochim. Acta.* 1993, *57*, 705–714.
228. Gratz, A.J.; Hillner, P.E.; Hansma, P.K. *Geochim. Cosmochim. Acta.* 1993, *57*, 491–495.
229. Hillner, P.E.; Gratz, A.J.; Manne, S.; Hansma, P.K. *Geology.* 1992, *20*, 359–362.
230. Blum, A.E.; Lasaga, A.C. *Aquatic Surface Chemistry: Chemical Processes at the Particle–Water Interface;* W. Stumm; Ed. John Wiley & Sons: New York, 1987; pp. 255–292.
231. Hachiya, K.; Sasaki, M.; Saruta, Y.; Mikami, N.; Yasumaga, T. *J. Phys. Chem.* 1984, *88*, 23–27.
232. Murphy, W.M.; Oelkers, E.H.; Lichtner, P.C. *Chem. Geol.* 1989, *78*, 357–380.
233. Ortoleva, P.; Merino, E.; Moore, C.; Chadam, J. *Am. J. Sci.* 1987, *287*, 979–1009.
234. Lasaga, A.C. *Am. J. Sci.* 1979, *279*, 324–346.
235. Li, Y.-H.; Gregory, S. *Geochim. Cosmochim. Acta.* 1974, *38*, 703–714.
236. Nielsen, A.E. *Croat. Chem. Acta.* 1980, *53*, 255–279.
237. Nielsen, A.E. *J. Crystal Growth.* 1984, *67*, 289–310.
238. Burton, W.K.; Cabrera, N.; Frank, F.C. *Philos. Trans. R. Soc. London.* 1951, *A243*, 299–358.
239. Söhnel, O. *J. Crystal Growth.* 1982, *57*, 101–108.
240. Reich, R.; Kahlweit M. *Ber. Bunsenges.* 1968, *72*, 66.
241. Ohara, M.; Reid, R.C. *Modeling Crystal Growth Rates from Solution.* Prentice-Hall: Englewood Cliffs, NJ, 1973.
242. Van Cappellen, P.; Berner, R.A. *Geochim. Cosmochim. Acta.* 1991, *55*, 1219–1234.
243. Brady, P.V. *Water–Rock Interaction WRI-7;* Y. Kharaka; A. Maest; Eds. Balkema: Rotterdam, The Netherlands, 1992; pp. 85–88.
244. Gardner, G.L.; Nancollas, G.H. *J. Dent. Res.* 1976, *55*, 342–352.
245. Doremus, R.H. *J. Phys. Chem.* 1958, *62*, 1068.
246. Nancollas, G.H.; Purdie, N. *Rev. Chem. Soc.* 1964, *18*, 1.
247. Nancollas, G.H. *DHEW Publ (NIH).* 1977, *NIH-77-1063*, 33–54.
248. Liu, S.T.; Nancollas, G.H. *J. Coll. Interf. Sci.* 1975, *52*, 593–601.
249. Inskeep, W.P.; Bloom, P.R. *Geochim. Cosmochim. Acta.* 1985, *49*, 2165–2180.
250. Reddy, M.M.; Nancollas, G.H. *Desalination.* 1973, *12*, 61–73.
251. Nielsen, A.E.; Toff, J.M. *J. Crystal Growth.* 1984, *67*, 278–288.
252. Reddy, M.M. *Studies in Diagenesis;* F.A. Mumpton; Ed. USGS Bull. 1578, 1986; pp. 169–182.
253. Mucci, A.; Morse, J.W. *Geochim. Cosmochim. Acta.* 1983, *47*, 217–233.
254. Mucci, A.; Morse, J.W. *Am. J. Sci.* 1985, *285*, 306–317.
255. Hirth, J.P.; Pound, G.M. *J. Chem. Phys.* 1958, *26*, 1216–1224.
256. Chiang, P.; Donohue, M.C.; Katz, J.L. *J. Coll. Interf. Sci.* 1988, *122*, 251–265.
257. Kleiner, J. *Water Res.* 1988, *22*, 1259–1265.
258. Raidt, H.; Koschel, R. *Limnologica.* 1993, *23*, 85–89.
259. House, W.A. *Water Res.* 1990, *24*, 1017–1023.
260. Walter, L.M.; Burton, E.A. *Chem. Geol.* 1986, *56*, 313–323.
261. Hinedi, Z.R.; Goldberg, S.; Chang, A.C.; Yesinowski, J.P. *J. Coll. Interf. Sci.* 1992, *152*, 141–160.
262. Davies, C.W.; Jones, A.L. *J. Chem. Soc. Trans. Faraday Soc.* 1955, *51*, 812–817.
263. House, W.A.; Shelley, N.; Fox, A.M. *Hydrobiologia.* 1988, *178*, 93–112.
264. Christoffersen, J.; Christoffersen, M.R. *J. Crystal Growth.* 1990, *100*, 203–211.
265. Christoffersen, J.; Christoffersen, M.R. *J. Crystal Growth.* 1981, *53*, 42–54.
266. Moutin, T.; Gal, J.Y.; Halouani, H.El.; Picot, B.; Bontoux, J. *Water Res.* 1990, *26*, 1445–1450.
267. van Kemenade, M.J.J.M.; de Bruyn, P.L. *J. Coll. Interf. Sci.* 1987, *118*, 564–585.
268. Christoffersen, M.R.; Christoffersen, J.; Kibalczyc, W. *J. Crystal Growth.* 1990, *106*, 349–354.
269. Christoffersen, J. *J. Crystal Growth.* 1980, *49*, 29–44.
270. Christoffersen, J.; Christoffersen, M.R. *J. Crystal Growth.* 1982, *57*, 21–26.
271. Van Leeuwen, C.; Blomen, L.J.M.J. *J. Crystal Growth.* 1979, *46*, 96–104.

272. Martens, C.S.; Harriss, R.C. *Geochim. Cosmochim. Acta.* 1970, *34*, 621–625.
273. Zawacki, S.J.; Koutsoukos, P.B.; Salimi, M.H.; Nancollas, G.H. *Geochemical Processes at Mineral Surfaces*; J.A. Davis; K.F. Hayes; Eds. ACS Symposium Series 323. American Chemical Society: Washington, D.C., 1990; pp. 650–662.
274. Christoffersen, J.; Christoffersen, M.R.; *J. Chem. Soc. Faraday Trans.* 1984, *77*, 235–242.
275. Ernst, W.G.; Calvert, S.E. *Am. J. Sci.* 1969, *267*, 114–133.
276. Stein, C.L.; Kirkpatrick, R.J. *J. Sediment. Petr.* 1976, *46*, 430–435.
277. Williams, L.A.; Crerar, D.A. *J. Sediment. Petr.* 1982, *55*, 312–321.
278. Gastuche, M.C.; Fripiat, J.J.; DeKempe, C. *Génese et Synthése dArgiles: C.N.R.S. Paris.* 1962, *105*, 57–65.
279. Lineares, J.; Huertas, F. *Science.* 1971, *171*, 896–897.
280. Siffert, B. *Mem. Ser. Carte. Geol. Alsace-Lorraine.* 1962, *21*, 1–86.
281. Siffert, B.; Wey, R. *C.R. Acad. Sci. Paris.* 1962, *254*, 549–581.
282. Wollast, R.; MacKenzie, F.T.; Bricker, O.P. *Am. Mineral.* 1968, *53*, 1645–1662.
283. Couture, R.A. Ph.D. dissertation, University of California: San Diego, 1977.
284. Kent, D.B.; Kastner, M. *Geochim. Cosmochim. Acta.* 1985, *49*, 1123–1136.
285. Bourcier, W.L.; Weed, H.C.; Nguyen, S.N.; Nielson, J.K.; Morgan, L.; Newton, L.; Knauss, K.G. *Water–Rock Interaction WRI-7*; Y. Kharaka; A. Maest; Eds. Balkema: Rotterdam, The Netherlands, 1992; pp. 81–84.
286. Berner, R.A. *Early Diagenesis: A Theoretical Approach*, Princeton University Press: Princeton, NJ, 1980.

Chapter 5

GEOCHEMICAL APPLICATIONS OF MINERAL SURFACE SCIENCE

Patrick V. Brady and John M. Zachara

CONTENTS

0-8493-8351-X/96/$0.00+$.50
© 1996 by CRC Press, Inc.

I. INTRODUCTION

A complete coverage of surface chemistry applications is impossible in one book, much less in a single chapter. The object of this chapter is to explore the link between chemical reactions occurring at mineral surfaces and macroscopic geochemical processes. Mineral surfaces greatly influence the chemistry of natural waters, both at the Earth–atmosphere interface and at depth. Adsorption and ion exchange reactions limit the movement of a large number of metals, nutrients, and organic molecules in soils and groundwaters. Particulate matter in surface waters, in turn, constitute mobile sinks for metals and many organic molecules. As a result, the biogeochemical cycling of most elements depends largely on surface chemistry and the fluxes of reactive mineral surfaces from soil to sea to sediment and back again. Surface-controlled chemical transport in soils is complex as soil surfaces are spatially heterogeneous, often organic, and seldom static. Soil surface chemistry can be reasonably generalized, and surface complexation models (see Chapter 3) can be used to explain and predict trends in solution–surface interactions. As a result, waste isolation and remediation techniques should rely more on an understanding of mineral–fluid interactions and will, as a result, become increasingly more involved and precise. The end result is that mineral surface chemistry may potentially provide more-effective pathways to waste isolation and environmental restoration.

Mineral–fluid interfaces influence biogeochemical cycling by catalyzing many, if not most, of the rate-determining chemical reactions occurring in soils, groundwaters, and sediments. Rates of weathering (mineral dissolution) and mineral growth depend on the chemistry of the respective crystal surface and on the stability of the underlying bonds (see Chapter 4). Some natural adsorbates (e.g., multifunctional carboxylic acids) accelerate corrosion of certain minerals, while others (e.g., phosphate) tend to inhibit dissolution and growth. Absolute rates of mineral growth and dissolution are important because weathering minerals are, to a large extent, the ultimate source of dissolved salts in natural waters. Likewise, growing minerals are important sinks.

The bulk of Earth's crust consists of silicates, mineral bases which titrate acidity out of surface waters and the atmosphere when they dissolve. When clays and Fe, Al (hydr)oxides grow, they return acidity to natural waters. Because CO_2 is the most common weak acid in the atmosphere, the mineral acid–base cycle connects directly to global climate through the greenhouse effect.[1] Over shorter, human time scales, surface-controlled dissolution of basic carbonates and silicates in soils limits the attenuation of acid rain inputs.[2] Surface-controlled weathering of metal sulfides from waste and mine dumps is an environmentally important source of acidity.

This chapter will proceed along a continuum, beginning with an examination of surface-controlled processes which operate over relatively short environmentally relevant time scales. The focus will be on mineral surface reactions (adsorption, desorption, dissolution, and growth) in soils and human tissue and, over longer time scale, during diagenesis. The linkage between surface-controlled mineral dissolution, geochemical cycling, and climate over geologic time will then be explored. The chapter will conclude with an examination of the role of surfaces in biomineralization and the origin of life nearly 4 billion years ago.

II. PARTICLE SURFACES, ION RETENTION, AND CHEMICAL TRANSPORT IN SOILS AND GROUNDWATER

The surfaces of particles in soils and subsurface geologic systems are dominated by mineral and organic material. Both of these contain reactive surface functional groups of different types that develop charge and form bonds of different character with a vast array of natural and xenobiotic ionic adsorbates. In natural systems, the functional groups are saturated to varying degrees by the electrolyte and minor ions in the bathing soil solution or groundwater, with surface speciation dependent on a complex array of factors, including interaction energies and concentrations of both aqueous ions and surface sites. These surface functional groups, and chemical reactions between them, adsorbates, and ions saturating their surface, play a preeminent role in the adsorptive retention and supply of nutrient ions; in the removal of ionic macrocontaminants from percolating waters; in the pH buffering of natural waters; in dissolution; in the nucleation of secondary, more stable products of weathering; and in the flocculation and aggregation of colloidal-sized materials. Surface chemical interactions have been implicated in economic-ore formation[3] and in the chemical evolution of groundwaters and are being intensively studied because of their implications to soil fertility, contaminant and waste management, and environmental restoration.

Ionic mineral solids common to soil and geologic environments contain two primary types of surface functional groups: (1) fixed charge sites arising from nonstoichiometric or isomorphic substitution in the bulk structure that yield a net charge imbalance and (2) variable charge sites arising from the interaction of solvent molecules or other dissolved aqueous species with partially coordinated structural ions on the crystallite surface (see, for example, References 4 and 5). Fixed charge sites are encountered on layer lattice silicates (e.g., smectites and vermiculites), zeolites, Mn(II,IV) oxides, and Ti(IV) substituted Fe(III) oxides. Variable charge sites, so termed because their charge varies with the concentrations of potential determining ions and other adsorbates in the aqueous phase, include hydroxylated metal ion centers on the surfaces of (hydr)oxides of Fe, Al, Si, and Mn; layer lattice silicate edges; and feldspars and the surfaces of carbonates, sulfides, and phosphates.[6] These different site types are summarized in Table 1. The intrinsic, or developed, charge on these sites propagates an electric double layer into solution where charge compensating ions of opposite sign are localized.

TABLE 1.

Common Types of Surface Sites in Natural Materials

Solid Phase	Surface Sites	Ref.
Variable Charge		
Al, Fe, and Mn (hydr)oxides	SOH[a,b]	7
Layer lattice silicate edges	AlOH, SiOH, or Al–OH–Si	8
Feldspars	AlOH, SiOH, Al–OH–Si	9,14
Carbonates (MeCO$_3$)	MeOH/MeO$^-$/MeOH$_2^+$/MeHCO$_3$/MeCO$_3^-$; CO$_3$H/CO$_3^-$/CO$_3$H$_2^+$/CO$_3$Me$^+$	10
Sulfides (MeS)	MeS/MeSH$^+$; SMeSMeOH$^-$	11
Phosphates	CaOH/CaOH$_2^+$; POH/PO$^-$	12
Organic matter	COOH/COO$^-$	
Fixed Charge		
Layer lattice silicate basal plane	X$^-$	13
Feldspars	X$^-$	14

[a] SOH is a generic term for a metal ion center, e.g., FeOH. Surface hydroxyls may be singly, doubly, or triply coordinated by convention; charge accounting assigns a charge of 0 or –1/2 to the SOH site.
[b] Develop charge by proton adsorption/desorption.

Surface functional groups on mineral solids undergo a wide variety of complexation type reactions with aqueous solutes (both inorganic and organic cations and anions) that have seen intense study using well-characterized phases. The results of these have been discussed in earlier chapters and other reviews.[15,16] The application of new generation surface spectroscopies has provided information on the structure and chemical characteristics of the adsorption complex.[17] Mass action models incorporating various double layer structures have been developed that can readily describe experimental adsorption data over wide ranges in pH and ionic strength, especially for Fe, Si, and Al oxides.[4,15,16] Similar models have also been used to describe charge development and sorbate binding to carbonates, sulfides, and phosphates.[10-12] Multisorbate, competitive equilibria are reasonably described by these models,[18,19] as well as other related phenomena such as coagulation.[20] These surface complexation models have been described in Chapter 3. Reactions between aqueous constituents and the surface sites in Table 1 yield adsorbed complexes of different structure and energetics, ranging from weak outer-sphere-type complexes, where the adsorbate retains its full hydration sphere and electrostatic forces predominate, to stronger inner-sphere complexes, where waters of hydration are shed and chemical forces predominate, and finally to surface precipitates containing a two- or three-dimensional structure and multiple attachment points. Some of the types of sorbate-binding reactions considered in SCM and adsorption models generally are tabulated in Table 2. These reactions are often functional in form, are based on the fitting of macroscopic chemical data, and may not be accurate at the molecular level.

TABLE 2.

Example Mass Equations Describing Sorbate — Surface Interactions

Surface Complexation to Hydroxylated Surface Sites on Oxides and Layer Silicates[a]	
$SOH + M^{m+}_{(aq)} = SOM^{(m-1)+} + H^+$ (inner sphere)	K_M
$SOH + M^{m+}_{(aq)} = SO^- - M^{m+} + H^+$ (outer sphere)	K_M
$SOH + L^{l-}_{(aq)} = SLM^{(l-1)-} + OH^-$ (inner sphere)	K_L
$SOH + H^+ + L^{l-}_{(aq)} = SOH^+_2 - L^{l-}$ (outer sphere)	K_L
Surface Precipitation/Solid Solution Formation on Hydrous Oxides[a]	
$S(OH)_3 \backslash SOMe^+ + Me^{2+}_{(aq)} + 2H_2O = \{S(OH)_3/Me(OH)_3\}/SOMe^+ + 2H^+$	K_{SS}
Ion Exchange to Fixed Charge Sites (X⁻) on Layer Silicates	
$mCX_u + uM^{m+}_{(aq)} = uMX_m + mC^{u+}_{(aq)}$	K_{MX}
Complexation to Carboxylate Sites (Lᵢ) on Organic Matter	
$HL_i + Me^{2+} = MeL^+_i + H^+$	$K_{Me}{}^{(i)}$
Surface Exchange of Structural Ions (Ca,co₃) on Calcite	
$CO_3Ca^+ + Me^{2+}_{(aq)} = CO_3Me^+ + Ca^{2+}_{(aq)}$	

[a] S is a metal ion center (e.g., Al^{3+} or Fe^{3+}) in or on a (hydr)oxide surface.

The energetics and reactivity of surface functional groups vary according to complex factors that are not fully understood. The strength of an adsorption complex on the ditrigonal cavity of a layer silicate (a fixed charge site), for example, depends on whether the structural charge deficit exists in the nearby siloxane tetrahedral layer or in the more distant gibbsitic octahedral layer.[21] The reactivity of inorganic surface hydroxyls on (hydr)oxides varies with the electronic properties of the metal ion center (i.e., SiOH vs. AlOH or FeOH) and with the number of coordinating metals to the hydroxyl (single, double, triple). Chemically distinct surface sites can exist on the same mineral phase as a result of varying coordination and structural environments on a given crystallographic face (as a result of defects), and on different crystallographic faces as well. For example, in the absence of surface defects, goethite has three different types of surface hydroxyls (A, B, and C), while gibbsite has at least two (singly coordinated hydroxyls on the edge and doubly coordinated hydroxyls on the basal plane). Crystallites of the same mineral type but with contrasting morphology or size may vary in reactivity as a result of different contributions of different crystallographic faces to the total surface area.[22,23] Recent advances in molecular scale microscopy (e.g., AFM and STM) have provided great insights on the relationships between surface structure and site reactivity and have shown that the number of reactive sites is often significantly lower than the total population of sites.[24-27]

The understanding of the interfacial surface chemistry and reactivity of particles in soils and geologic environments has lagged significantly behind that of single mineral/water systems because of the chemical and mineralogic heterogeneity of natural particles. In natural environments, particle surfaces are typically a complex mosaic of different phases. The surfaces of primary mineral grains are changed by chemical weathering and biotic activity, resulting in the formation of coatings and alteration rinds, discrete surface precipitates, and secondary mineral phases all of complex composition and properties.[4] The chemical properties of surfaces may change seasonally as a result of redox, temperature, pore water compositional variations, or biologic activity.

The transformation of a relatively clean, unaltered mineral composite into a physically complex aggregated media occurs during weathering, soil formation, and diagenesis. These transformations may have profound implications for the rate and magnitude of surface chemical reactions that occur at the grain scale in soil and subsurface porous media. Aggregation occurs as mineral particles of unlike charge associate,[28-30] organic materials and secondary mineral precipitates bind particles together,[31-33] and pressure induces more compact grain arrangements. The aggregation process, combined with chemical weathering, yields a multiphase physical association of mineral material that may contain appreciable volumes of intra- and intergranular porosity.[34,35] Such porosity complicates the understanding of chemical processes in soils and groundwaters, in that mass transfer limitations to reactive surface sites and phases may impart complex kinetics to reactions that achieve equilibrium rapidly in single-phase mineral suspensions. Grain-scale photographs of several representative subsurface materials are shown in Figure 1.

The presence of organic matter distinguishes soils from deeper geologic materials. In soils, organic matter derived from plant debris functions as a substrate for diverse and copious microbiologic activity. It also acts to aggregate mineral particles forming organomineral associations with complex physical and chemical properties.[32] The organic fraction is highly varied in composition and in its constituent components, i.e., humic substances, organic acids, amino acids, polysaccharides, etc.[36-38] These components interact with the mineral fraction and surfaces in different ways and with different effects and impacts.[32,39] Humic substances adsorb strongly to oxide surfaces modifying their surface properties and charge.[40-42] The surface properties of the mineral fraction of surface soil may be dominated by adsorbed and particle-associated organic matter. Indeed, the cation

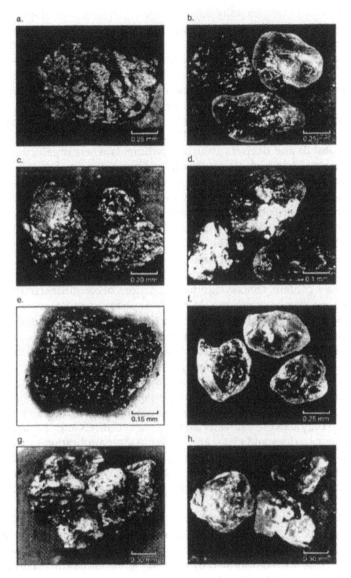

Figure 1
Optical micrographs of subsurface sediments containing: (a) grain put fillings of iron oxy-hydroxides, (c) interparticle cements of iron oxyhydroxides, (e) discrete mixed surface pre-cipitates of Fe and Mn oxides, and (g) interparticle cements of Mn oxides. Panels b, d, f, and h show matrix mineralogy for sediments to their left.

exchange capacity of surface soils correlates with organic matter content.[43] Organic substances impact secondary mineral formation; they block forma-tion of goethite and hematite in favor of ferrihydrite precipitation[44] and inhibit the crystallization of Al-oxides such as boehmite and gibbsite.[45]

Organic matter generally carries negative charge as a result of carboxylate ionization and is an ion exchanger/complexant of cations and a strong complexer of certain metals (e.g., Fe^{3+}, Al^{3+}, Cu^{2+}, Am^{3+}, and Pu^{4+}) and a reductant of others.[46] The charge density of organic matter per unit mass is high, and it impacts the surface chemistry/reactivity of natural particles at low mass fractions.[43] Acidity and metal-binding constants for organic matter span a broad range with surface saturation (i.e., by H^+ or a given metal) as a result of its heterogeneous chemical composition and electrostatic effects (see Reference 47 and associated references).

In subsoils and deeper subsurface sediments where organic matter is negligible, particle surfaces are dominated by inorganic phases resulting from the weathering of primary mineral grains and secondary precipitation reactions. These surface phases are most commonly iron, aluminum, and Mn oxyhydroxides, aluminosilicates, and layer silicates of various forms that exhibit different surface properties and reactivity from the substrate mineralogy.[33,48-52] The physical, mineralogic, and chemical properties of these phases are complex. Their mineralogic characterization is difficult as the most reactive, high surface area phases are often present in low mass content, contain significant fractions of constituent impurities, and may be poorly crystalline. Not only is phase identification difficult, but so is the quantification of their reactive properties, such as site concentration, and their affinity for protons and electrolyte ions. These properties are not readily or uniquely determined by measurement because reactive phases may (1) be intimately comingled, (2) be strongly associated with substrate mineral grains in ways that compromise properties measurement, or (3) show a range of properties as a result of compositional impurities and variation. For example, potentiometric titrations, which are commonly used to characterize site concentrations and the acid–base properties of oxides and clays, are limited in their application to soils and geologic materials because of (1) competing reactions such as Al dissolution from soluble Al-containing solid phases and (2) inability to determine the proton adsorption density on the individual phases contributing to acid or base consumption.

X-ray absorption spectroscopy (XAS) and other spectroscopic and specialized beam techniques that have advanced understanding of the identity and structure of surface complexes on single mineral phases (see Chapter 3) are also complicated by the heterogeneous nature of composite mineral materials. The EXAFS spectra of an adsorbed constituent in a soil or subsurface material may reflect spectral contributions from surface complexes in various and multiple chemical environments, all in different unknown concentrations, preventing unambiguous interpretation of the integrated spectra. XAS measurements, however, have proved particularly useful in characterizing more-general or semiquantitative aspects of metal speciation in soil, such as the valence of sorbed Cr or U.[53] Natural materials present themselves as horribly irregular, aggregated, intergrown mineral materials that currently challenge such useful microscopic techniques as AFM and STM, except in the most select of instances.

The net result of the characterizational difficulties noted above is that the identity of reactive phases in soils and subsurface sediments and their properties are often established by inference and by comparison with controlled studies of single phase materials of known composition and properties. Additionally, interpretation and modeling of ion retention measurements with these complex natural materials require semiempirical or empirical approaches because fundamental surface chemical properties and mechanisms cannot easily be established.[47,54]

It is clear, however, that the same types of surface chemical reactions observed in single-phase mineral suspensions also occur in geologic materials containing these phases. Differences are often noted because natural materials contain multiple reactive phases, and their properties, such as site concentration, sorbate bonding energies, and solubility, may differ from comparable materials synthesized in the laboratory. Examples are shown in Figures 2 and 3, and the relevant surface reactions for these are summarized in Tables 3 and 4. Generally, however, metal partitioning to composite mineral materials reflects the integrated response of reactions on multiple reactive phases (Figure 4 is an example), although one phase may dominate. Figure 4, for example, shows the effect of mineral-bound humic substances on Co^{2+} adsorption by the kaolinite and subsurface material shown in Figure 2. While the sorptive effect of the humic substance was, in this case, shown to be additive,[52] the contributions of the different reactive phases are generally not additive because of particle interactions and other complex phenomena, but sum in nonlinear ways. Thus, adsorption isotherms may exhibit greater nonlinearity, and adsorption edges may slope more gently than comparable single-phase mineral materials because of the overlapping contributions of different sorbing phases or because the properties of the sorbing phase(s) span a range of values.

Modeling these ion retention reactions can be accomplished with the surface complexation approach described for discrete mineral solids in Chapter 3, but invariably a number of assumptions are required about the nature of the sorbing phase and its properties.[15,54] The basic information requirements for modeling are the number of surface sites and their concentrations and equilibrium constants between these sites and aqueous solutes (i.e., components including H^+ and relevant electrolyte ions) leading to surface species or complexes. The surface area(s) of the adsorbing phases are required for double-layer-corrected SCMs (i.e., the constant capacitance model or the triple layer model). The surface area of the sorbing phase in a composite mineral material is difficult to measure and has been estimated from isotherm sorption maxima and solute molecular dimensions.[55] The total site concentration is also difficult to measure, and Davis and Kent[15] argued for the adoption of a common site density of

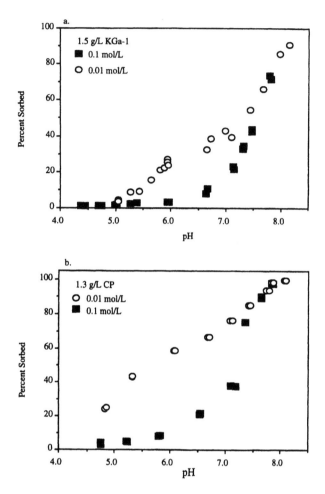

Figure 2

Co^{2+} (10^{-5} mol/L) adsorption to specimen kaolinite (KGa-1) and a clay-sized mineral isolate (<2.0 μm, CP) containing kaolinite, Al, and Fe-oxides and some 2:1 layer silicates as reactive phases. The reduction in sorption on both materials with increasing ionic strength results from the suppression of ion exchange reactions to fixed charge sites on layer silicates. The increase in adsorption with increasing pH results from complexation to hydroxylated sites on the edges of the layer silicates and on the oxide fraction. The CP has a larger concentration of fixed charge sites but comparable concentrations of hydroxylated sites to KGa.

3.84 μmol/m² for mineral material to standardize modeling and bypass characterization difficulties. Furthermore, because of the complex particle geometries of natural materials, many investigators use SCMs with simple or no electrostatic corrections. An example is given in Figure 5 for the sorption of Zn^{2+} by an aquifer material that contains grain coatings of a mixed Fe/Al/Si oxide phase.[51,56] The titration data were well modeled with a single surface complexation/exchange reaction of the form:

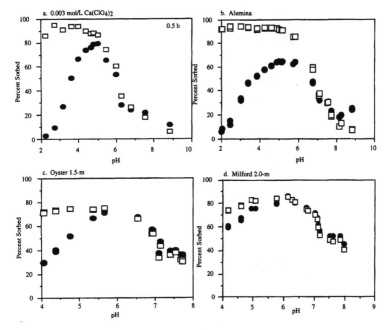

Figure 3
Sorption of Co^{2+} (circles) and $EDTA^{4-}$ (squares) added as the preformed 1:1 complex
$[Co(II)EDTA^{2-}$ at 10^{-5} mol/L to suspensions of (a) goethite (25 m^2/L), (b) alumina (25 m^2/L),
(c) the Oyster 1.5-m sediment (500 g/L), and (d) the Milford 2.0-m sediment (500 g/L).
Equivalence in percent sorption of Co^{2+} and $EDTA^{4-}$ (e.g., at higher pHs) signifies adsorption
of the intact complex. The disparity in sorption of Co^{2+} and $EDTA^{4-}$ at lower pH's indicates
that the complex has partially dissociated to a complex suite of sorbates — $[Co^{2+}$,
$Co(II)EDTA^{2-}$, and either or both $Al(III)EDTA^-$ and $Fe(III)EDTA^-]$ — that each show varying
affinity for surface hydroxyl sites on the reference solids; the subsurface materials suggests
that the same reaction network is responsible for the sorption process.

$$SOH + Zn^{2+} = SOZn^+ + H^+ \quad K_{SOZn} \qquad (1)$$

where SOH is a generic surface hydroxyl site. No electrostatic correction
(see Chapter 3) was applied and the data were modeled by (1) fixing the
total site concentration ($SOH_T = SOH + SOZn^+$) at a value equal to $N_s \cdot A \cdot S$,
where N_s is the standardized site density noted above (3.84 $\mu mol/m^2$), A
is the solid-to-solution ratio of the experiment (g/L), and S is the surface
area (m^2/g) and (2) fitting an equilibrium constant for the surface reaction
(K_{SOZn}).

Two general approaches have been taken to model adsorption reactions with the surface complexation approach:[15,54] (1) assume a given
sorbent is dominant (e.g., Fe oxide), that analog-phase (e.g., goethite)
properties apply, and that its site concentration is proportional in some
way to its mass content or (2) assume that the chemical behavior of the
material is an intrinsic property that can be fitted, like a single mineral

TABLE 3.

Aqueous and Surface Reactions Affecting
Co^{2+} and LHA Distribution in Suspensions of
Kaolinite and Kaolinite-Containing
Subsurface Material

Aqueous Reactions

1. $Co^{2+} + H_2O = CoOH + H^+$
2. $nL^- + Co^{2+} = L_nCo^{(2-n)a}$

Reactions in Mineral Material (Figure 2)

3. $FeO^-_{(H,G)} + Co^{2+} = FeOCo^+_{(H,G)}$ [b]
4. $AlO^-_{(G,K)} + Co^{2+} = AlOCo^+_{(G,K)}$ [c]
5. $2X^-_{(K,S)} + Co^{2+} = CoX_{2(K,S)}$ [d]

Sorption Reactions of LHA (Figure 4)

6. $xSOH_2^+ + yL^- = (SL)_x^- (L^-)_{y-x}$ [e]
7. $-nL^- + Co^{2+} = -L_nCo^{(2-n)f}$

---

[a]　nL^- = n equivalents of dissociated carboxylate
　　or phenolate groups on the water soluble humic
　　substance; n = 1 or 2.

[b]　$FeO^-_{(H,G)}$ = deprotonated surface site on hematite
　　or goethite.

[c]　$AlO^-_{(G,K)}$ = deprotonated surface site on gibbsite
　　or the kaolinite edge.

[d]　$X^-_{(K,S)}$ = fixed charge sites on kaolinite or smec-
　　tite.

[e]　Sorption reaction of humic substance (HS) to
　　protonated AlOH or FeOH surface sites
　　(SOH_2^+). HS represented as a dissociated ligand
　　(L^-). Sorption consumes surface sites (SOH) by
　　ligand exchange (SL), leaving a portion of HS
　　sites $[(L^-)_{y-x}]$ available for metal complexation.

[f]　$-nL^-$ = n dissociated HS sites associated with the
　　mineral-bound HS.

phase, using selected global characterization measurements (e.g., Figure 5) . These two approaches have been termed *pseudomechanistic* and *empirical*, respectively, and are discussed by Zachara and Westall.[54] The two approaches differ significantly in characterization needs. In the pseudomechanistic approach, one attempts to identify the controlling sorbent, its crystalline structure if inorganic, and its concentration of surface sites by direct or surrogate measurement. Identifying these items may require mineralogic, electrolyte-binding, and surface chemical/compositional measurements of significant detail and sophistication. The surface properties of the sorbing phase may also be interrogated, if feasible, through size fractionation experiments or direct-beam/spectroscopic measurement. An example of this type of modeling is shown in Figure 6 using

TABLE 4.

Important Reactions Controlling $Co(II)EDTA^{2-}$ Distribution in FeOOH, Al_2O_3, and Subsurface Material Suspensions

Solubility

$Fe(OH)_{3(s)}$ or $FeOOH + 3H^+ = Fe^{3+} + 3$ or $2H_2O$
$Al(OH)_{3(s)} + 3H^+ = Al^{3+} + 3H_2O$

Aqueous Complexation

$EDTA^{4-} + pH^+ = HpEDTA^{(4-p)}$
$Co^{2+} + EDTA^{4-} = CoEDTA^{2-}$
$Fe^{3+} + EDTA^{4-} = FeEDTA^-$
$Fe^{3+} + EDTA^{4-} + H_2O = FeOHEDTA^{2-} + H^+$
$Al^{3+} + EDTA^{4-} = AlEDTA^-$
$Al^{3+} + EDTA^{4-} + H_2O = AlOHEDTA^{2-} + H^+$

Dissociation

$Co(II)EDTA^{2-}_{(aq\ or\ surface)} + Fe^{3+}/Al^{3+}_{(aq\ or\ surface)} = FeEDTA^-_{(aq)}$ or $AlEDTA^-_{(aq)} + Co^{2+}_{(aq)}$

Adsorption

$SOH + H^+ = SOH_2^+$
$SOH_2^+ + Co(II)EDTA^{2-} = SOH_2 - Co(II)EDTA^-$
$SOH_2^+ + FeEDTA^- = SOH_2 - FeEDTA$
$SOH_2^+ + AlEDTA^- = SOH_2 - AlEDTA$
$SOH = SO^- + H^+$
$SO^- + Co^{2+} = SOCo^+$

the data in Figure 2b. Cobalt adsorption is modeled as the sum of three surface species resulting from reactions 3, 4, and 5 in Table 3. The change in the adsorbed Co concentration with ionic strength is modeled to result from an ion exchange reaction on fixed charge sites. The reactions were parameterized using measurements from pure sorbent phases and the subsurface material itself as described in Table 5.

The empirical approach requires less characterization information, such as the solute sorption maxima, because the sorption data are fit to yield soil-specific reaction parameters. Many variants of this approach have been taken, ranging from the calculations in Figure 5 to the application of more elaborate pK spectrum models.[47,54,58] The latter offer real advantages for the description of heterogeneous natural materials and also provide the ability to extrapolate sorption data over wide ranges in intensive variables such as pH, ionic strength, and sorbate concentration.[54]

The surface reactions described above concentrate adsorbates at the mineral–organic water interface and retard the migration velocity of these solutes relative to nonreactive solutes in soils and aquifer materials (see, for example, References 59 and 60). The adsorptive retardation of a reactive solute, i, can be described using the retardation factor:

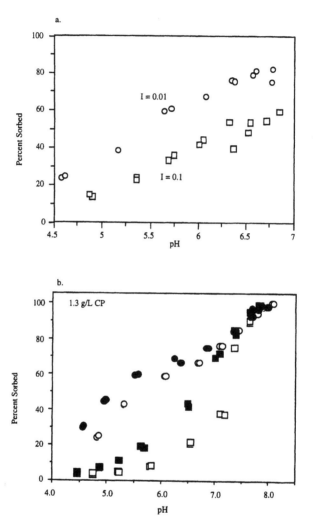

Figure 4

In aqueous solution the leonardite humic acid (LHA) complexes Co^{2+} (a). When the LHA is contacted with the CP material shown in Figure 2 it is strongly adsorbed by hydroxylated surface sites between pH 4.5 and 7.0. The adsorbed LHA incrementally enhances Co^{2+} adsorption by the mineral phase.

$$R_{f,i} = t_i/t_{nr} \qquad (2)$$

which is the ratio of the arrival time of the solute of interest, i, to the travel time of a nonreactive tracer, nr, following the same transport path. The R_f is a simple, yet fundamental macroscopic measurement of solute transport velocity often measured in the field,[60,61] and used in some chemical transport models.[62] If it is assumed that chemical conditions are constant throughout

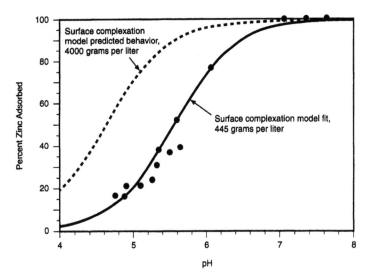

Figure 5
Modeling Zn^{2+} adsorption on aquifer material using a nonelectrostatic surface complexation model. Solid circles are experimental results from a batch experiment: 445 g/L aquifer material, 5×10^{-6} mol/L Zn^{2+}, and I = 0.1. Solid curve is model fit to the data. The dashed curve is predicted adsorption at the solid:liquid ratio in the aquifer (4000 g/L). (From Rea, B.A. et al. *U.S. Geological Survey Toxic Substances Hydrology Program — Proceedings of the Technical Meeting*; Monterey, CA, March 11–15, 1991. G.E. Mallard; D.A. Aronson; Eds. U.S. Geological Survey, 91–4034, 1991.)

the flow path (i.e., pH, ion composition, reactive surface area), that the amount adsorbed is a linear function of the adsorbate aqueous concentrations, and that adsorption achieves local equilibrium on the time scale of transport, then the R_f can be related to, or predicted by, sorption data such as that shown in Figures 2 through 4 by the following relationship:

$$R_f = 1 + p_s(1 - n/n)V/W(F/1 - F) \qquad (3)$$

where p_s is the density of the solid, n is porosity, F is the fraction adsorbed, and V/W is the ratio of the solution volume to weight of solid (cm^3/g). Models that can be used to simulate the adsorption data in Figures 2 through 4, such as applied in Figure 5 and 6, can be used to estimate F and consequently R_f, under the constraints of the assumptions noted above. Existing chemical transport codes for soil and subsurface environments describe adsorptive retardation (1) using an R_f, (2) through coupling to a generalized equilibrium speciation and SCM, or (3) by direct linkage to a minimal reaction set (that may be either kinetically controlled or at equilibrium) describing speciation and adsorption.[62,63] Therefore, the SCMs that were described in Chapter 3 have direct application in the modeling of ion transport in soil and subsurface systems.

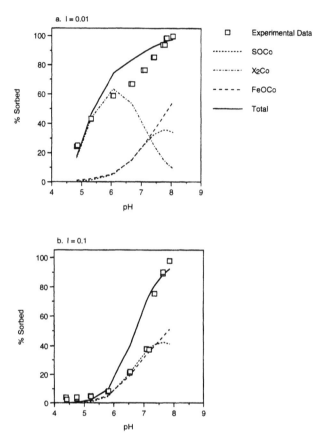

Figure 6
Nonelectrostatic surface complexation modeling of Co^{2+} on the subsurface isolate (CP) in Figure 2. Electrolyte was $NaClO_4$ at (a) I = 0.01 mol/L and (b) I = 0.01 mol/L. Symbols are experimental data. Solid and dashed lines are calculations with the nonelectrostatic model.

While it was noted above that heterogeneity exists at the grain scale in soils and subsurface materials, complicating identification and quantification of individual surface chemical reactions, heterogeneity also exists at the meter scale and above and has profound impacts on the rates and extent of surface chemical reactions controlling contaminant and ion concentrations in larger systems such as aquifers. Indeed, understanding and quantification of the effects of these macroscopic heterogeneities is required to appreciate and predict the effects of surface reactions at the field scale. Geologic depositional processes and postdepositional geochemical phenomena create deposits of porous media with great spatial heterogeneity in grain size distribution, structure, hydraulic conductivity, and reactive surface area and mineralogy. Reactive phases may be correlated or inversely correlated with hydraulic conductivity, with contrasting implications to

TABLE 5.

Parameters for Multisite Adsorption Modeling

Site	Concentration	K' (log K)[a]	K'' (log K)[b]	K_{Na} [c]	K_{Co} (log K)
X^-	CEC @ pH[d]	NA	NA	Fixed at reference value	Fitted to DCP (4.37)
AlOH	$S_{(N2-DCP)} \cdot N_{s(K)}$[e]	From KGa-1 (8.18)	From KGa-1 (5.38)	NA	From KGa-1 (−2.53)
FeOH	$\Delta S_{(N2)} \cdot N_{s(G)}$[f]	From Al-goethite (11.6)	From Al-goethite (3.16)	NA	From Al-goethite (−3.49)

a $SO^- + H^+ = SOH$ K'; K' in mol L^{-1}.

b $SOH + H^+ = SOH_2^+$ K''; K'' in mol L^{-1}.

c $X^- + Na = NaX$ K_{Na}; K' in mol L^{-1} fixed at high value to assure cation saturation of exchange sites.

d CEC @ pH = cation exchange capacity measured at pH of calculation.

e $S_{(N2-DCP)}$ = surface area of the deferrated isolate (DCP) measured by multipoint nitrogen adsorption. $N_{s(K)}$ = surface site density for kaolinite reported by Sposito.[4]

f ΔS = difference in N_2 surface area before and after DCB extraction (23 m^2/g). $N_{s(G)}$ = surface site density of goethite (10 sites/nm^2 from Hayes et al.[57]).

g NA = not applicable.

reactive transport. Stringers or lenses of lower conductivity, highly reactive materials, may be bypassed by a significant fraction of the reactive solutes, but may function as long-term diffusive sources and sinks. Ions of contrasting chemistry may show different reactive effects to heterogeneities if the phases responsible for retardation are not the same and are distributed unequally throughout the aquifer material.

III. INTERFACES, SEDIMENTS, AND DIAGENESIS

Roughly 18% of the particulate matter produced in the surface layers of the ocean sinks into the deep ocean.[64] By weight, this material is 60% biogenic $CaCO_3$, 10% biogenic SiO_2, 25% organic material, and 5% detrital mineral phases. What actually reaches the sediment–water interface on the ocean floor is notably less rich in organic matter and represents less than 1% of the organic matter which left the ocean surface. In areas where the ocean floor is below the carbonate compensation depth, the fraction of calcium carbonate reaching the floor is much diminished, as well, and clays predominate. Particulate matter in the water column is an important control on trace element cycling in the ocean.[65-67] Comparisons of trace element residence times with independently measured trace metal–metal (hydr)oxide binding constants indicate that the latter are too weak to account for the observed residence times.[68] Instead, it appears that the surfaces of organisms and organically coated inorganic particulates control trace element movement.

Microporous mineral surfaces may limit net rates of organic carbon accumulation in sediments on a global scale by preventing degradation by hydrolytic enzymes[69,70] and by fostering the condensation of labile organic compounds into macromolecules.[71] Historically, the petroleum precursor kerogen has been envisioned to form primarily through condensation of small organic molecules in solution, i.e., by way of a homogeneous process. As will be argued at the end of the chapter, thermodynamics does not favor homogeneous polymerization from most natural waters. Recent work has argued for mineral surface–directed preservation and condensation of organic material. There is a linear relationship between sediment surface area and organic carbon content (OC)[69,71] of 0.8 mg OC m^{-2}. Whereas major element ocean chemistry is reasonably constant and static through time, the specific surface areas of mineral grains can vary by orders of magnitude in natural systems.[72] The linear relation between the latter and OC points to surface interaction as being potentially the critical determinant of preservation. Adsorption decreases the degradation rate of organic molecules by shielding them (Kiel et al.[73] measured a 10^5-fold decrease in degradation rate from the dissolved state). There is considerable evidence pointing to a surface role in the subsequent polymerization of adsorbed organic matter (see review of Collins et al.[71]), and a symbiotic relation probably exists between sorbed organic matter and host mineral throughout diagenesis. Defunctionalization will lead to increasingly hydrophobic surface coatings, which should, in turn, passivate and protect the mineral against dissolution in the face of changes in pore water saturation state.

Kerogen is transformed through the combined influences of temperature and time, and apparently mineral surfaces, into crude petroleum. Experimental efforts (laboratory pyrolysis of organic material) to duplicate the formation of petroleum indicate that the process depends on the specific catalytic properties of the adjacent minerals, primarily clays. (For reviews of the pyrolysis literature see References 74 and 75.) In general, methane and C_2 hydrocarbons are generated from organic matter without the aid of surfaces. Branched C_4–C_6 hydrocarbons and minor amounts of olefins are produced from kerogen and bitumen (an early by-product of organic matter alteration) by clays such as illite and montmorillonite.[74] Calcite is a much less effective catalyst. Acid Al sites on the clay edges are thought to be the active sites for catalysis.[76] The type and amount of hydrocarbon produced during burial and thermal alteration depend on the particular mineral catalyst present during burial,[74,77] as well as the nature of the organic source.

The economic value of a particular hydrocarbon reservoir depends to a large extent on the porosity of the rock matrix in which it resides. Primary porosity is the void space that a particular rock begins with. Secondary porosity is the void space which is formed during subsequent burial through a combination of mineral dissolution and growth. Recent work on diagenesis of siliciclastic reservoirs suggests that secondary porosity may depend, in part, on a complex set of interactions between kerogen and mineral surfaces. Production of carboxylic acids has been ascribed to surface-mediated redox reactions.[78,79] Carboxylates and phenolates also may be generated when ferric groups at (hydr)oxide or clay surfaces (or polysulfides) accept electrons from kerogen in petroleum source rocks.[80] Carboxylate groups are then able to contribute to aluminosilicate dissolution, hence, porosity enhancement, by elevating the solubility of Al-bearing phases.[81] At the same time, adsorbed carboxylate groups accelerate the dissolution rates of the same phases, again driving the enhancement of secondary porosity.

The interaction between petroleum hydrocarbons and the surrounding host rocks largely determines overall recovery (see review of Morrow[82]). Oil–brine–rock systems form when migrating oil displaces brine. A layer of brine (10 to 1000 Å) is thought to remain between the mineral surface and the oil (e.g., Buckley et al.[83]). When the layer is relatively thin, the contact angle between the mineral–oil and the oil–water interfaces approaches zero, and the system is termed *oil-wet*. When the brine layer is thick, the contact angle between the mineral–oil and oil–water interface is high, and the system is termed *water-wet* (for extensive reviews of wetting phenomenon see those of Israelachvili[84] and Berg[85]). Petroleum recovery from oil-wet systems is greatly diminished relative to the water-wet case, and a mechanistic understanding of *in situ* wettability is important in economic terms.

The net of the competing attractive and repulsive forces in the thin brine film ultimately determines whether an oil–brine–rock system is water-wet or oil-wet. Attractive, van der Waals forces thin the film, whereas short-range structural forces at the mineral surface, arising from strongly held waters of hydration associated with adsorbed ions,[86] repel oil. An additional force, which can be attractive or repulsive, comes from electrostatic interaction between charged groups at the mineral surface and ionized functional groups exposed at the oil surface. Layer thinning and oil wetting is favored when the oil–water interface and the mineral–water interface are oppositely charged. Repulsion and water wetting are favored when the two interfaces possess like charge.

The electrostatic stabilization of water films plays a very large role in determining adhesion potential, in general,[87] and wettability in oil–brine–rock systems, in particular.[86-91] In theory, one should be able to predict the electrostatic effect, and wettability trends in general, given a knowledge of the electrostatic behavior of the oil–water and mineral–water interfaces as a function of temperature and solution composition.[92] The oil side of the oil–water interface is zwitterionic and modeled as being made up of carboxylate groups, which are deprotonated for the most part above ~pH 4, and nitrogen bases, which are protonated and positively charged below pH ~ 5.[94] Thus, in mildy acidic settings the oil surface will have a positive charge, while at neutral pH and above the charge will become progressively more negative. Historically, ζ-potential measurements of crude oil droplets have been fit using the ionizable site group model of Healy and White,[93] which is analogous to mineral SCMs (see Chapter 3), in that surface site ionization (in this case at the oil–water interface) depends on the composition of the fluid (e.g., pH, ionic strength) and the degree to which adjacent sites are ionized.

Laboratory measurements of electrostatic film stabilization in oil–brine–mineral systems have been done primarily using quartz slides (or soft glass) as the mineral interface. Surface charging of quartz as a function of pH and ionic strength is often described using the Gouy–Chapman model. Recall from the previous chapter that quartz surface charge is negative in most dilute solutions, particularly above pH 7, and the net negative charge becomes greater with pH and ionic strength. The same is qualitatively true for the oil surface as well. One result of this is that oil-brine-quartz systems should be increasingly water-wet at high pH and ionic strength. At pH < 7, electrostatic repulsion between the oil and quartz surface should be lessened and oil wetting more attainable, because of increasing protonation of nitrogen bases at the oil-water interface. The pH and ionic strength dependency of wettability is shown in Figure 7

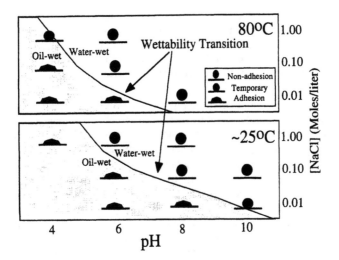

Figure 7
Adhesion map of Alaska-93 crude oil on quartz as a function of pH, NaCl content and temperature. (From Liu, Y.; Buckley, J.S. Paper presented at the 3rd Int. Symp. Evaluation of Reservoir Wettability and Its Effect on Oil Recovery. Sept. 21–23; Laramie, WY, 1994. With permission.)

using wettability maps of Alaska-93 crude oil on quartz[94] as a function of temperature, pH, and ionic strength. Note that the transition between water-wet and oil-wet conditions shifts to lower pH by 1 to 2 pH units between 25 and 80°C. Following the analysis above, the wettability shift occurs because, at a given pH, either (1) the oil–water interface becomes more negatively charged with increasing temperature because of temperature-dependent shifts in the carboxylate or base equilibria or (2) the quartz surface becomes more negatively charged because of an increase in surface acidity with temperature. Because carboxylic acid pK values are relatively insensitive to temperature, it would appear that the wettability shift arises because of the effect of temperature on mineral surface acidity or the basicity of nitrogen groups exposed at the oil–water interface. As the effect of temperature on wettability is different for different oils on the same substrate,[95] both probably contribute. At the same time, the nature and number of functional groups exposed from the oil side of the oil–water interface may change in response to changes in temperature as well.

Obvious avenues for extending the link between surface chemistry and wettability to consider *in situ* conditions in more detail are to examine:

1. *Other mineral interfaces.* Calcite, dolomite, and kaolinite are routinely the dominant component of the integrated mineral surface area exposed to subsurface fluids. Even if only a small fraction of the rock, by coating pore walls, calcite cements and clays would tend to have an inordinately large effect on wettability.[82] The electric structures of

carbonate and clay surfaces are quite different from that of quartz. The isoelectric points of each mineral are substantially higher than quartz (see Chapters 3 and 4). In the case of calcite and dolomite, surface charging depends explicitly on solution alkalinity,[96] unlike quartz. Wettability trends are therefore certain to differ among the various minerals.

2. *Temperature. In situ* reservoir temperatures are always greater than 25°C, and the effects of temperature on surface equilibria are nonnegligible.[97-99] The isoelectric points of most minerals decrease substantially with temperature, favoring water-wetting conditions, assuming constant pH.

3. *Effects of multivalent, charge-reversing species on* in situ *surface charge.* When cations bond chemically to mineral surface sites they compress the double layer, decrease the ζ potential, and, therefore, favor adhesion. Very high concentrations of Na are observed to produce a wettability reversal for quartz at high pH, which can be explained by specific binding of Na.[100] Multivalent cations such as Ca^{++} are more troublesome as they can potentially interact both with mineral surfaces and with carboxylate groups at the oil–water interface,[95] making it difficult to predict their net effect on wettability. Specific binding of Al drastically alters the electric double layer at mineral surfaces[101] (see Figure 8). At the same time, specific adsorption of multivalent anions, such as SO_4^{-2}, at near neutral pH might potentially extend regions of water wetting for minerals having positively charged sites (e.g., Al on clay edges and Ca on carbonates).

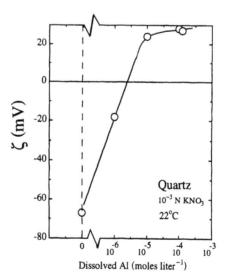

Figure 8
ζ potential (mV) of quartz as a function of solution Al (mol L[-1]) content. (From Ishido, T.; Mizutani, M. *J. Geophys. Res.* 1981, *86*, 1763–1775. With permission.)

IV. WEATHERING

As noted in the previous chapter, the dissolution of crustal minerals (the majority of which are sparingly soluble) is largely surface controlled. Rates vary orders of magnitude in response to seemingly trivial changes in the chemistry of their surfaces. Temperature and composition of the dissolving fluid (pH, salt content, concentration of adsorbing ligands) largely determine the latter. Microscope studies of minerals weathered in the field[102-106] emphasize the complicated and heterogeneous nature of weathering. Weathering is clearly an extremely complex process which can, at best, only be approximated by laboratory experiments.

The most obvious drawback to applying the laboratory rates outlined in Chapter 4 to understand global geochemical and environmental processes is that laboratory rates are a great deal higher than those measured in the field. The latter have historically been calculated by monitoring solute budgets of mineral components (e.g., Si, K, HCO_3^-). This involves determining the concentrations of these components dissolved in surface runoff, subtracting out the starting concentrations of each in the rainwater, and correcting for any reactions, such as biotic uptake[107] and ion exchange, which might occur along the way.[108] The whole calculation is best done using small, closed catchment basins, where sources and sinks are well known. To convert this mass balance approach to a rate comparable to laboratory values requires an assumption of the amount of mineral surface area exposed to weathering solutions, as well as an estimate of the amount of time it takes for precipitation to pass out of the basin. Obviously, it is very difficult to get a precise number for the amount of weathering surface area on a basin scale, but order-of-magnitude estimates can be made which are sufficient to make the point. Even with the most optimistic estimates of the reactive surface area, field weathering rates of primary silicates are found to be 100 to 1000 times less than those measured in the laboratory.[72,109-111]. A number of explanations have been proposed to account for the difference. To begin with, most soil solutions are sufficiently far from saturation with dissolving silicates that there is little chance for appreciable back reaction (see Chapter 4) which would cause field rates to be slower.[103] Temperature differences between laboratory setups at 25°C and natural soils, which are often 10 to 20°C less, can probably account for around a factor of five of the difference.[112] On the other hand, the surface areas of natural materials are often geometrically measured, and hence underestimated, whereas laboratory rates are measured by the BET method (see Chapter 4). This would tend to exacerbate the laboratory–field disparity by roughly a factor of three to five.

The temperature and surface area differences between laboratory and field rates tend to cancel, leaving a large absolute difference between the two. Recent results, however, indicate that much of the discrepancy between the laboratory and field comes from hydrologic, not crystallographic factors.

Drever and Swoboda-Colberg[111] and Swoboda-Colberg and Drever[113] showed that minerals taken from natural soils dissolve in the laboratory at essentially the same rate as those prepared for laboratory dissolution experiments. Differences in laboratory surface pretreatment and grinding can account for very little of the difference between laboratory and field rates.[114] The only remaining explanation is that the diminished degree of fluid exposure to soil minerals during percolation through the unsaturated zone accounts for the residual difference between laboratory and field rates. Discontinuous and heterogeneous flow of fluids through soil pores lessens the overall impact of weathering in the field. At the same time, reactive sites at soil mineral surfaces may be blocked by adsorbed poisons, e.g., HCO_3^- at high pH[115] or humic substances.[116]

An important point is that the factors which control fluid movement in soils are not those which directly affect mineral dissolution and growth. Temperature and pH affect field and laboratory rates similarly.[112,117,118] Moreover, reactivity trends observed in the laboratory between various minerals also are quantitatively the same in the field.[9] For a given watershed, the ratio of weathering rate for two different silicate minerals is identical to the same ratio calculated from laboratory data.[118] In sum, this means that there is a reasonable physical explanation which relates rates measured in the laboratory with those measured in the field, and laboratory weathering rates can be linked to field rates with a "lab/field adjustment coefficient"[118] which primarily monitors soil physical processes (hydrologic regime, soil wetting, etc.). In other words, given the appropriate hydrologic model, surface science can be applied to understand weathering.

A. Acid Rain Neutralization

A number of natural processes produce acidity in watersheds. Vegetation takes up base cations (K^+, Ca^{++}, Mg^{++}, etc.) from soil solutions, in effect, satisfying charge balance by leaving protons in their place. Roots exude organic acids. Decaying organic matter raises soil CO_2 levels and adds to organic acid budgets as well. Oxidation of ammonia to N_2 and nitrogen oxides (nitrification) adds acidity to soil waters. Base exchange reactions occur wherein cations displace protons from mineral surfaces. Formation of new minerals (reverse weathering) also contributes to proton budgets. Each of those processes that affect the proton budget is outlined in Table 6. The movement of protons and base cations through soil systems is shown schematically in Figure 9. Often the magnitude of the combined natural proton sources rivals that from artificial sources, the most important of the latter coming from the burning of fossil fuels.

Acidity production in atmospheric precipitation is an unavoidable side effect when fossil fuels containing C, N, and S are burned. The degree to which this acidity affects natural waters depends on the magnitude of

TABLE 6.

Proton-Consuming Processes in Terrestrial and Aquatic Ecosystems

Proton-Producing Processes		Proton-Consuming Processes
Atmospheric input	$(H^+)_{rain} \leftrightarrow (H+)_{soil\ solution}$	Drainage
Assimilation of cations	$M^{n+} + nR\text{–}OH_{(org)} \leftrightarrow (R\text{–}O)_n M_{(org)} + nH^+$	Mineralization of cations
Mineralization of anions	$nH_2O\text{–}R_n - A_{(org)} \leftrightarrow nR\text{–}OH + A^{n-} + nH^+$	Assimilation of anions
Dissociation of acids	$HnA \leftrightarrow A^{n-} + nH^+$	Protonation of anions
Oxidations	$Red + mO_2 + nH_2O \leftrightarrow Ox^{r-} + rH^+$	Reductions
Reverse weathering of cations	$M^{n+} + 0.5H_2O \leftrightarrow 0.5nM_{2/n}O + nH^+$	Weathering of cations
Weathering of anionic components	$NO_{(m+n)} + mH_2O \leftrightarrow NO^{2m-}_{(2m+n)} + 2mH^+$	Reverse weathering of anions

From van Breeman, N.; Driscoll, C.T.; Mulders, J. *Nature*. 1984, *307*, 559–604. With permission.

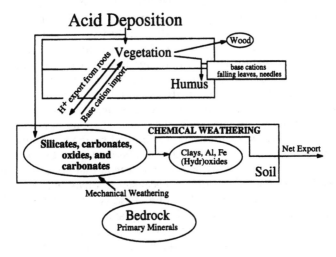

Figure 9
Schematic of base cation cycling in soils. (Adapted from Reference 119.)

the offsetting proton-consuming reactions in the respective soil. The list of natural proton-consuming reactions is often the reverse of the proton-producing reactions (see Table 6). Important proton-consuming reactions include reduction of oxidized nitrogen (e.g., denitrification) and sulfur (sulfate reduction), photosynthesis, exchange of protons for base cations at mineral surfaces, and the assimilation of anions by organic matter. The opposite of proton addition from the atmosphere is proton export through drainage. The most important net consumer of protons in soils, though, is the weathering of base-containing minerals, carbonates and silicates.[2,119-124] Base fluxes are large enough to exhaust the capacity of rapidly exchangeable soil minerals in a few decades, emphasizing the importance of replenishment from weathering of slower-reacting, bedrock minerals in the soil.

All minerals are not equal in their ability to mitigate environmental acidity. The acid-neutralizing capacity, ANC, of the inorganic component of a soil is the sum of the basic component oxides minus the strongly acid components:[2]

$$ANC = 6[Al_2O_3] + 2[CaO] + 2[K_2O] + 2[Na_2O] + 4[MnO_2]$$

$$+ 2[MnO] + 2[FeO] -2[SO_3] -2[P_2O_5] - [HCl] \qquad (4)$$

[HCl] and [SO_3] are the molal concentration of the respective input acids; the bracketed oxides represent the molal concentration of available bases as minerals. CO_2, H_2S, and SiO_2^{aq} are neglected in the calculation because from pH 2 to 5 they neither accept nor donate protons to solution. By the same token, Fe_2O_3 is neglected because over the range of soil pH ferric iron can be considered to be insoluble. The ANC measures the degree to which a given rock mineralogy can buffer external and internal inputs of acidity. Soils rich in SiO_2 but poor in Mg, Ca, etc. (e.g., granites, weathered gneisses, quartzites, etc.) have low ANCs. Lakes atop such rocks will be sensitive to atmospheric acidity inputs, tending to acidify rapidly. Rocks rich in Mg and Ca (limestones, basic igneous rocks, young volcanics) tend to have high ANCs and are relatively resistant to sudden increases in soil acidity. Lakes set in limestone, in particular, are able to resist external proton inputs. ANC measures only the potential ability of a rock to neutralize acidity. That is, neutralization is limited by the availability of the base cation, hence the rate at which it is liberated from the rock matrix through dissolution reactions. Part of the reason that lakes in limestone terranes do not generally acidify is because carbonate minerals (calcite, aragonite, and, to some extent, dolomite) dissolve very quickly, orders of magnitude faster than the fastest dissolving silicates (see Chapter 4).

Those silicates which dissolve slowest (quartz, clays) are also those with the lowest mineral ANC. In effect, waters in contact with Si-rich rocks are doubly doomed by acid rain. The rocks possess very little inherent acid-buffering ability and that which they do possess is only available over very long periods of time. Obviously, absolute acid neutralization also depends upon the amount of time acid inputs from the atmosphere are in contact with dissolving mineral bases.[125] Lakes that receive most of their water directly from the atmosphere (precipitation-dominated lakes) and lose it through seepage will understandably be strongly affected if that input is acidic. Mineralogy will have little to do with the final pH. The same is true for lakes in basins with shallow soil and abbreviated hydraulic residence times.[119] Lakes in basins with extended soil residence times, and/or containing mineralogically high ANCs, will be appreciably less sensitive to atmospheric acidity inputs.

Because the neutralization rate is tied closely to the dissolution rates of base-containing minerals, the dependencies of the latter (see Chapter 4) are important inputs in forward modeling of acid neutralization. Early models[126-129] varied weathering rates to fit field data. More-recent models[119,130-134] have used laboratory rates, and their dependencies, and accounted for field surface areas and hydrologic factors separately (see above). One of the more advanced acid rain response models is the PRO-FILE model of Sverdrup and Warfvinge.[130,131,133]

Some important conclusions are that:

1. Field weathering rates appear to be independently calculable (to within ±8%) given an initial knowledge of mineralogy, soil physics, etc.,

2. Laboratory rates are consistent with field rates, once the hydrologic regime of a given basin is fully accounted for, and

3. Plagioclase feldspar, epidote, hornblende, and, to an extent, apatite, are the most important reactive phases which must be considered in carbonate-poor terrains.

In line with the latter, Sverdrup and Warfvinge[131] and Warfvinge and Sverdrup[133] have suggested that future advances in the modeling of soil acidification depend largely on the accumulation of additional mineral dissolution rates. Dissolution rates for minerals, such as epidote, hornblende, apatite, and plagioclase feldspar, are the most important of the latter. Hornblende dissolution appears to be a particularly important sink for protons.[135,136]

Although the more recent models appear to roughly approximate the response of dissolving soil silicates to acid inputs, a number of uncertainties remain. Warfvinge and Sverdrup[133] and many previous authors have argued for a soil CO_2 effect on rates, an effect that presently appears doubtful (see Chapter 4). Moreover, while the temperature dependence of weathering has been extensively studied in the laboratory (see Chapter 4), its effect in the field has remained controversial. The primary chemical effect of organic activity is thought to be an acceleration of weathering relative to the abiotic case, because of organic acid production.[137-140] Microbial activity in general appears to accelerate silicate weathering.[141,142] At the same time, many naturally occurring organic polymers (e.g., extracellular polysaccharides) inhibit dissolution.[143] Plants also accelerate weathering because of the physical action of roots.[144]

B. Global Carbon Cycle and Climate

The observed effects of temperature, CO_2, and organic activity (chemical and physical effects) on silicate weathering are routinely applied to understand the carbon cycle and global climate. Over geologic time the surface-controlled dissolution of silicates dominates the geochemical cycling of CO_2, globally the most abundant acid in the atmosphere. In Figure 10 is shown a schematic of the carbonate–silicate cycle. Carbon dioxide is produced by mantle outgassing (volcanic exhalations can be greater that 10% CO_2), metamorphic decarbonation, and, to a lesser extent, by oxidation of organic matter. Over time spans greater than the average stay of HCO_3^- in the ocean (>10^5 years), weathering of Ca and Mg silicates is the primary sink for atmospheric CO_2. Over time spans this long, the carbon dioxide produced through photosynthesis must be balanced by biologic respiration.[145] The biotic and atmospheric reservoirs for carbon are very small, though the fluxes between the two are very large and assumed to be equal. The reservoir of organic carbon buried in sediments has varied with time,[146] but is a second-order feature in terms of long-term climate. Carbonate weathering plays a minor role in the silicate–climate link. While weathering of carbonates consumes acidity, the subsequent transport and deposition of calcium and bicarbonate in the ocean as calcite releases CO_2 again to the atmosphere. Calcite weathering is, therefore, equivalent to the mechanical transport of calcite from the continents to the oceans and has no net long-term effect on atmospheric CO_2.

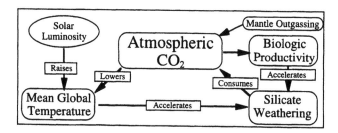

Figure 10
Feedbacks between Ca/Mg-silicate weathering and the long-term global carbon cycle.

The balance between mantle outgassing and surface-controlled silicate weathering is one of the geochemical feedbacks most critical to life on Earth because it determines the amount of CO_2 in the atmosphere. Atmospheric CO_2, by adsorbing and reradiating back to Earth terrestrial radiation, largely determines mean global temperatures through the greenhouse effect. The negative feedback between weathering and climate works as follows. When global temperatures are raised because of positioning of the continents,[147] increases in solar luminosity,[1] or any of a number of other geophysical events, silicates weather faster, consuming larger quantities of CO_2 in the process. This causes CO_2 levels to decrease and global temperature, through the greenhouse effect, to drop. At low temperatures, silicate weathering, and CO_2 consumption, is minimal, causing CO_2 to accumulate in the atmosphere. In this way the feedback mechanism between weathering and CO_2 resists large increases and large decreases in global temperature.

Over the lifetime of Earth the geochemical feedback between weathering and mantle outgassing has also compensated for increases in solar luminosity, ultimately keeping surface temperatures relatively constant. Solar radiation has increased by ~30% over the life span of Earth. Nevertheless, the fossil record suggests that global temperatures have varied relatively little. There certainly hasn't been an unbroken record of global warming over geologic time. The weathering feedback has over the past 4 billion years effected a net transfer of atmospheric CO_2 from the atmosphere through the dissolution of silicates.[148]

The strength of the climate control depends directly on the degree to which global silicate weathering responds to temperature and atmospheric CO_2. A strong temperature dependency of weathering leads to a tightly controlled global thermostat, in that the system will respond quickly to moderate changes in temperature, subtracting CO_2 from the atmosphere as needed. Conversely, a very weak temperature dependency for silicate weathering would result in poor system control over atmospheric CO_2 and climate. Cometary impacts,[149] changes in mid-ocean ridge hydrothermal activity and tectonic regime,[150,151] the rise of plant life,[152,153] and the ultimate demise of the latter,[154] all would test the resilience of the weathering–climate feedback.

Note that the temperature, CO_2, and organic activity–weathering link operative over geologic time is probably different from that which controls solute fluxes and acid rain attenuation over human time scales. Over short time, scales, temperature is linked to differences in latitude and elevation.[155] Over geologic time, temperature-induced changes in weathering rate are linked to latitudinal position, elevation, and changes in global mean temperature. Over short time spans, soil CO_2 is linked directly to soil organic productivity. Over geologic time, soil CO_2 will vary with atmospheric CO_2 as well, though levels are likely to be functions of soil thickness and wetting, which are likely to vary depending on plant evolution.[156] Lastly, the physical component of the organic activity control over weathering probably assumes greater importance on geologic time scales, because of the sequential evolution of plants having vastly different potentials for soil working and stabilization (e.g., fungi vs. angiosperms).[152,156-159]

The conclusions derived from models of the global carbon cycle and long-term climate reflect the choice of input weathering dependencies to an inordinate degree,[159,160] and the input parameters are, therefore, areas of intense study. Most models of the long-term global carbon cycle have used as input temperature dependencies measured in the laboratory.[156,159,160,162] Although the continental crust, taken as a whole, approximates the composition of an alkali feldspar, it is the dissolution of Ca- and Mg-containing silicates which consume the largest fraction of CO_2.[163] Activation energies for dissolution of Ca/Mg-silicates in organic-free solutions at low temperature generally fall in the range of 10 to 20 kcal/mol[160] (see Chapter 4). Laboratory temperature dependencies are generally in accord with weathering temperature dependencies derived from solute-based field studies. Velbel[155] measured an increase in weathering in the Blue Ridge at lower elevations, hence higher mean temperatures, which corresponds to a plagioclase weathering activation energy of 18.4 kcal/mol. Drever[164] measured a slightly smaller activation energy from solute budgets in the Swiss Alps. White and Blum[165] examined watershed records from a large number of basins and derived activation energies of 14.2 and 14.9 kcal mol^{-1} for, respectively, Na and Si fluxes. One of the critical conclusions of the latter study is that watershed topography, physical erosion rates, glacial history, and soil age are all second-order effects. In other words, it is precipitation and temperature, not other, hydrologic factors which control global CO_2 fluxes due to weathering over long periods of time. Dorn and Brady[166] measured the temperature dependency of plagioclase weathering using backscattered electron imaging of pit formation on exposed basalt surfaces to determine a plagioclase activation energy of ~26 kcal/mol (see Figure 11). This was done by imaging 2000-year-old basalts taken along lines of equal rainfall, but increasing mean temperature, on Hualalai Volcano in Hawaii. Because the basalts were mineralogically similar, and had been weathered for long, but well-constrained periods of time (through ^{14}C dating of buried vegetation), the derived activation energy is one of the only measures of temperature-dependent long-term dissolution derived from the rocks themselves, as opposed to solute budgets. Nevertheless, any amplifying effects of temperature-dependent organic activity are not included in the number, because the measurements were done on bare rock, free of fungal colonies, etc.

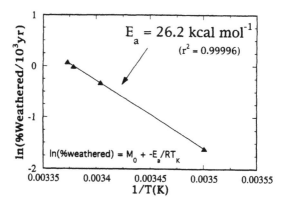

Figure 11

Rock-based activation energy for plagioclase weathering trends along the 500 mm isohyet on ^{14}C-dated basalt flows on Hualalai Hawaii.[166] %Weathered is the percent of the plagioclase surface weathered as measured by backscattered imaging. (From Dorn, R.I.; Brady, P.V. *Geochim. Cosmochim. Acta.* 1995, *59*, 2847–2852. With permission.)

Figure 12 shows mean global temperatures, calculated for the past 80 million years using the carbonate–silicate model of Berner.[153] Berner's model considers the effects of runoff, continental elevation, tectonic activity, temperature, and soil organic activity on weathering, atmospheric CO_2, and global mean temperature. The values in Figure 12 were calculated by retaining all of the tectonic and climatic input of Berner[153] and only allowing the input activation energy of silicate dissolution to vary (Berner's soil organic activity–weathering relation was deleted as well — see below). Note that paleo-P_{CO_2} values can vary by nearly a factor of two depending on the weathering T dependency which is used, highlighting the sensitivity of global climate to the temperature dependency of mineral weathering.

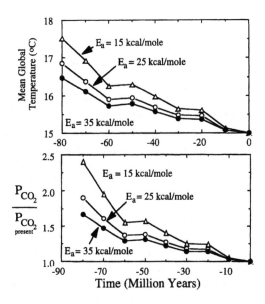

Figure 12
Dependence of P_{CO_2} and global mean temperature in the geologic past as a function of weathering activation energy, calculated using the model of Berner.[153] See text for further explanation. (From Dorn, R.I.; Brady, P.V. *Geochim. Cosmochim. Acta*, 1995, 59, 2847–2852. With permission.)

Two large uncertainties in the weathering sink are:

1. The integrated effect of organic activity on weathering globally is unclear. Drever[164] has suggested that organic acid levels may, in fact, be negligibly small in all but the most organically rich soils. If this turns out to be the case, this, and the fact that activation energies for inorganic-mediated dissolution are only slightly different from those for organic-mediated dissolution, would indicate effectively no biotic enhancement of the weathering feedback through chemical means. The

root–soil interface is estimated globally to be between 40 and 200 times that of the total land surface area,[167] which emphasizes the ability of plants to amplify weathering by increasing the surface area of soils,[156] extending the residence time of water in soil profiles, and anchoring fine particles.[164]

2. A worldwide activation energy for CO_2 consumption by weathering over geologic time is needed. Activation energies for silicates vary widely, giving substantially different model paleoclimate results. The actual weathering response to temperature will presumably be a sum of mineral activation energies weighted to reflect areal exposure, latitudinal distribution, and rainfall, etc. There is the possibility that changing lithologies due to shifts in tectonic regime in the past may have given rise to distinctly different weathering-temperature dependencies than exist at present.

V. BIOMINERALIZATION

Organisms grow minerals to encase and protect themselves (e.g., shells), as magnetic sensors (Fe_3O_4 in magnetotactic bacteria), to store nutrients, and to better consume nutrients (e.g., teeth). Examples are listed in Table 7. Salts of calcium (Ca-phosphates, carbonates, oxalate, and sulfate) dominate the list, acting as exoskeletons, teeth, Ca stores, and as gravity devices. Calcium is the third most abundant cation in the ocean after Na and Mg and has a role in a number of important metabolic functions (glycogen breakdown, neural transmission, control of intracellular transport). Metal sulfates are primarily used as gravity devices, iron oxides as teeth or for storage. Silica is primarily utilized as an exoskeleton or to give rigidity to plants. Bamboo is stiff because of its high silica content. Stinging nettles don't possess the rigidity to sting unless they have first deposited silica in the thorns.[169] Not only are biominerals intricately detailed, they are volumetrically imposing on a global scale.[168]

The minerals formed during biomineralization are typically of very specific morphology (crystal orientation, size, etc.) indicating that the mechanistic steps involved in biomineralization are far from random. Indeed, it is the precision, order, and efficiency of the biomineralization process that has drawn interest from both a biological and an industrial perspective. The steps involved in seeded crystal growth were outlined in the preceding chapter. For a mineral seed to nucleate from supersaturated solution, be it the ocean or cellular fluids, crystal embryos of a critical size must form. The free energy for formation of these nucleii, ΔG_n (kJ mol^{-1}) is:

$$\Delta G_n = \frac{16\pi\gamma^3}{(kT \ln S)^3} \tag{5}$$

TABLE 7.

Types and Functions of Inorganic Solids Found in Biologic Systems

Mineral	Formula	Organism	Function
Calcite	$CaCO_3$	Algae	Exoskeletons
		Trilobites	Eye lense
Aragonite	$CaCO_3$	Fish	Gravity device
		Molluscs	Exoskeleton
Vaterite	$CaCO_3$	Ascidians	Spicules
Amorphous	$CaCO_3 \cdot nH_2O$	Plants	Ca store
Hydroxyapatite	$Ca_5(PO_4)_3(OH)$	Vertebrates	Endoskeletons/teeth/Ca store
Octa-calcium phosphate	$Ca_8H_2(PO_4)_6$	Vertebrates	Precursor phase in bones
Amorphous Ca/P		Mussels	Ca store
		Vertebrates	Ca store
Calcium oxalate	$CaC_2O_4 \cdot H_2O$	Plants	Ca store
Gypsum	$CaSO_4 \cdot 2H_2O$	Jellyfish larvae	Gravity device
Barite	$BaSO_4$	Algae	Gravity device
Celestite	$SrSO_4$	Acantharia	Cellular support
Magnetite	Fe_3O_4	Bacteria	Magnetotaxis
		Chitons	Teeth
Goethite	α-FeOOH	Limpets	Teeth
Lepidocrocite	γ-FeOOH	Chitons	Teeth
Ferrihydrite	$5Fe_2O_3 \cdot 9H_2O$	Animals and plants	Fe storage proteins
Silica	$SiO_2 \cdot nH_2O$	Algae	Exoskeletons
		Diatoms, Sponges	Exoskeletons
		Sponges	Cellular support

From Mann, S. *Nature*. 1988, *332*, 119–124. With permission.

γ is the edge free energy for growth (kJ mol^{-1}), k is Boltzmann's constant, and S is saturation state (see Chapter 4). Note that ΔG_n can be lowered (and nucleation favored) by decreasing γ or increasing S. Organisms direct mineral nucleation by doing both. Nucleation is often onto an organic surface where γ reflects the local bonding environment at the substrate–nuclei interface. At the same time, the saturation state of nucleating fluids are controlled in biologic systems by the ability of biologic membranes to selectively transport ions to surface sites and/or water away from surface sites. Note that there must be a distinct minimum free energy for each organic-directed nucleation process, not a range. If the latter were true, crystallochemical specificity could not result.[168]

Biominerals, once nucleated, grow according to one of the heterogeneous crystal growth rate laws outlined in the previous chapter. Hence, rates will depend on γ and S. γ can be modified by selective adsorption of organic molecules onto active sites and S again controlled by membrane-directed ion transport. The net effect of biologically directed nucleation, growth, and inhibition is to determine the identity, morphology, and abundance of biominerals, often by very different pathways than abiotically grown minerals.

The mechanisms of biomineralization can be simplistically illustrated using the example of biomineral–biopolymer interactions. Biocomposites (biominerals + biopolymers) such as teeth and bones consist of a polymeric matrix strengthened by a biomineral. Calvert and Mann[170] categorize biomineral–polymer interactions as being one of four types, depending on the degree to which the organic matrix interacts with the mineral growing *in situ*. Type I biocomposites are passively engineered, in that only the saturation state of the growing mineral is controlled. Type II biocomposites possess minerals which are selectively nucleated and uniformly oriented by specific sites on the organic template. The formation of type III biocomposites involves the growth of amorphous minerals due to the inhibition of crystal nucleation. Type IV biocomposites form morphologically precise minerals through directed crystal nucleation and controlled inhibition of specific growth sites. To a large extent, biocomposites end up in one class or another depending on the thermodynamic function $(S, \gamma, $ etc.$)$ the organic matrix controls during mineral growth. The latter obviously reflects the chemical state (bonding geometry, electrostatics, etc.) of the mineral–polymer interface.

One of the more effective approaches to unraveling the relative contribution of the various factors involved in template-directed crystal growth at biologic surfaces has been through the use of Langmuir monolayers having ionic and/or polar sites exposed to supersaturated solutions.[171,172] This approach provides a nucleating surface which might broadly mimic those seen by nucleating biominerals. By compressing the monolayer, one is able to vary the density, hence separation of sites, of the film, exposed to a supersaturated solution. By doing the latter, and by varying the charge, hydrophobicity, etc. of the headgroups on the solution side of the film, one can get a rough idea of the relative importance of such factors as geometry, charge, and hydrogen bonding on biologically directed mineral growth. Mann et al.[172] studied nucleation of calcite, and its metastable polymorph vaterite, on compressed Langmuir monolayer interfaces consisting of exposed carboxylate or amine groups. Films with an anionic charge accumulated calcium at the carboxylate sites, and the calcite crystals subsequently nucleated were oriented with the $[1\bar{1}.0]$ axis perpendicular to the monolayer. At lower levels of dissolved calcium, vaterite formed with the [00.1] axis normal to the monolayer. The identity of the nucleated phase depends explicitly on the extent of adsorption at the interface, which in turn can depend on dissolved calcium levels, as well as cations such as Na^+ and Li^+ which may compete with Ca^{++} for carboxylate sites.[173] One implication is that nucleation appears to be a three-dimensional process which is largely controlled by ion interactions in the double layer adjacent to the template.[173]

Only vaterite nucleated under the cationic octadecylamine monolayer. The identical polarity of the interface and calcium suggests that binding of the latter is not required for vaterite nucleation. The fact that nucleated crystals of calcite and vaterite were nonrandomly oriented points to a geometric component of the process. Apparently, a lattice match at the monolayer–nucleii interface facilitates growth and provides a mechanism for the transfer of geometric information into the growing crystal.[174] The argument for molecular recognition of monolayer headgroups by nucleating crystals has subsequently been used to selectively grow the calcite (00.1) face using sulfate as the headgroup and aragonite crystals with the [001] axes perpendicular to the monolayer, again using sulfate as the headgroup.[173] Nonionic, polar templates with exposed alcohol groups are able to nucleate hydrated crystals such as gypsum[175] and ice.[176] Hydrogen bonding apparently orders water molecules at the monolayer–solution interface.

Historically, work in biomineralization has dealt primarily with Ca-bearing minerals.[177] More recently, the role of organisms in the nucleation and growth of elemental gold,[178,179] and in particular that of iron (hydr)oxides and sulfides, has been argued. Directed growth of minerals such as magnetite is obviously of industrial interest. At the same time iron mineral surface chemistry may be a link to organic–mineral interactions which occurred soon after Earth cooled and life began.[180,181] Carbon fixation (in this case, conversion of inorganic carbon to organic acids) is a thermodynamically favored side effect of pyrite growth.[182] Wächtershäuser[182] has argued that pyrite surfaces catalyzed the first accumulation and assembly of abiotically produced organic compounds to produce life in the far geologic past (see below). The details will be examined more fully below, but it is important to note that biomineralization probably began at almost the same time as life itself.

Processes occurring at the mineral–fluid interface are sometimes found to induce disease, particularly when the mineral surface is that of quartz, chrysotile asbestos, asbestiform amphibole, or the zeolite erionite, to name the most widely studied minerals. The link between quartz surface behavior and silicosis is perhaps the best understood process. More-recent attention has focused on the pathogenetic propertes of chrysotile asbestos and amphiboles. Thorough examinations of the quartz–disease link are those of Iler[169] and Ross et al.[183] upon which the following summary is based. Inhalation of quartz dusts leads to formation of fibrotic (scar) tissue, loss of lung function, hardening of tissues, and a greater susceptibility to tuberculosis. Early theories attributed silicosis to the interaction of dissolved silica with nucleic acids or to protein denaturation by silica polymers. In either case, minerals were thought to be important only as a source of dissolved silica. One of the primary drawbacks to this "solubility-based" view is that the cytotoxicity of the various SiO_2 polymorphs appears to be inversely related to their solubility. Quartz is relatively insoluble, but quite toxic. Silica is much more soluble, but biologically inert. Note that the inverse should be true if the ingested minerals are active only by providing aqueous silica.

Current theories identify the quartz surface as the primary cause of silicosis. Phagocytes in the lungs are envisioned to coat quartz particles upon ingestion. Interactions at the quartz–phagocyte interface result in products (cytokines, proteins, and growth factors), which stimulate production of hydroxyproline by fibroblasts and subsequent formation of fibrotic tissue.[169] It is the specific geometric arrangement and chemical reactivity of the quartz surface which causes organic molecules to behave abnormally. Unlike chrysotile asbestos and asbestiform amphiboles (see below), particle physical dimension is not a controlling factor in quartz–phagocyte interaction.

Chrysotile asbestos and asbestiform amphiboles, when ingested, cause pulmonary disease (see the review of Guthrie and Mossman[184]). Nevertheless, the work of Stanton et al.[185] correlating the probability of tumor formation with asbestos mineral fiber dimension pointed to a geometric control over pathogenesis, at least for fibrous minerals. Apparently, outside of the minerals causing silicosis and a few others, reactions at the mineral–biofluid interface are non-mineral-specific. Unlike quartz in lung tissue, the identity of the mineral surface is less important, suggesting that the chemical behavior of specific mineral surfaces is a secondary factor in pathogenesis. Surface chemistry is relevant to the toxicity of asbestiform minerals primarily as it controls the rate of mineral dissolution, that is, the residence time of a given mineral in the body.[186] The same approach applies to dissolution of Pb-containing solids.[187] To apply laboratory dissolution rates to assess *in vivo* durability requires knowledge of the effects of temperature (most laboratory rates are done at 10 to 15°C less than metabolic temperatures) and of the integrated effect of the individual rate-affecting components in metabolic fluids.

Minerals that thermodynamics favor to form at Earth's surface generally grow fast (with or without the aid of biota) or seemingly not at all, due to the constant recycling of sedimentary materials. Dolomite is an intriguing exception to the rule because it appears to have grown readily in the geologic past, but only rarely does so today. Abundant field evidence suggests that the dolomite which once grew from seawater, or modified versions thereof, in substantial quantities from the Precambrian at least until the late Paleozoic (see summaries of Zenger[188] and Holland[189]) provided a destination for riverine fluxes of magnesium, calcium, and carbon and a subsequent resting place for much of world petroleum.[190] At present, dolomite seldom grows in marine environments,[191] and, rarely, in laboratory beakers.[190] However, dolomite seems to grow very rapidly as an alteration product on some marble building stones.[192] There is no universally accepted explanation for its on-again, off-again pattern of growth.[193] The latter remains one of the fundamental riddles of geochemistry and a glaring exception to the notion of uniformitarianism, i.e., that the geologic present is the key to the geologic past.

Much of what is known about dolomite growth is derived from field relations of ancient dolomites[191] and examinations of the growth of recent,

Holocene dolomites formed in evaporative environments.[194-196] As field evidence is often amenable to a variety of interpretations,[191] attention has also focused on laboratory synthesis of dolomite to unravel the chemical controls on growth. To achieve discernible reaction in the laboratory, high temperatures (200 to 300°C)[197-199] and/or extreme supersaturation are required. The combination potentially makes it difficult to unravel the chemical steps involved in the growth of dolomite at Earth's surface, where temperatures are low and supersaturation moderate. There is also the distinct possibility that reaction mechanisms operative at high temperatures are not those which control dolomite growth at low temperatures.

To add a unit cell of dolomite onto a dolomite seed crystal (either through dissolution–reprecipitation or *de novo* growth) requires the stoichiometric adsorption of Ca and Mg onto their respective planes followed by the dehydration and carbonation of each ion. Note that there is a continuum between metal sorption and metal carbonate mineral growth.[200-202] If adsorbate incorporation is relatively rapid, growth rates will depend largely on the equilibrium sorption of mineral components. For this reason, sorption measurements should give important clues to the controls on growth. Ca and Mg adsorption to stoichiometric dolomite is roughly one to one over wide ranges in solution composition[203,204] (see Figure 13), suggesting that metal dehydration and carbonation must be the slow, rate-determining step for dolomite growth.[205] As calcite ($CaCO_3$) grows rapidly at low temperatures, dehydration and carbonation of adsorbed Ca is probably rapid relative to that of Mg (magnesite, $MgCO_3$, is notoriously difficult to grow in the laboratory at low temperatures[206]). The rate-limiting steps for dolomite growth are therefore thought to be the hydration and carbonation of the adsorbed Mg ion.[205] Because dolomite grows in a variety of geologic settings, a number of pathways to growth probably exist. Organic acids which complex Mg are able to remove the tight hydration sheath of the Mg atom, a critical step for the incorporation of magnesium into carbonate lattices.[207-208] Sulfate is a more common ligand than organic acids in marine and evaporative environments. Sulfate binds strongly to Mg in solution (K_{assoc} = $10^{2.23}$)[209] and, through adsorption, might provide a pathway for bringing dehydrated Mg to the mineral surface. Sulfate appears to amplify metal sorption to dolomite when sulfate levels are greater than ~5 mM in 0.05 M NaCl solutions, but inhibits sorption at levels less than this (see Figure 14). Once Mg-sulfate complexes are adsorbed to the surface, the problem then becomes one of exchanging carbonate for coadsorbed sulfate. Note that the Mg–sulfate complex is appreciably weaker than is the $MgCO_3$ complex (K_{assoc} = $10^{2.98}$).[210] Although carbonate occurs at low levels in marine environments, it is thermodynamically favored to displace sulfate from Mg ions when present. One potential pathway for dolomite growth might then be:

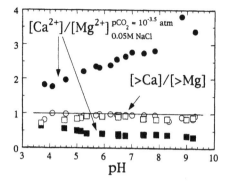

Figure 13
Ca and Mg adsorption (mol L^{-1}) onto dolomite (10 g L^{-1}) as a function of $[Ca^{2+}]/[Mg^{2+}]$ (I = 0.05 M NaCl; P_{CO_2} = $10^{-3.5}$ atm).[204]

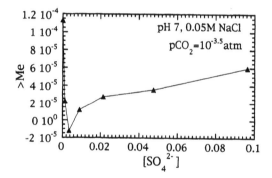

Figure 14
Ca + Mg adsorption (mol L^{-1}) onto dolomite as a function of $[SO_4^{2-}]$ (mol L^{-1}) (pH 7; I = 0.05 M NaCl; P_{CO_2} = $10^{-3.5}$ atm).[204]

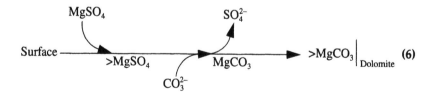

$>MgCO_3$ is a dehydrated, carbonated Mg at the dolomite surface. It should be emphasized that the above scenario is at this point largely speculation. Nevertheless, a fair bit of circumstantial evidence supports it. Siegel[211] showed that sulfate is a catalyst for protodolomite growth at low temperatures.[188,191] At the same time, dolomite grows in contact with thenardite (Na_2SO_4)-saturated and epsomite $(MgSO_4.7H_2O)$-saturated solutions at Deep

Springs and Basque Lakes approaching 2 to 70 (30,000 to 200,000 ppm) times marine sulfate levels.[212] Although adsorbed sulfate may foster growth in some environments at low temperatures, sulfate has been shown to inhibit growth at T > 200°C.[198] This may be due to the low levels of sulfate used in the latter experiments — recall that, at least at 25°C, sulfate levels this low cause Mg to desorb from dolomite. Another explanation for dolomite inhibition is that some trace substance in the present-day ocean inhibits growth and/or replacive dissolution of calcite. If this were the case, however, it still should be possible to grow dolomite in the laboratory under controlled experiments at low temperatures. Unraveling the growth of dolomite is a very critical test of the surface complexation approach to kinetics that is, at the same time, useful in a practical sense. The absence of even one rate law for dolomite growth makes it very difficult to test the linkages between tectonics, ocean chemistry, evaporite formation, and Mg cycling over geologic time.[189]

VI. ORIGINS OF LIFE

RNA sequencing of present-day organisms indicates that the earliest living creature on Earth was a very primitive microorganism which evolved in relatively high-temperature fluids (50°C < T < 150°C).[213] All life that has appeared in the intervening 4 billion years, or so, is thought to have evolved from this ancient ancestor.[214] The universality of the genetic code, and the fact that no organisms with even slightly different codes are known to exist, points to a singular origin.[215] Moreover, precellular evolution is thought to have occurred quite rapidly. The accepted age for Earth is ~4.6 billion years, the age of the oldest rocks ~4 billion years, and the age of the oldest fossils less than a few hundred million years younger.

In the prebiotic world there were a limited number of environments with consistently high temperatures for thermophilic bacteria to evolve in. The surface oceans may not have been warmer than the present ones.[189] If the prebiotic atmosphere lacked free oxygen and ozone, high inputs of ultraviolet radiation would have made it very difficult for any biologic compounds formed at the surface to survive.[216] The most reasonable physical setting for the advent of life has, therefore, been argued to be geothermal zones located along the ocean floor.[217] The original, abiotic source of primitive organic compounds in the Archean has been argued to be electric discharges in the presence of the primitive atmosphere.[218] Miller and co-workers showed that electric discharges in an ammonia-rich atmosphere produce amino acids, nitrogen bases, and pentose sugars, in essence, many of the building blocks for life. Because recent evidence points to a prebiotic atmosphere which was not reducing,[219,220] the original chemosynthetic reactions probably involved reduction of CO_2 and/or CO.[221] Organic compounds are also produced by coronal discharges, ultraviolet radiation, and may have arrived on Earth in the form of interplanetary dust particles.[222] Abiotic production of organic material by exposed, reduced mantle minerals in mid-ocean ridge hydrothermal systems was probably important in the Archean as well.[223] Abiotic synthesis of reduced organic compounds occurs today on basalt surfaces in contact with volcanic gases.[224] Even with the most optimistic production rates from the various sources, organic material would still have been only a trace constituent of world oceans[225] and present at levels much less than that needed to polymerize homogeneously. Obviously, some mechanism for concentrating organic material was required. Said differently, the trick was the removal of water from an aqueous medium.[226] And this is where mineral surfaces come in.

Mineral surfaces are thought to have provided the original template for the precursors of life to begin.[216,227,228] Chemical evidence for this is the reduced water activity at mineral surfaces and the ability of mineral surfaces to catalyze many important organic processes. Geologic evidence for the mineral surface role is that biologic compounds do not spontaneously self-assemble in the oceans today. If the present is any guide, biologic compounds didn't self-assemble 4 billion years ago either. Minerals do spontaneously form from natural waters and, once formed, are quite handy at adsorbing and concentrating organic matter at their surfaces. In the distant past they might have, in this fashion, provided the requisite scaffolding for the construction of Earth's first biologic compounds.[229] Put most succinctly, "the question of the origin of life is assumed to be related to solid–liquid interfacial activity, and the answer may be associated with sediment–water interaction rather than with solution chemistry."[225]

Two approaches have historically been used to explore the sediment–water answer to the origin of life. The first is the crystallographic approach (see, for example, Cairns-Smith[229]) where structures of proteins, DNA, etc. are compared with the repeating structures of clay minerals to establish points of physical similarity. By making enough links between the two, one might then advance the argument that the life was given its start by minerals. Obviously, one has to assume that the imprint of the original template has not been obscured irrevocably by subsequent evolutionary change. The chemical approach, which will largely be emphasized here, instead asks the question: what did a surface need to do to start life? and then proceeds to test various minerals for their ability to perform the assumed tasks (see Wächtershäuser[182]). Some tests which have been proposed thus far are:

1. **Surface-controlled enantiomeric segregation** — Optical activity in organic molecules is one of the primary criteria used for the recognition of life.[230] A number of minerals adsorb and preferentially polymerize enantiomeric amino acids from racemic mixtures. Both L- and D-type exist in nature, yet all proteins in present-day biologic compounds are L-type. Peptide chains composed of random L- and D-type isomers would be

sterically hindered from forming the highly organized, complex structures typical of biologic activity.[231] The preference of present-day biologic compounds for L-types is thought to be a relict of surface adsorption processes operative in the prebiotic world.[232-234] The ability to polymerize enantiomerically segregated, racemic molecules is therefore one of the strongest tests of mineral-directed evolution.[231,233,235-237]

2. **Phosphate adsorption** — Phosphorus is a limiting nutrient in much of the ocean today and probably was in the prebiotic world as well. Phosphorus is critical for the formation of phospholipid membranes and is the metabolic currency of energy through ATP–ADP cycles in all present-day organisms.[238] It is difficult to imagine life at any level in the absence of phosphorylation reactions. Phosphate concentrations were in all likelihood very low in the prebiotic ocean.[239] Griffith et al.[239] calculate that 3 billion years were required for the oceans to become saturated with phosphate, derived primarily from the surface-controlled weathering of primary apatite. To avoid phosphate starvation, precellular organisms needed a mechanism for concentrating any available phosphorus out of seawater. This means that the surface which grew life had to strongly adsorb phosphate from seawater-like solutions.

3. **Trace metal adsorption** — The chemical composition of present-day biota reflects the chemical conditions that their ancestors contended with.[215,240,241] Biologic utilization of elements correlates imperfectly with marine and crustal abundances,[242] suggesting that an important component of biologic specificity exists. For this reason, the chemical compositions of present-day organisms provide a test of hypotheses about the evolution of the universal ancestor.[215,243] The metabolic function of metals is probably the best example of this. Divalent metals can be categorized as being either universal nutrients (e.g., Ca, Mg, Co, Fe), necessary for life, or as universal toxins (e.g., Cd, Hg, Pb, Sn), which perform no metabolic function.[182] Nutrient metals exist in large concentrations in living organisms; toxins do not. The sharp divide between nutrients and toxins is thought to exist because there was a selection process occurring at the earliest stage of evolution. If precellular evolution was surface mediated, the segregation of metals into nutrient and toxin classes should also be a surface-controlled process.[182] This constitutes a powerful test for any prebiotic template. Specifically, if a mineral adsorbs nutrients and toxins equally, it is less likely that precellular evolution occurred on that particular surface. If a mineral can differentiate between nutrient and toxic metals, it is a more reasonable site for precellular evolution.[182] The metal–surface connection may involve interactions between coadsorbates. Lawless and Levi[243] showed that yields for surface-controlled condensation of glycine and alanine depended upon the identity of the metal ions present. Nucleotide adsorption onto bentonites also is metal co-ion dependent.[244]

4. **Relative abundance** — A mineral template had to be present in sufficient quantities in the appropriate geologic setting for precellular

evolution to occur. Moreover, it had to grow abiotically in a sedimentary or hydrothermal regime.

These tests are by no means monolithic, but rather a distillation of the criteria which biologists, geochemists, and the like, have historically used to approach the problem over the past four decades. Obviously, any answers are likely to be inferential at best because of uncertainties about the chemical state of primitive Earth and the vast expanse of elapsed time involved. A number of mineral surfaces have been examined for their ability to accelerate life's first steps (or whatever), including kaolinite and/or Al_2O_3,[245,246] montmorillonite,[247] Ca-phosphates,[248] bentonites,[243,244] quartz,[233,249,250] and pyrite.[251,252] Some of the positive and negative aspects of each mineral are summarized below.

> **Kaolinite.** (+ montmorillonites and bentonites). *Positive* — Clays are abundant and grow in an abiotic fashion rapidly. At the same time, they possess high reactive surface areas. Kaolinite and bentonite adsorb and polymerize enantiomerically segregated organic molecules.[231,235,253,254] Bentonites adsorb nucleotides.[244] Catalytic activity seems to be most important at exposed Al sites.[253,255] *Negative* — Scenarios involving aluminosilicate clays generally involve cyclic wetting and drying,[246] hence a geologic scenario of prebiotic evolution in a near-surface environment where ultraviolet radiation should have countered the accumulation of abiotically produced organic compounds. Also, kaolinite and montmorillonite are scarce in Precambrian sediments.[256]
>
> **Ca-Phosphates.** *Positive* — Nucleotides are readily formed at Ca-phosphate surfaces at 65 to 100°C from urea and ammonium chloride solutions.[248] *Negative* — Abundance is a problem, due to the probable scarcity of authigenic phosphates in the early oceans (see above).
>
> **Quartz.** *Positive* — Quartz is optically active and able to segregate amino acids enantiomerially.[234] Quartz adsorbs amino acids and phosphate from dilute solutions and adsorbs poisons and nutrient metals (see above) by different mechanisms.[250] By adsorbing multivalent metals in a unidentate, charge-reversing fashion, quartz has the ability to adsorb anionic organic groups such as carboxylates. *Negative* — Quartz specific surface areas are generally low and were probably not all that abundant in seafloor hydrothermal systems, relative to other candidates such as pyrite.
>
> **Pyrite, Iron Sulfides.** Based primarily on the work of Wächtershäuser, much of the recent attention has focused on pyrite as the original template for life. Recent work[257,258] has pointed to hydrothermally precipitated iron sulfide membranes as the

catalyst. Fe sulfides are abundant in and near seafloor hydrothermal systems. It has been theorized that the pyrite surface is
optically active,[251] hence potentially able to enantiomerically
segregate organic molecules. Central to the pyrite as incubator
theory is the nature of bonding at the pyrite–solution interface.
For self-selection of biologic molecules a number of conditions
must exist. Adsorbed constituents must be bound to the surface
but, at the same time, be free to slowly migrate in a two-dimensional sense. The latter is needed for association and polymerization of progressively larger organic molecules to occur and
suggests that outer sphere complexes (see Chapter 3) are the
basis for subsequent growth. The requirement of relatively
weak surface bonding also plays a role in the method by which
pyrite surfaces are argued to differentiate between nutrient and
poison metals.[182] Metal sulfides of the poisons Cd^{++}, Pb^{++}, Cu^{++},
Ag^+, Hg^{++}, Sn^{++}, As^{+++}, Sb^{+++}, and Bi^{+++} are highly insoluble, and
they generally sorb very strongly. Metal sulfides of the nutri-
ents Mn^{++}, Fe^{++}, Ni^{++}, Co^{++}, and Zn^{++} are less soluble and can be
expected to be less strongly bound. The first group are envi-
sioned to be rapidly scavenged in seafloor hydrothermal systems, whereas the latter are predicted to remain appreciably in
solution, hence able to participate in surface metabolism.[182]
Outer-sphere metal complexes and weakly sorbed, long-chain
organic molecules are thermodynamically favored to associate
further at high temperatures,[182] the surface-bonded configuration working against thermal degradation and depolymeriza-
tion. As noted above, pyrite has the added advantage of being
a potential electron donor and, hence, has the ability to stabilize
reducing conditions at the surface.[182]

Many of the critical features of the surface metabolism theory are
experimentally testable.[182] Two of the more obvious assumptions to be
verified are those of phosphate adsorption and nitrogen fixation. Phosphate is assumed to adsorb to pyrite from solutions similar to seawater,
and to then participate in surface metabolism. A more telling test is whether nitrogen fixation is possible at the pyrite surface. Nitrogen fixation is
thought to be one of the most ancient biochemical processes. In the model
of Wächtershäuser[182] nitrogen fixation is assumed to be controlled by
organic carbon adsorbed at the pyrite surface.

VII. ENDNOTE

A complex series of interlocking chemical fluxes determines the hab-
itability of Earth's surface. Their collective sensitivity to human activities

is a critical unknown. Many of the underlying processes, such as carbon cycling, climate change, and chemical transport in soils, depend specifically on the reactions occurring at mineral surfaces, such as adsorption, mineral dissolution, and growth. This chapter has sought to (1) point out a number of the more important biogeochemical processes and (2) highlight the mechanistic link between these macroscopic processes and microscopic mineral surface reactions.

ACKNOWLEDGMENTS

PVB — Many thanks to those who have reviewed sections of this chapter: Jill Buckley, surface–petroleum interactions; Michael Velbel, weathering; Everett Shock, origins of life; Leslie Melim, dolomite growth; Jim Krumhansl and Diane Marozas, the whole thing. Funding for PVB has been graciously supplied by the National Science Foundation, the Petroleum Research Fund of the American Chemical Society, DOE/BES-Geosciences, and the Nuclear Regulatory Commission. JMZ's review was funded, in part, by the Subsurface Science Program, Office of Health and Environmental Research, U.S. Department of Energy.

REFERENCES

1. Walker, J.C.G.; Hays, P.B.; Kasting, J.F. *J. Geophys. Res.* 1981, *86*, 9776–9782.
2. Van Breemen, N.; Driscoll, C.T.; Mulders, J. *Nature.* 1984, *307*, 599–604.
3. Samama, J.C. *Ore Fields and Continental Weathering.* Van Nostrand Reinhold: New York, 1988.
4. Sposito, G. *The Surface Chemistry of Soils.* Oxford University Press: New York, 1984.
5. McBride, M.B. *Environmental Chemistry of Soils.* Oxford University Press: New York, 1994.
6. Stumm, W. *Chemistry of the Solid-Water Interface.* John Wiley: New York, 1992.
7. Schindler, P.W. In *Adsorption of Inorganics at Solid-Liquid Interfaces.* Ann Arbor Science: Ann Arbor, MI, 1981; pp. 1–47.
8. White, G.W.; Zelazny, L.W. *Clays Clay Min.* 1988, *36*, 141–146.
9. Brady, P.V.; Walther, J.V. *Geochim. Cosmochim. Acta.* 1989, *53*, 2823–2830.
10. Van Cappellen, P.; Charlet, L.; Stumm, W.; Wersin, P. *Geochim. Cosmochim. Acta.* 1993, *57*, 3505–3518.
11. Rönngren, L.; Sjöberg, S.; Zhonxi, S.; Forsling W.; Schindler, P.W. *J. Coll. Interf. Sci.* 1991, *145*(20), 396–404.
12. Wu, L.; Forsling, W.; Schnidler, P.W. *J. Coll. Interf. Sci.* 1991, *147*, 178–185.
13. Sposito, G. *Chemical Equilibria and Kinetics in Soils.* Oxford University Press: New York, 1994.
14. Blum, A.E.; Lasaga, A.C. *Geochim. Cosmochim. Acta.* 1991, *55*, 2193–2201.
15. Davis, J.A.; Kent, D.B. *Mineral-Water Interface Geochemistry.* M.F. Hochella; A.F. White; Eds. Mineralogical Society of America: Washington, D.C., 1990; pp. 177–248.

16. Goldberg, S. *Adv. Agr.* 1991, *47*, 233.
17. Brown, G.E. *Mineral-Water Interface Geochemistry.* M.F. Hochella; A.F. White; Eds. Mineralogical Society of America: Washington, D.C., 1990; pp. 309–353.
18. Balistrieri, L.S.; Murray, J.W. *Geochim. Cosmochim. Acta.* 1985, *46*, 1253–1265.
19. Zachara, J.M.; Girvin, D.C.; Schmidt, R.L.; Resch, C.T. *Environ. Sci. Tech.* 1987, *21*, 589.
20. Liang, L.; Morgan, J.J. *Chemical Modeling of Aqueous Systems II.* D.C. Melchior; R.L. Basset; Eds. American Chemical Society: Washington, D.C., 1993; p. 293.
21. Xu, S.; Harsh, J.B. *Clays Clay Miner.* 1992, *40*, 567–574.
22. Barron, V.; Herruzo, M.; Torrent, J. *Soil Sci. Soc. Am. J.* 1988, *52*, 647–651.
23. Colombo, C.; Barron, V.; Torrent, J. *Geochim. Cosmochim. Acta.* 1994, *58*, 1261–1269.
24. Hochella, M.F.; Eggleston, C.M.; Elings, V.B.; Thompson, M.S. *Am. Mineral.* 1990, *75*, 723–730.
25. Dove, P.M.; Hochella, M.F. *Geochim. Cosmochim. Acta.* 1993, *57*, 705–714.
26. Eggleston, C.M.; Stumm, W. *Geochim. Cosmochim. Acta.* 1993, *57*, 4843–4850.
27. Maurice, P.A.; Hochella, M.F.; Parks, G.A.; Sposito, G.; Schwertmann, U. *Clays Clay Miner.* 1995, *43*, 29–38.
28. Greenland, D.J. *Clay Miner.* 1975, *10*, 407–416.
29. Golden, D.C.; Dixon, J.B. *Soil Sci. Soc. Am. J.* 1985, *49*, 1568–1576.
30. Ryan, J.N.; Gschwend, P.M. *Environ. Sci. Tech.* 1994, *28*, 1717–1726.
31. Deshpande, T.L.; Greenland, D.J.; Quirk, J.P. *Nature.* 1964, *201*, 107–108.
32. Theng, B.K.G. *Formation and Properties of Clay-Polymer Complexes.* Elsevier Scientific Publishing: New York, 1979.
33. Ryan, J.N.; Gschwend, P.M. *Geochim. Cosmochim. Acta.* 1992, *56*, 1507–1521.
34. Wood, W.W.; Kraemer, T.F.; Hearn, P.P., Jr. *Science.* 1990, *247*, 1569–1572.
35. Brusseau, M.L. *Am. Geophys. Union.* 1994, *32*, 285–313.
36. Stevenson, F.J. *Humus Chemistry.* John Wiley & Sons: New York, 1982.
37. Thurman, E.M. *Organic Geochemistry of Natural Waters.* Martinus Nijhoff: The Hague, Netherlands, 1985.
38. Aiken, G.R.; McKnight, D.M.; Wershaw, R.L.; MacCarthy, P. *Humic Substances in Soil, Sediment, and Water.* John Wiley & Sons: New York, 1985.
39. Huang, P.M.; Schnitzer, M. *Interactions of Soil Minerals with Natural Organics and Microbes.* Soil Science Society of America: Madison, Wisconsin, 1986.
40. Davis, J.A. *Geochim. Cosmochim. Acta.* 1982, *46*, 2381–2393.
41. Tipping, E.; Cooke, D. *Geochim. Cosmochim. Acta.* 1982, *46*, 75–80.
42. Zachara, J.M.; Resch, C.T; Smith, S.C. *Geochim. Cosmochim. Acta.* 1994, *58*, 553–566.
43. Sposito, G. *The Chemistry of Soils.* Oxford University Press: New York, 1989.
44. Cornell, R.M.; Schwertmann, U. *Clays Clay Miner.* 1979, *27*, 402–410.
45. Huang, P.M.; Violante, A. *Interactions of Soil Minerals with Natural Organics and Microbes*; P.M. Huang; A.F. White; Eds. Soil Science Society of America: Madison, Wisconsin, 1986; p. 159.
46. Nakashima, S. *Adv. Org. Geochem.* 1992, *19*, 421–430.
47. Westall, J.C.; Jones, J.D.; Turner, G.D.; Zachara, J.M. *Environ. Sci. Tech.* 1995, *29*, 951–959.
48. Jones, R.C.; Uehara, G. *Soil Sci. Soc. Am. Proc.* 1973, *37*, 792–795.
49. Postma, D.; Brockenhaus-Schack, B.S. *J. Sediment. Petrol.* 1987, *57*, 1040–1053.
50. Banfield, J.F.; Eggleton, R.A. *Clays Clay Miner.* 1988, *36*, 47–60.
51. Coston, J.A.; Fuller, C.C.; Davis, J.A. *Geochim. Cosmochim. Acta.* 1995. In press.
52. Zachara, J.M.; Smith, S.C.; Kuzel, L.S. *Geochim. Cosmochim. Acta.* 1995. In press.
53. Bertsch, P.M.; Hunter, D.B.; Sutton, S.R.; Bajt, S.; Rivers, M.L. *Environ. Sci. Tech.* 1994, *28*, 980–984.
54. Zachara, J.M.; Westall, J.C. *Soil Physical Chemistry; 2nd Edition*; D.L. Sparks; Ed. CRC Press: Boca Raton, FL, 1995. In press.
55. Goldberg, S.; Sposito, G. *Soil Sci. Soc. Am. J.* 1984, *48*, 772–783.

56. Rea, B.A.; Kent, D.B.; LeBlanc, D.R.; Davis, J.A. *U.S. Geological Survey Toxic Substances Hydrology Program — Proceedings of the Technical Meeting:* Monterey, CA, March 11–15, 1991; G.E. Mallard; D.A. Aronson; Eds. U.S. Geological Survey 91–4034, 1991.
57. Hayes, K.F.; Redden, G.; Ela, W.; Leckie, J.O. *J. Coll. Interf. Sci.* 1991, *142*, 448–460.
58. Wagner, J.; Zachara, J.M.; Westall, J.C. *Environ. Sci. Tech.* 1995. Submitted.
59. Kent, D.B.; Davis, J.A.; Anderson, L.D.D.; Rea, B.A. *Water Resour. Res.* 1995, *31*, 1041–1050.
60. Burr, D.T.; Sudicky, E.A.; Naff, R.L. *Water Resour. Res.* 1994, *30*, 791–815.
61. Kent, D.B.; Davis, J.A.; Anderson, L.C.D.; Rea, B.A.; Waite, T.D. *Water Resour. Res.* 1995, *30*, 1099–1114.
62. Mangold, D.C.; Tsang, C.F. *Am. Geophys. Union.* 1991, *29*, 51–79.
63. Yeh, G.T.; Tripathi, V.S. *Water Resour. Res.* 1989, *27*, 3075.
64. Collier, R.W.; Edmond, J.M. *Progr. Oceanogr.* 1984, *13*, 113–199.
65. Broecker, W.S.; Peng, T.H. *Tracers in the Sea;* Lamont-Doherty Geological Observatory: New York, 1982.
66. Whitfield, M.; Turner, D.R. *Aquatic Surface Chemistry.* W. Stumm; Ed. Wiley-Interscience: New York, 1987.
67. Erel, Y.; Morgan, J.J. *Geochim. Cosmochim. Acta.* 1991, *57*, 513–518.
68. Balistrieri, L.; Brewer, P.G.; Murray, J.W. *Deep Sea Res.* 1981, *28A*, 101–121.
69. Mayer, L.M. *Chem. Geol.* 1994, *114*, 347–363.
70. Kiel, R.G.; Tsamakis, E.; Fuh, C.B.; Giddings, J.C.; Hedges, J.I. *Geochim. Cosmochim. Acta.* 1994, *58*, 879–893.
71. Collins, M.W.; Bishop, A.N.; Farrimond, P. *Geochim. Cosmochim. Acta.* 1995, *59*, 2387–2391.
72. White, A.F., Peterson, M.L. *Chemical Modeling in Aqueous Systems II;* D.C. Melchior; R.L. Bassett; Eds. ACS Symposium Series 416; American Chemical Society: Washington, D.C., 1990, Chapter 18.
73. Kiel, R.G.; Montluçon, D.B.; Prahl, F.G.; Hedges, J.I. *Nature.* 1994, *370*, 549–552.
74. Tannenbaum, E.; Kaplan, I.R. *Geochim. Cosmochim. Acta.* 1985, *49*, 2589–2604.
75. Lewan, M.D. *Philos. Trans. R. Soc. London.* 1985, *A312*, 123–134.
76. Theng, B.K.G. *The Chemistry of Clay-Organic Reactions.* Adam Hilger: London, 1974.
77. Ivanov, V.V.; Scherban, O.V. *Org. Geochim.* 1983, *4*, 185–194.
78. Surdam, R.C.; Crossey, L.J.; Hagen, E.S.; Heasler, H.P. *Am. Assoc. Pet. Geol. J.* 1989, *73*, 1–23.
79. Almon, W.R. Ph.D. Thesis. University of Missouri: Columbia, MO, 1974.
80. Crossey, L.J.; Surdam, R.C.; Lahann, R.W. *Roles of Organic Matter in Sediment Diagenesis;* D. Gautier; Ed. SEPM, Tulsa, OK, special publication 38, pp. 147–156.
81. MacGowan, D.; Surdam, R.C. *Org. Geochem.* 1988, *12*, 245–259.
82. Morrow, N. *J. Pet. Tech.* 1990, Dec., 1476–1484.
83. Buckley, J.S.; Takamura, K.; Morrow, N.R. *Soc. Petrol. Eng. Res. Eng.* 1989, Aug., 332–340.
84. Israelachvili, J. *Intermolecular and Surface Forces; 2nd Edition.* Academic Press: New York, 1992.
85. Berg, J.C.; Ed. *Wettability.* Marcel Dekker, Inc.: New York, 1993.
86. Pashley, R.M.; Israelachvili, J. *J. Coll. Interf. Sci.* 1984, *97*, 446–455.
87. Bolger, J.C.; Michaels, A.S. *Interface Conversion;* P. Weiss; G.D. Cheevers; Eds. Elsevier: New York, 1969, 53–60.
88. Brown, C.E.; Neustadter, E.L. *J. Can. Pet. Tech.* 1980, *19*, 100–110.
89. Takamura, K.; Chow, R.S. *Coll. Surf.* 1983, *15*, 34–48.
90. Takamura, K.; Chow, R.S. *J. Can. Pet. Tech.* 1983, *22*, 22–30.
91. Hall, A.C.; Collins, S.H.; Melrose, J.C. *Soc. Pet. Eng. J.* 1983, Apr., 249–258.
92. Buckley, J.S. Chemistry of the crude oil/brine interface. Paper presented at the 3rd International Symposium on Evaluation of Reservoir Wettability and Its Effect on Oil Recovery; Sept. 21–23, Laramie, WY, 1994.

93. Healy, T.W.; White, L.R. *Adv. Coll. Interf. Sci.* 1978, 9, 303–345.
94. Liu, Y.; Buckley, J.S. Wetting alteration by adsorption from crude oil. Paper presented at the 3rd International Symposium on Evaluation of Reservoir Wettability and Its Effect on Oil Recovery; Sept. 21–23, Laramie, WY, 1994.
95. Buckley, J.S.; Morrow, N.R. An overview of crude oil adhesion phenomena. IFP Research Conference on Exploration-Production, Saint Raphaël, France, September 4–6, 1991.
96. Thompson, D.W.; Pownall, P.G. *J. Coll. Interf. Sci.* 1989, 131, 74–82.
97. Machesky, M.L. *Chemical Modeling in Aqueous Systems II*; D.C. Melchior; R.L. Bassett; Eds. ACS Symposium Series 416. American Chemical Society: Washington, D.C., 1990; pp. 262–274.
98. Brady, P.V. *Geochim. Cosmochim. Acta.* 1992, 56, 2941–2946.
99. Schoonen, M.A.A. *Geochim. Cosmochim. Acta.* 1994, 58, 2845–2852.
100. Doe, P.H. Salinity dependence in the wetting of silica by oils. Paper presented at the 3rd International Symposium on Evaluation of Reservoir Wettability and Its Effect on Oil Recovery. Sept. 21–23, 1994, Laramie, WY.
101. Ishido, T.; Mizutani, M. *J. Geophys. Res.* 1981, 86, 1763–1775.
102. Banfield, J.F.; Eggleton, R.A. *Clays Clay Miner.* 1990, 38, 77–89.
103. Velbel, M.A. *Clays Clay Miner.* 1989, 6, 515–524.
104. Banfield, J.F.; Veblen, D.R.; Jones, B.F. *Contr. Min. Pet.* 1990, 106, 110–123.
105. Pettit, J.-C.; Della Mea, G.; Dran, J.-C.; Schott, J.; Berner, R.A. *Nature.* 1987, 325, 705–707.
106. Nahon, D.; Colin, F.; Tardy, Y. *Clay Mineral.* 1982, 17, 339–348.
107. Taylor, A.B.; Velbel, M.A. *Geod.* 1991, 51, 29–50.
108. Garrels, R.M.; MacKenzie, F.T. *Natural Water Systems*; W. Stumm, Ed. American Chemical Society: Washington, D.C., 1967; pp. 222–242.
109. Paĉes, T. *Geochim. Cosmochim. Acta.* 1983, 47, 1855–1863.
110. Velbel, M.A. *Am. J. Sci.* 1985, 285, 904–930.
111. Drever, J.I.; Swoboda-Colberg, N.G. *Water-Rock Interaction WRI-7*; Y. Kharaka; A. Maest; Eds. Balkema: Rotterdam, The Netherlands, 1992; pp. 211–214.
112. Velbel, M.A. *Chem. Geol.* 1989, 78, 245–253.
113. Swoboda-Colberg, N.G.; Drever, J.I. *Water-Rock Interaction WRI-7*; Y. Kharaka; A. Maest; Eds. Balkema: Rotterdam, The Netherlands, 1992; pp. 585–590.
114. Velbel, M.A. *Geochemical Processes at Mineral Surfaces*; J.A. Davis; K.F. Hayes; Eds. American Chemical Society Symposium. 1986, 323, 615–634.
115. Wogelius, R.A.; Walther, J.V. *Geochim. Cosmochim. Acta.* 1991, 55, 943–954.
116. Hering, J.; Stumm, W. *Mineral-Water Interface Geochemistry*; M.F. Hochella; A.F. White; Eds. Mineralogical Society of America: Washington, D.C., 1990; pp. 427–465.
117. Schnoor, J. *Aquatic Chemical Kinetics*; W. Stumm; Ed. Wiley Interscience: New York, 1990; pp. 475–504.
118. Velbel, M.A. *Chem. Geol.* 1993, 107, 337–339.
119. Schnoor, J.L.; Stumm, W. *Schweiz. Z. Hydrol.* 1986, 48, 171–195.
120. Johnson, N.M. *Science.* 1979, 204, 497–499.
121. Johnson, N.M.; Reynolds, R.C.; Likens, G.E. *Science.* 1972, 177, 514–516.
122. Reuss, J.O.; Cosby, B.J.; Wright, R.F. *Nature.* 1987, 329, 27–32.
123. Giovanoli, R.; Schnoor, J.L.; Sigg, L.; Stumm, W.; Zobrist, J. *Clays Clay Min.* 1989, 36, 521–529.
124. Bricker, O.P.; Rice, K.C. *Environ. Sci. Tech.* 1989, 23, 379–385.
125. Anderson, M.P.; and Bowser, C.J. *Water Resour. Res.* 1986, 22, 1101–1108.
126. Cosby, B.J.; Hornberger, G.M.; Galloway, J.N.; Wright, R.F. *Water Resour. Res.* 1985, 21, 51–63.
127. Cosby, B.J.; Wright, R.F.; Hornberger, G.M.; Galloway, J.N. *Water Resour. Res.* 1985, 21, 1591–1601.
128. Christoffersen, N.; Siep, H.; Wright, R.F. *Water Resour. Res.* 1982, 18, 977–996.

129. Gherini, S.A.; Mok, L.; Hudson, R.J.M.; Davis, G.F.; Chen, C.W.; Goldstein, R.A. *Water Air Soil Pollut.* 1985, *26*, 425–459.

130. Sverdrup, H.U.; Warfvinge, P. *Water Air Soil. Pollut.* 1988, *38*, 387–408.

131. Sverdrup, H.U.; Warfvinge, P. *Water-Rock Interaction WRI-7;* Y. Kharaka; A. Maest; Eds. Balkema: Rotterdam, The Netherlands, 1992; pp. 585–590.

132. Furrer, G.; Sollins, P.; Westall, J.C. *Geochim. Cosmochim. Acta.* 1990, *54*, 2363–2374.

133. Warfvinge, P.; Sverdrup, H.U., *Water-Rock Interaction WRI-7;* Y. Kharaka; A. Maest; Eds. Balkema: Rotterdam, The Netherlands, 1992; pp. 603–606.

134. Drever, J.I.; Hurcomb, D.R. *Geology,* 1986, *14*, 221–224.

135. April, R.A.; Newton, R.M.; Coles, L.T. *Geol. Soc. Am. Bull.* 1986, *97*, 1232–1238.

136. Sverdrup, H.U. *The Kinetics of Base Cation Release Due to Chemical Weathering.* Lund Univerity Press: Lund, 1990.

137. Huang, W.H.; Kiang, W.C. *Am. Min.* 1972, *57*, 1849–1859.

138. Manley, E.P.; Evans, L.J. *Soil Sci.* 1986, *41*, 359–369.

139. Amrhein, C.; Suarez, D. *Geochim. Cosmochim. Acta.* 1988, *52*, 2785–2794.

140. Welch, S.A.; Ullman, W.J. *Geochim. Cosmochim. Acta.* 1993, *57*, 2725–2736.

141. Berthelin, J. *Physical and Chemical Weathering in Geochemical Cycles;* A. Lerman; M. Meybeck; Eds. Kluwer Academic: Norwell, MA, 1988; pp. 33–59.

142. Heber, F.K.; Bennett, P.C. *Science.* 1992, *258*, 278–281.

143. Welch, S.A.; Vandevivere, P. *Geomicrobiol. J.* 1994, *12*, 227–238.

144. Mast, M.A.; Drever, J.I. *Geochim. Cosmochim. Acta.* 1987, *51*, 2259–2568.

145. Berner, R.A.; Lasaga, A.C.; Garrels, R.M. *Am. J. Sci.* 1983, *283*, 641–683.

146. Lasaga, A.C.; Berner, R.A.; Garrels, R.M. *The Carbon Cycle and Atmospheric CO_2: Natural Variations, Archean to Present;* E.T. Sundquist; W.S. Broecker; Eds. Geophys. Monogr. Ser. 1985, *32*, 397–411.

147. Barron, E.J.; Washington, W.M. *The Carbon Cycle and Atmospheric CO_2: Natural Variations, Archean to Present;* E.T. Sundquist; W.S. Broecker; Eds. Geophys. Monogr. Ser. 1985, *32*, 546–553.

148. Lovelock, J.E.; Whitfield, M. *Nature.* 1982, *296*, 561–563.

149. Kasting, J.F.; Richardson, S.M.; Pollack, J.B.; Toon, O.B. *Am. J. Sci.* 1986, *286*, 361–389.

150. Kasting, J.F.; Richardson, S.M. *Geochim. Cosmochim. Acta.* 1985, *49*, 2541–2544.

151. Caldeira, K.; Rampino, M.R. *Geophys. Res. Lett.* 1991, *18*, 987–990.

152. Volk, T. *Geology.* 1989, *17*, 107–110.

153. Berner, R.A. *Am. J. Sci.* 1991, *291*, 339–376.

154. Caldeira, K.; Kasting, J.F. *Nature.* 1992, *360*, 721–723.

155. Velbel, M.A. *Geology.* 1993, *21*, 1059–1062.

156. Schwartzman, D.; Volk, T. *Nature.* 1989, *340*, 457–460.

157. Knoll, M.A.; James, W.C. *Geology.* 1987, *15*, 1099–1102.

158. Cochran, M.F.; Berner, R.A. *Chem. Geol.* 1993, *107*, 213–215.

159. Volk, T. *Am. J. Sci.* 1987, *287*, 763–779.

160. Brady, P.V. *J. Geophys. Res.* 1991, *96*, 18101–18106.

161. Marshall, H.G.; Walker, J.C.G.; Kuhn, W.R. *J. Geophys. Res.* 1988, *93*, 791–801.

162. Berner, R.A. *Am. J. Sci.* 1994, *294*, 56–91.

163. Berner, E.K.; Berner, R.A. *The Global Water Cycle: Geochemistry and Environment.* Prentice-Hall: Engelwood Cliffs, NJ, 1987.

164. Drever, J.I. *Geochim. Cosmochim. Acta.* 1994, *58*, 2325–2332.

165. White, A.F.; Blum, A.E. *Geochim. Cosmochim. Acta.* 1995, *59*, 1729–1747.

166. Dorn, R.I.; Brady, P.V. *Geochim. Cosmochim. Acta.* 1995, *59*, 2847–2852.

167. Nobel, P.S. *Biophysical Plant Physiology and Ecology,* W.H. Freeman: New York, 1983.

168. Mann, S. *Nature.* 1988, *332*, 119–124.

169. Iler, R.K. *The Chemistry of Silica.* Wiley-Interscience: New York, 1979.

170. Calvert, P.D.; Mann, S. *J. Mater. Sci.* 1987, *5*, 309–314.

171. Landau, E.M.; Grayer Wolf, S.; Levanon, M.; Leiserowitz, L.; Lahav, M.; Sagiv, J. *J. Am. Chem. Soc.* 1989, *111*, 1436.

172. Mann, S.; Heywood, B.R.; Rajam, S.; Walker, J.B.A. *J. Phys. D: Appl. Phys.* 1991, 24, 154–164.
173. Heywood, B.R.; Mann, S. Adv. Mater. 1994, 6, 9–20.
174. Weissbuch, I.; Addadi, L.; Lahav, M.; Leiserowitz, L. *Science.* 1991, 253, 637–645.
175. Douglas, T.; Mann, S. *Mat. Sci. Eng.* 1994, 1994, 193–199.
176. Riecke, P.C.; Calvert, P.D.; Alper, M. Eds. *Materials Synthesis Utilizing Biological Processes, Materials Research Society Proceedings,* 1990.
177. Lowenstam, H.A.; Weiner, S. *On Biomineralization.* Oxford University Press: New York, 1979.
178. Watterston, J.R. *Geology.* 1992, 20, 315–318.
179. Southam, G.; Beveridge, T.J. *Geochim. Cosmochim. Acta.* 1994, 58, 4527–4530.
180. Hartman, H. *J. Mol. Evol.* 1975, 4, 359–370.
181. Williams, Q.; Williams, K.J. *Lancet.* 1991, 337, 791.
182. Wächtershäuser, G. *Microb. Rev.* 1988, 52, 452–484.
183. Ross M.; Nolan, R.P.; Langer, A.M.; Cooper, W.C. *Health Effects of Mineral Dusts;* G.D. Guthrie, Jr.; B.T. Mossman; Eds. Mineralogical Society of America: Washington, D.C., 1993.
184. Guthrie, G.D., Jr.; Mossman, B.T., Eds. *Health Effects of Mineral Dusts.* Mineralogical Society of America: Washington, D.C., 1993.
185. Stanton, M.F.; Layard, M.; Tegeris, A.; Miller, E.; May, M.; Morgan, E.; Smith, A. *J. Natl. Cancer Inst.* 1981, 67, 965–975.
186. Hume, L.A.; Rimstidt, L.D. *Am. Mineral.* 1992, 77, 1125–1128.
187. Ruby, M.V.; Davis, A.; Kempton, J.H.; Drexler, J.W.; Bergstrom, P.D. *Environ. Sci. Tech.* 1992, 26, 1242–1248.
188. Zenger, D.H. *J. Geol. Ed.* 1972, 20, 107–124.
189. Holland, H.D. *The Chemical Evolution of the Atmosphere and Oceans.* Princeton University Press: Princeton, NJ, 1984.
190. Morrow, D.W. *Geosc. Can.* 1972, 9, 5–13.
191. Hardie, L.A. *J. Sed. Pet.* 1987, 57, 166–183.
192. Del Monte, M.; Sabbioni, C. *Nature.* 1980, 288, 350–351.
193. Land, L.S. *Concepts and Models of Dolomitization;* D.H. Zenger; J.B. Dunham; L. Ethington; Eds. Soc. Econ. Paleo. Min. Spec. Pub. 28, 1980, pp. 87–110.
194. Clayton, R.N.; Jones, B.F.; Berner, R.A. *Geochim. Cosmochim. Acta.* 1968, 32, 415–432.
195. Badiozamani, K. *J. Sed. Petrol.* 1973, 43, 965–984.
196. Folk, R.L.; Land, L.S. *AAPG Bull.* 1975, 59, 60–68.
197. Katz, A.; Matthews, A. *Geochim. Cosmochim. Acta.* 1977, 41, 297–308.
198. Baker, P.; Kastner, M. *Science.* 1981, 213, 214–216.
199. Sibley, D.F.; Nordeng, S.H.; Borkwoski, M.L. *J. Sed. Pet.* 1994, A64, 630–637.
200. McBride, M.B. *Soil Sci. Soc. Am. J.* 1979, 44, 26–28.
201. Davis, J.A.; Fuller, C.C.; Cook, A.D. *Geochim. Cosmochim. Acta.* 1987, 51, 1477–1490.
202. Wersin, P.; Charlet, P.; Karthein, R.; Stumm, W. *Geochim. Cosmochim. Acta.* 1989, 53, 2787–2796.
203. Brätter, P.; Möller, P.; Rösick, U. *Earth Plan. Sci. Lett.* 1972, 14, 50–54.
204. Brady, P.V.; Krumhansl, J.L.; Papenguth, H.W. *Geochim. Cosmochim. Acta.* Submitted.
205. Lippmann, F. *Sedimentary Carbonates.* Springer: New York, 1973.
206. Sayles, F.L.; Fyfe, W.S. *Geochim. Cosmochim. Acta.* 1973, 37, 87–99.
207. Kitano, Y.; Kanamori, N. *Geochem. J.* 1966, 1, 1–10.
208. Murata, K.J.; Friedman, I., Madsen, B.M. *USGS Prof. Pap.* 614-B. 1972.
209. Smith, R.M.; Martell, A.E. *Critical Stability Constants. 4 - Inorganic Complexes.* Plenum Press: New York, 1974.
210. Ball, J.W.; Nordstrom, D.K.; Jenne, E.A. *USGS Wat. Res. Invest.* WRI 78–116, 1980.
211. Siegel, F.R. *Kansas Geol. Surv. Bull.* 1961, 152, 127–158.
212. Jones, B.F. *USGS Prof. Pap.* 502-A, 1965.
213. Pace, N.R. *Cell.* 1991, 65, 531–533.

214. Woese, C.R. *Microb. Rev.* 1987, *51*, 221–271.
215. Crick, F.H.C.; Orgel, L.E. *Icarus.* 1973, *19*, 341–346.
216. Bernal, J.D. *Proc. Phys. Soc.* 1949, *62*, 537–558.
217. Corliss, J.B. *Nature.* 1990, *347*, 624.
218. Miller, S.L. *Science.* 1953, *117*, 528–529.
219. Levine, J. *Mol. Evol.* 1982, *18*, 161.
220. Kasting, J.F.; Zahnle, K.J.; Walker, J.C.G. *Precamb. Res.* 1983, *20*, 121.
221. Shock, E.L.; McCollom, T.; Schulte, M.D.; *Origins Life Evol. Biosphere.* 1995, *25*, 141–159.
222. Chyba, C.F.; Sagan, C. *Nature.* 1992, *355*, 125–132.
223. Shock, E.L. *Origins Life Evol. Biosphere.* 1992, *22*, 67–107.
224. Tingle, T.N.; Hochella, M.F., Jr. *Geochim. Cosmochim. Acta.* 1993, *57*, 3245–3249.
225. Nissenbaum, A. *Origins Life.* 1976, *7*, 413–416.
226. Katchalsky, A. *Naturwissenschaften.* 1973, *60*, 215–220.
227. Goldschmidt, V.M. *New Biol.* 1952, *12*, 97–105.
228. Oparin, A.I. *The Origin of Life on Earth.* Academic Press: New York, 1957.
229. Cairns-Smith, A.G. *Genetic Takeover and the Mineral Origins of Life.* Cambridge: New York, 1982.
230. Gause, G.F. Optical Activity and Living Matter. Biodynamica: Normandy, MO, 1941; pp. 19–34.
231. Jackson, T.A. *Chem. Geol.* 1971, *7*, 295–306.
232. Bonner, W.A. *Exobiology;* C. Ponnamperuma; Ed. North-Holland: Amsterdam, 1972; pp. 170–274.
233. Bonner, W.A.; Kavasmaneck, P.R.; Martin, F.S.; Flores, J.S. *Science.* 1974, *186*, 143–144.
234. Bonner, W.A.; Kavasmaneck, P.R.; Martin, F.S.; Flores, J.S. *Origins Life.* 1975, *6*, 367–376.
235. Degens, E.T., Matheja, J.; Jackson, T.A. *Nature.* 1970, *227*, 492–493.
236. Jackson, T.A. *Experientia.* 1971, *27*, 242–243.
237. Tranter, G.E. *Nature.* 1985, *318*, 172–173.
238. Stryer, L. *Biochemistry;* 2nd edition; Freeman: New York, 1981.
239. Griffith, E.J.; Ponnamperuma, C.; Gabel, N. *Origins Life.* 1977, *8*, 71–85.
240. Banin, A.; Navrot, J. *Science.* 1975, *189*, 350–351.
241. Egami, F. *J. Mol. Evol.* 1974, *4*, 113–120.
242. McClendon, J.H. *J. Mol. Evol.* 1976, *8*, 175–195.
243. Lawless, J.G., Levi, N. *Mol. Evol.* 1979, *13*, 281–286.
244. Odom, D.G.; Rao, M.; Lawless, J.G.; Oro, J. *J. Mol. Evol.* 1979, *12*, 365–367.
245. Gabel, N.W.; Ponnamperuma, C. *Nature.* 1967, *216*, 453–455.
246. Lahav, N., White, D.; Chang, S. *Science.* 1978, *201*, 67–69.
247. Paecht-Horowitz, M.; Berger, J.; Katchalsky, A. *Nature.* 1970, *228*, 636–639.
248. Lohrmann, R.; Orgel, L. *Science.* 1971, *171*, 490–494.
249. Kavasmaneck, P.R.; Bonner, W.A. *J. Am. Chem. Soc.* 1977, *99*, 44–50.
250. Brady, P.V. Paper presented at the American Geophysical Union Meeting, San Francisco, CA, December, 1992.
251. Wächtershäuser, G. *Med. Hypoth.* 1991, *36*, 307–311.
252. Wächtershäuser, G. *Prog. Biophys. Mol. Biol.* 1992, *58*, 85–201.
253. Harvey, G.R.; Mopper, K.; Degens, E.T. *Chem. Geol.* 1972, *9*, 79–87.
254. Lahav, N.; Chang, S. *J. Mol. Evol.* 1976, *8*, 357–380.
255. Theng, B.K.G.; Walker, G.F. *Isr. J. Chem.* 1970, *8*, 417–424.
256. Weaver, C.E. *Geochim. Cosmochim. Acta.* 1967, *31*, 2181–2196.
257. Russell, M.J.; Daniel, R.M.; Hall, A.J. *Terra Nova*, 1993, *5*, 343.
258. Russell, M.J.; Daniel, R.M.; Hall, A.J.; Sherringham, J.A. *J. Mol. Evol.*, 1994, *39*, 231.

INDEX

N

O

P

Milton Keynes UK
Ingram Content Group UK Ltd.
UKHW020016071024
449327UK00031B/2809